Turbulence: Numerical Analysis, Modelling and Simulation

Special Issue Editor

William Layton

MDPI • Basel • Beijing • Wuhan • Barcelona • Belgrade

Special Issue Editor
William Layton
University of Pittsburgh
USA

Editorial Office
MDPI
St. Alban-Anlage 66
Basel, Switzerland

This edition is a reprint of the Special Issue published online in the open access journal *Fluids* (ISSN 2311-5521) from 2017–2018 (available at: http://www.mdpi.com/journal/fluids/special_issues/turbulence).

For citation purposes, cite each article independently as indicated on the article page online and as indicated below:

Lastname, F.M.; Lastname, F.M. Article title. *Journal Name* **Year**, *Article number*, page range.

First Edition 2018

ISBN 978-3-03842-809-1 (Pbk)
ISBN 978-3-03842-810-7 (PDF)

Table of Contents

About the Special Issue Editor

William Layton, Professor of Mathematics, University of Pittsburgh, performs research regarding numerical analysis, a field focusing on understanding what is predictable and how to predict within time and resource constraints. His research reflects his continued fascination with the detailed mathematics of fluid motion, which, to paraphrase, is an area of mathematics in which "a gnat may bathe and an elephant may drown." He has authored approximately 200 papers and four books; guided 38 PhD students, who have since become accomplished researchers and who are now successfully supervising PhD students themselves. Outside of mathematics, he was the chess champion of Georgia in 1976 and is currently an avid mid-level whitewater kayaker, getting many mathematical ideas from observations of turbulent flows in nature.

fluids

MDPI

Editorial

Turbulence: Numerical Analysis, Modeling, and Simulation

William Layton

Department of Mathematics, University of Pittsburgh, Pittsburgh, PA 15260, USA; wjl@pitt.edu

Received: 13 February 2018; Accepted: 14 February 2018; Published: 18 February 2018

The problem of accurate and reliable prediction of turbulent flows is a central and intractable challenge that crosses disciplinary boundaries. As the needs for accuracy increase and the applications expand beyond flows where extensive data is available for calibration, the importance of a sound mathematical foundation that addresses the needs of practical computing increases. This special issue is directed at this crossroads of rigorous numerical analysis, the physics of turbulence, and the practical needs of turbulent flow simulations. It contains papers providing a broad understanding of the status of the problem considered and open problems that comprise further steps. It consists of papers covering fundamentals, applications, theory, simulations, experiments, and reviews. The papers cover the general topics summarized below.

Kubacki and Tran [1] present a modern, efficient approach for uncoupling groundwater–surface water flows governed by the fully evolutionary Stokes-Darcy equations. These algorithms treat the coupling terms explicitly and at each time level require only one sub-physics, sub-domain solve that can be done by codes highly optimized for individual processes. Obviously, such methods have greater accuracy and efficiency per time step than non-optimized, fully coupled, monolithic methods. Thus, the key to their utility is whether a price in stability must be paid. This paper presents algorithms with unconditional stability and high accuracy.

Nguyen et al. [2] study the simulation and modeling of the dispersion from an instantaneous source of heat or mass located at the center of a turbulent flow channel. This work is at the intersection of high impact in applications and the leading edge of the understanding of turbulence modulation by transport effects.

Bowers and Rebholz [3] present a review of recent results for the reduced Navier-Stokes-α (rNS-α) model of incompressible flow. The model was recently developed as a numerical approximation of the well-known Navier-Stokes-α model. Numerical simulations are far more efficient with the reduced model. Those simulations have revealed interesting features of the reduced model as an independent fluids model.

Basse [4] presents a comparison of turbulence intensity profiles for smooth and rough wall pipe flow measurements made in the Princeton Superpipe. The profile development in the transition from smooth to rough wall flow is analyzed from the data. In this paper, the highly difficult problem wherein maximum insight must be obtained from available data is addressed.

Chen and Lo [5] present a numerical study of coherent structure evolution in boundary layer transition flow using high order compact difference schemes with non-uniform grids in the wall-normal direction. Efficient solutions and high accuracy are provided in this interesting study.

Maulik and San [6] present the results of a study solving two-dimensional (2D), compressible turbulence. Their paper compares two promising computational approaches and draws valuable conclusions.

Brkić and Ćojbašić [7] present evolutionary optimization for approximations of the Colebrook's equation for the turbulent friction factor. This calculation is used for the calculation of turbulent hydraulic resistance in hydraulically smooth and rough pipes including the transient zone.

Breckling et al. [8] present an overview of time relaxation models. To date, these have been one of the few models for LES where the model solution is proven to converge to the true averages of the turbulent flow. This approach is related to data assimilation by a technique called nudging. It has proven to be effective in the regularization of Navier-Stokes equations, and as such this summary of completed theory, necessary algorithms, and new directions is very welcome.

Dunca [9] studies a very promising family of alpha-deconvolution models. It is widely observed that high-order models and methods always outperform low-order, ones even for problems with rough solutions when the available theory indicates no advantage. This paper is one of the first theoretical studies to explain this advantage. The paper provides a theoretical analysis of model accuracy with many complicating factors that would suggest no advantage of high-order models to compensate for their greater cost. This paper is a landmark theoretical result.

In the final paper [10], Ries et al. present a database generated by numerical and physical experiments of a database of near-wall turbulence properties of a highly turbulent jet impinging on a solid surface under different inclination angles. The dilemma of resolving near-wall turbulence or employing error-prone near-wall models is inescapable in LES. Their database will be useful, even essential, for the development of accurate near-wall models in large-eddy simulations (LES).

I thank all of the contributors for submitting high quality papers for the special issue. I also thank the reviewers for their dedicated time and help supporting the quality of the papers.

Conflicts of Interest: The authors declare no conflicts of interests.

References

1. Kubacki, M.; Tran, H. Non-Iterative Partitioned Methods for Uncoupling Evolutionary Groundwater–Surface Water Flows. *Fluids* **2017**, 2, 47. [CrossRef]
2. Nguyen, Q.; Feher, S.E.; Papavassiliou, D.V. Lagrangian Modeling of Turbulent Dispersion from Instantaneous Point Sources at the Center of a Turbulent Flow Channel. *Fluids* **2017**, 2, 46. [CrossRef]
3. Bowers, A.L.; Rebholz, L.G. The Reduced NS-α Model for Incompressible Flow: A Review of Recent Progress. *Fluids* **2017**, 2, 38. [CrossRef]
4. Basse, N.T. Turbulence Intensity and the Friction Factor for Smooth- and Rough-Wall Pipe Flow. *Fluids* **2017**, 2, 30. [CrossRef]
5. Chen, W.; Lo, E.Y. High Wavenumber Coherent Structures in Low Re APG-Boundary-Layer Transition Flow—A Numerical Study. *Fluids* **2017**, 2, 21. [CrossRef]
6. Maulik, R.; San, O. Energy Dissipation Characteristics of Implicit LES and Explicit Filtering Models for Compressible Turbulence. *Fluids* **2017**, 2, 14. [CrossRef]
7. Brkić, D.; Ćojbašić, Z. Evolutionary Optimization of Colebrook's Turbulent Flow Friction Approximations. *Fluids* **2017**, 2, 15. [CrossRef]
8. Breckling, S.; Neda, M.; Hill, T. A Review of Time Relaxation Methods. *Fluids* **2017**, 2, 40. [CrossRef]
9. Dunca, A.A. Improving Accuracy in α-Models of Turbulence through Approximate Deconvolution. *Fluids* **2017**, 2, 58. [CrossRef]
10. Ries, F.; Li, Y.; Rißmann, M.; Klingenberg, D.; Nishad, K.; Böhm, B.; Dreizler, A.; Janicka, J.; Sadiki, A. Database of Near-Wall Turbulent Flow Properties of a Jet Impinging on a Solid Surface under Different Inclination Angles. *Fluids* **2018**, 3, 5. [CrossRef]

fluids

MDPI

Article

Non-Iterative Partitioned Methods for Uncoupling Evolutionary Groundwater–Surface Water Flows

Michaela Kubacki [1] and Hoang Tran [2,*]

1 Department of Mathematics, Middlebury College, Middlebury, VT 05753, USA; mkubacki@middlebury.edu
2 Computer Science and Mathematics Division, Oak Ridge National Laboratory, Oak Ridge, TN 37831, USA
* Correspondence: tranha@ornl.gov; Tel.: +1-865-574-1283

Received: 1 July 2017; Accepted: 26 August 2017; Published: 10 September 2017

Abstract: We present an overview of a modern, efficient approach for uncoupling groundwater–surface water flows governed by the fully evolutionary Stokes–Darcy equations. Referred to as non-iterative partitioned methods, these algorithms treat the coupling terms explicitly and at each time level require only one Stokes and one Darcy sub-physics solve, thus taking advantage of existing solvers optimized for each sub-flow. This strategy often results in a time-step condition for stability. Furthermore, small problem parameters, specifically those related to the physical characteristics of the porous media domain, can render certain time-step conditions impractical. Despite these obstacles, researchers have made significant progress towards efficient, stable, and accurate partitioned methods. Herein, we provide a comprehensive survey and comparison of recent developments utilizing these non-iterative numerical schemes.

Keywords: implicit-explicit schemes; finite difference methods; Stokes–Darcy equations

---

1. Introduction

Access to the clean freshwater is absolutely imperative for the continued survival of humankind. As a necessity for our agricultural, industrial and domestic practices, water constitutes an integral part of all civilizations. However, only 2.5% of the water present on Earth is freshwater, and the majority of this amount is either frozen or inaccessible. Furthermore, 96% of accessible freshwater comes from aquifers underground. Because of the scarcity of this resource, we must prioritize the protection and conservation of groundwater sources. Too often, human and natural processes threaten groundwater quality, sometimes irreversibly. For example, in hydro-fracturing, companies inject a mixture of water with sand and chemicals at high pressure into a well to create fractures to allow for the collection of shale gas. Companies do not recover the majority of the chemicals in this mixture and many fear that eventually these pollutants will leave the well to contaminate the local groundwater supply. Pesticide application in agriculture can have devastating effects on surrounding freshwater resources due to chemical run-off into nearby rivers, lakes, and streams, and seepage deep into the soil. Furthermore, many storage facilities for radioactive materials exist underground for both safety and convenience. Over time, as storage containers become compromised, nuclear waste can migrate into nearby freshwater aquifers. Even natural processes may result in contaminated freshwater, as evident in the devastation of forests growing above coastal aquifers from salt-water intrusion.

Tracking these contaminants necessitates accurate numerical models for this coupled flow. Scientists have thoroughly studied the individual groundwater and surface water flows (see, for example, Pinder and Celia [1], Watson and Burnett [2], or Bear [3] for an extensive study on subsurface flows, and Kundu, Cohan and Dowling [4] for surface water flows). As a result, many accurate and efficient solvers for the independent flow processes exist. Modeling the interaction of groundwater and surface water, however, presents additional difficulties as we must preserve the physical processes in each sub-flow while accurately describing the activity occurring along the interface.

An attractive and practical strategy, which is the main focus of this survey paper, is to make use of the existing solvers for separate fluid and porous media flow by investigating methods that uncouple the flow equations in time so that the individual flow problems may be solved separately. Called partitioned methods (also domain decomposition methods), these methods allow us to utilize, in a black-box manner, solvers already optimized for the separate flow problems. It is important that the partitioned methods maintain stability and accuracy along the interface where the two flows meet. In addition, potentially small physical parameters create an additional challenge for stability. We are concerned with methods that are efficient for the time-dependent models, in particular, the ones that are stable over long-time intervals, since groundwater moves slowly and numerical simulations may span long-time periods. Along these same lines, we want methods that converge within a reasonable amount of time to be of practical use, making higher-order convergence a desirable property.

In recent years, several researchers have made substantial progress in the development of non-iterative, partitioned methods applied to the evolutionary groundwater–surface water flow problem. Based on an implicit discretization in time for each subproblem, these methods, however, make use of results from previous time steps to predict the values on the interface at the current time step, thus requiring only one solve for groundwater and one solve for surface water flow at each time level (thus non-iterative). In this work, we will review and discuss several such methods so as to illustrate the current status of this important problem. The modeling of coupled fluid-porous media flow begins with the coupling of the Stokes or Navier–Stokes equations describing the flow in the fluid region, along with the Darcy or Brinkman equations for the flow in the aquifer, or porous media region containing the groundwater. This survey focuses on the Stokes–Darcy coupling that is suitable for slow moving flows over large domains.

Studies on the continuum surface water-groundwater model have been performed in [5–8]. The literature on numerical analysis of methods for the coupled Stokes–Darcy problem has grown extensively since [9,10] (see, for example, [11–13] for analysis of the steady-state problem). There exist many effective and efficient domain decomposition techniques for decoupling the Stokes–Darcy system in the stationary case [14–24]. To solve the fully evolutionary Stokes–Darcy problem, one approach is monolithic discretization by an implicit scheme (see, e.g., [25,26]). These schemes can also be solved by an iterative domain decomposition method at each time step. In general, any decoupling technique for stationary Stokes–Darcy (many cited above) may be applied to find the solution at each time level in the time-dependent case.

Non-iterative partitioned methods, an alternative approach, are advantageous in that they allow uncoupling into only one (SPD) Stokes and one (SPD) Darcy system per time step. Mu and Zhu presented the first non-iterative partitioned scheme in [27], proposing employing Backward Euler discretization for each subproblem while treating the coupling term explicitly by Forward Euler. Layton, Tran and Trenchea revisited this method in [28], with an improved analysis showing long-time stability. In that work, the authors also developed and tested for efficiency a second first-order scheme, Backward Euler–Leap Frog. Following these methods, others proposed several other implicit-explicit (IMEX) methods of high order, such as Crank–Nicolson–Leap Frog [29], second-order backward-differentiation with Gear's extrapolation [30], and Adam–Moulton–Bashforth [30,31]. Although these methods use explicit discretizations for the coupling terms, all are now known to be long-time stable and optimally convergent uniformly in time (possibly under a small time-step constraint). With the addition of suitable stabilization terms, it is possible to further enhance the stability property, for instance, a stabilized Crank–Nicolson–Leap Frog, developed in [32,33], requires no time-step restriction for the long-time stability and convergence. Another way for uncoupling groundwater–surface water systems is using splitting schemes. Unlike the aforementioned IMEX schemes that solve for separate sub-flows in parallel, splitting methods require sequential sub-problem solves at each time step. In [34], the authors proposed four first and second-order splitting schemes. Theoretical and numerical evidence provided therein suggests that these methods are stable for larger time steps than the first order IMEX schemes and, in particular, a good option in case of small physical parameters. Finally, asynchronous (aka, multiple-time-step, multi-rate) partitioned methods allow for different time steps in the two subregions, motivated by the observation that the flow in fluid region occurs with higher velocities compared to flow in porous media region. Such methods may be more efficient, as we may apply two different time steps to separately solve the fast and slow flows. Developed in [35,36], these asynchronous techniques utilize the Backward Euler-Forward Euler time discretization, with long-time stability acquired in the latter work.

We organize this paper as follows. Section 2 reviews the preliminaries of the Stokes–Darcy equation, including interface conditions, variational formulation and semi-discrete approximations. We briefly discuss the implicit time discretization, together with the iterative domain-decomposition approach. Section 3 focuses on first-order partitioned methods. We will survey several different approaches including first-order IMEX schemes and splitting schemes. We review high-order methods in Section 4 and asynchronous partitioned techniques in Section 5. Finally, we provide some conclusions and outlooks in Section 6.

2. The Stokes–Darcy Equation

Let the fluid region be denoted by Ω_f and the porous media region by Ω_p. Assume both domains are bounded and regular. Let I represent the interface between the two domains. We assume the time-dependent Stokes flow in Ω_f and the time-dependent groundwater flow along with Darcy's law in Ω_p. The Stokes–Darcy equation, describing the fluid velocity field $u = u(x,t)$ and pressure $p = p(x,t)$ on Ω_f and the porous media hydraulic head $\phi = \phi(x,t)$ on Ω_p, can be written as follows:

$$\begin{aligned} u_t - \nu\Delta u + \nabla p = \mathbf{f}_f, \nabla \cdot u = 0, &\quad \text{in } \Omega_f, \\ S_0\phi_t - \nabla\cdot(\mathcal{K}\nabla\phi) = f_p, &\quad \text{in } \Omega_p, \\ u(x,0) = u_0, \text{ in } \Omega_f \text{ and } \phi(x,0) = \phi_0, &\quad \text{in } \Omega_p, \\ u(x,t) = 0, \text{ in } \partial\Omega_f\backslash I \text{ and } \phi(x,t) = 0, &\quad \text{in } \partial\Omega_p\backslash I, \\ &+ \text{coupling conditions across } I. \end{aligned} \tag{1}$$

Here, $\mathbf{f}_f$ denotes the body force in the fluid region, f_p is the sink or source in the porous media region, $\nu > 0$ is the kinematic viscosity of the fluid, S_0 is the specific mass storativity coefficient and $\mathcal{K}$ is the hydraulic conductivity tensor, assumed to be symmetric, positive definite with spectrum$(\mathcal{K}) \in [k_{\min}, k_{\max}]$.

It is worth noting that values of S_0 and the smallest eigenvector $k_{\min}$ of $\mathcal{K}$ can be very small (see Tables 1 and 2 for the values of S_0 and $k_{\min}$ for different materials). As we shall see, this poses a major challenge in designing partitioned methods with good stability. Indeed, partitioning often induces time-step restrictions for long-time stability, which may become severe in the case of small system parameters.

Table 1. Specific storage (S_0) values for different materials [37,38].

Material	S_0 (m^{-1})
Plastic clay	$2.6 \times 10^{-3} - 2.0 \times 10^{-2}$
Stiff clay	$1.3 \times 10^{-3} - 2.6 \times 10^{-3}$
Medium hard clay	$9.2 \times 10^{-4} - 1.3 \times 10^{-3}$
Loose sand	$4.9 \times 10^{-4} - 1.0 \times 10^{-3}$
Dense sand	$1.3 \times 10^{-4} - 2.0 \times 10^{-4}$
Dense sandy gravel	$4.9 \times 10^{-5} - 1.0 \times 10^{-4}$
Rock, fissured jointed	$3.3 \times 10^{-6} - 6.9 \times 10^{-5}$
Rock, sound	less than 3.3×10^{-6}

Table 2. Hydraulic conductivity ($k_{\min}$) values for different materials [3].

Material	$k_{\min}$ **(m/s)**
Well sorted gravel	$10^{-1} - 10^{0}$
Highly fractured rocks	$10^{-3} - 10^{0}$
Well sorted sand or sand and gravel	$10^{-4} - 10^{-2}$
Oil reservoir rocks	$10^{-6} - 10^{-4}$
Very fine sand, silt, loess, loam	$10^{-8} - 10^{-5}$
Layered clay	$10^{-8} - 10^{-6}$
Fresh sandstone, limestone, dolomite, granite	$10^{-12} - 10^{-7}$
Fat/Unweathered clay	$10^{-12} - 10^{-9}$

2.1. Interface Conditions

To close the coupled problem formulation, a set of conditions has to be defined on the interface. Let $\widehat{n}_{f/p}$ denote the indicated, outward pointing, unit normal vector on I. The first two coupling conditions involve the conservation of mass and balance of forces on I:

$$u \cdot \widehat{n}_f - \mathcal{K}\nabla\phi \cdot \widehat{n}_p = 0, \text{ on } I,$$
$$p - \nu\, \widehat{n}_f \cdot \nabla u \cdot \widehat{n}_f = g\phi \text{ on } I.$$

In addition, we need a tangential condition on the fluid region's velocity along the interface. In [5], Beavers and Joseph proposed the following slip–flow condition, expressing that slip velocity along I is proportional to the shear stresses along I

$$-\nu\, \widehat{\tau}_i \cdot \nabla u \cdot \widehat{n}_f = \alpha_{BJ}\sqrt{\frac{\nu g}{\widehat{\tau}_i \cdot \mathcal{K} \cdot \widehat{\tau}_i}}(u - \mathbf{u}_p) \cdot \widehat{\tau}_i, \text{ on } I \text{ for any } \widehat{\tau}_i \text{ tangent vector on } I,$$

where α_{BJ} is a dimensionless constant depending solely on the porous media properties and ranges from 0.01 to 5, g is the gravitational acceleration constant, and $\mathbf{u}_p$ is the average velocity in the porous media region. The validity of Beavers–Joseph interface condition has been supported by abundant empirical evidence; however, one challenge in adopting this condition is that the bilinear form in the weak formulation is not coercive. Several simplifications have been considered. In [6], Saffman proposed a modification to the Beavers–Joseph coupling condition by dropping the porous media averaged velocity $\mathbf{u}_p$, based on observations that the term $\mathbf{u}_p \cdot \widehat{\tau}_i$ is negligible compared to the fluid velocity $u \cdot \widehat{\tau}_i$. This simplified condition was mathematically justified in [39] and has been shown satisfactory for many fluid-porous media systems. Known as Beavers–Joseph–Saffman(–Jones) coupling condition, this is the third and final condition we use in this article:

$$-\nu\, \widehat{\tau}_i \cdot \nabla u \cdot \widehat{n}_f = \alpha_{BJ}\sqrt{\frac{\nu g}{\widehat{\tau}_i \cdot \mathcal{K} \cdot \widehat{\tau}_i}}\, u \cdot \widehat{\tau}_i, \text{ on } I \text{ for any } \widehat{\tau}_i \text{ tangent vector on } I.$$

For the analysis and numerical methods for Stokes–Darcy systems with Beavers–Joseph condition, we refer to [21,26,40].

2.2. Variational Formulation and Semi-Discrete Approximations Using Finite Element Method

We denote the $L^2(I)$ norm by $\|\cdot\|_I$ and the $L^2(\Omega_{f/p})$ norms by $\|\cdot\|_{f/p}$, respectively, and the corresponding inner products are denoted by $(\cdot,\cdot)_{f/p}$. In addition, define the $H_{\text{div}}(\Omega_f)$ and $H^1(\Omega_{f/p})$ norms

$$\|u\|_{\text{div},f} := \sqrt{\|u\|_f^2 + \|\nabla \cdot u\|_f^2}, \quad \|u\|_{1,f/p} = \sqrt{\|u\|_{f/p}^2 + \|\nabla u\|_{f/p}^2},$$

the functional spaces

$$\begin{aligned} X_f &= \{v \in \left(H^1(\Omega_f)\right)^d : v = 0 \text{ on } \partial\Omega_f \backslash I\},\ Q_f = L^2(\Omega_f), \\ X_p &= \{\psi \in H^1(\Omega_p) : \psi = 0 \text{ on } \partial\Omega_p \backslash I\}, \end{aligned}$$

and the bilinear forms

$$\begin{aligned} a_f(u,v) &= (\nu \nabla u, \nabla v)_f + \sum_i \int_I \alpha_{BJ} \sqrt{\frac{\nu g}{\widehat{\tau}_i \cdot \mathcal{K} \cdot \widehat{\tau}_i}} (u \cdot \widehat{\tau}_i)(v \cdot \widehat{\tau}_i) ds, \\ a_p(\phi, \psi) &= g(\mathcal{K} \nabla \phi, \nabla \psi)_p, \text{ and } c_I(u,\phi) = g \int_I \phi u \cdot \widehat{n}_f ds. \end{aligned}$$

It can be shown that $a_{f/p}(\cdot,\cdot)$ are continuous and coercive.

A (monolithic) variational formulation of the coupled problem is to find (u,p,ϕ) : $[0,\infty) \to X_f \times Q_f \times X_p$ satisfying the given initial conditions and, for all $v \in X_f, q \in Q_f, \psi \in X_p$,

$$\begin{gathered} (u_t, v)_f + a_f(u,v) - (p, \nabla \cdot v)_f + c_I(v,\phi) = (\mathbf{f}_f, v)_f, \\ (q, \nabla \cdot u)_f = 0, \\ gS_0(\phi_t, \psi)_p + a_p(\phi,\psi) - c_I(u,\psi) = g(f_p,\psi)_p. \end{gathered}$$

Note that, setting $v = u, \psi = \phi$ and adding, the coupling terms exactly cancel out in the monolithic sum yielding the energy estimate for the coupled system.

To discretize the Stokes–Darcy problem in space by the finite element method (FEM), we select finite element spaces

$$\text{velocity: } X_f^h \subset X_f,\ \text{Darcy pressure: } X_p^h \subset X_p,\ \text{Stokes pressure: } Q_f^h \subset Q_f$$

based on a conforming FEM triangulation with maximum triangle diameter denoted by "h". We do not assume mesh compatibility or interdomain continuity at the interface I between the FEM meshes in the two subdomains. The Stokes velocity-pressure FEM spaces are assumed to satisfy the usual discrete inf-sup condition for stability of the discrete pressure (see, e.g., [41])

$$\exists \beta_h > 0 \text{ such that } \inf_{q_h \in Q_f^h, q_h \neq 0} \sup_{v_h \in X_f^h, v_h \neq 0} \frac{(q_h, \nabla \cdot v_h)_f}{\|\nabla v_h\|_f \|q_h\|_f} > \beta_h. \tag{2}$$

Assume X_f^h, X_p^h, Q_f^h satisfy approximation properties of piecewise polynomials on quasi-uniform meshes of local degrees k, k, $k-1$, respectively, that is,

$$\begin{aligned} \inf_{v_h \in X_f^h} \|u - v_h\|_f &\le Ch^{k+1} \|u\|_{H^{k+1}(\Omega_f)}, \quad \forall u \in H^{k+1}(\Omega_f), \\ \inf_{v_h \in X_f^h} \|\nabla(u - v_h)\|_f &\le Ch^{k} \|u\|_{H^{k+1}(\Omega_f)}, \quad \forall u \in H^{k+1}(\Omega_f), \\ \inf_{\psi_h \in X_p^h} \|\phi - \psi_h\|_p &\le Ch^{k+1} \|\phi\|_{H^{k+1}(\Omega_p)}, \quad \forall \phi \in H^{k+1}(\Omega_p), \\ \inf_{\psi_h \in X_p^h} \|\nabla(\phi - \psi_h)\|_p &\le Ch^{k} \|\phi\|_{H^{k+1}(\Omega_p)}, \quad \forall \phi \in H^{k+1}(\Omega_p), \\ \inf_{q_h \in Q_f^h} \|p - q_h\|_f &\le Ch^{k} \|p\|_{H^{k}(\Omega_f)}, \quad \forall p \in H^{k}(\Omega_f). \end{aligned}$$

The semi-discretization for the time-dependent Stokes–Darcy problem is as follows: find $(u_h, p_h, \phi_h) : [0, \infty) \to X_f^h \times Q_f^h \times X_p^h$ satisfying the given initial conditions and, for all $v_h \in X_f^h$, $q_h \in Q_f^h, \psi_h \in X_p^h$,

$$
\begin{aligned}
(u_{h,t}, v_h)_f + a_f(u_h, v_h) - (p_h, \nabla \cdot v_h)_f + c_I(v_h, \phi_h) &= (\mathbf{f}_f, v_h)_f, \\
(q_h, \nabla \cdot u_h)_f &= 0, \\
gS_0(\phi_{h,t}, \psi_h)_p + a_p(\phi_h, \psi_h) - c_I(u_h, \psi_h) &= g(f_p, \psi_h)_p.
\end{aligned}
$$

It is worth noting that the coupling between the Stokes and the Darcy sub-problems is exactly skew symmetric.

2.3. Fully-Discrete Approximations with Fully Implicit Temporal Schemes

Let $t^n := n\Delta t$ and $w^n := w(x, t^n)$ for any function $w(x, t)$. For V being a Banach space with norm $\|\cdot\|_V$, we denote the following discrete norms

$$
|||w|||_{L^2(0,T;V)} := \left(\Delta t \sum_{n=0}^{N} \|w^n\|_V^2\right)^{1/2}, \quad |||w|||_{L^\infty(0,T;V)} := \sup_{0 \le n \le N} \|w^n\|_V,
$$

where $N = T/\Delta t$ and T can be ∞. For fixed $T > 0$, the discrete norm $|||\cdot|||_{L^\infty(0,T;V)}$ is bounded by the continuous norm $\|\cdot\|_{L^\infty(0,T;V)}$. The discrete norm $|||\cdot|||_{L^2(0,T;V)}$, on the other hand, depends on the time step Δt. However, for functions smooth in time, this norm converges to the continuous norm $\|\cdot\|_{L^2(0,T;V)}$ as $\Delta t \to 0$. In those cases, one can reasonably assume the uniform bound of $|||\cdot|||_{L^2(0,T;V)}$, independent of Δt.

The most natural time discretization for the Stokes–Darcy equation is perhaps the first-order backward Euler scheme, which, in combination with the aforementioned finite element Galerkin method for the spatial discretization, leads to the following fully implicit, coupled problem.

Algorithm 1 Backward Euler

Given $(u_h^n, p_h^n, \phi_h^n) \in X_f^h \times Q_f^h \times X_p^h$, find $(u_h^{n+1}, p_h^{n+1}, \phi_h^{n+1}) \in X_f^h \times Q_f^h \times X_p^h$ such that for all $v_h \in X_f^h, q_h \in Q_f^h, \psi_h \in X_p^h$,

$$
\begin{aligned}
&(\frac{u_h^{n+1} - u_h^n}{\Delta t}, v_h)_f + a_f(u_h^{n+1}, v_h) - (p_h^{n+1}, \nabla \cdot v_h)_f + c_I(v_h, \phi_h^{n+1}) = (\mathbf{f}_f^{n+1}, v_h)_f, \\
&(q_h, \nabla \cdot u_h^{n+1})_f = 0, \\
&gS_0(\frac{\phi_h^{n+1} - \phi_h^n}{\Delta t}, \psi_h)_p + a_p(\phi_h^{n+1}, \psi_h) - c_I(u_h^{n+1}, \psi_h) = g(f_p^{n+1}, \psi_h)_p.
\end{aligned}
$$

Stability and convergence analysis of this scheme were conducted in [25–27], for both Beavers–Joseph and Beavers–Joseph-Saffman–Jones interface conditions. Higher order fully implicit schemes, such as the Crank–Nicolson, can also be considered. In general, fully implicit methods possess superior stability compared to IMEX or splitting temporal schemes. The major concern here is that this approach must solve a coupled problem at each time level. Partitioning the coupled problem at each time step is possible, but involves an iterative procedure with additional cost. In principle, any decoupled methods developed for the stationary model can be used in iteration at each time level.

3. First Order Partitioned Schemes

An attractive alternative to fully implicit, fully coupled discretization is exploiting information obtained in previous time steps to construct a non-iterative uncoupling scheme, which only need a single Stokes solve and a single Darcy solve at each time step. This approach allows the use of legacy subproblems' codes and obviously requires less programming effort (compared to solving coupled Stokes–Darcy system directly) as well as less computation cost (compared to iterative domain decomposition approach). As the interface values are obtained in an explicit manner, the main challenge here is how to obtain optimal accuracy and good stability properties. Many non-iterative partitioned methods have been developed in the literature recently [27–36,42], whose stability and accuracy have been proved (over a long time or without time-step condition) and numerically tested. Several of them maintain good performance even in the case of small parameters. The rest of this paper represents an overview of these developments. Our discussion will be divided into three parts: in Section 3, we survey first order schemes; in Section 4, high order schemes will be discussed; Section 5 is devoted to asynchronous partitioned methods. Unless otherwise stated, C denotes a generic positive constant whose value may be different from place to place but which is independent of mesh size, time step, and final time. For all the methods surveyed, approximations are needed at the first few (one or more) time steps to begin, and we always assume these are computed to sufficient accuracy.

3.1. Backward Euler-Forward Euler

The first non-iterative uncoupling scheme is Backward Euler-Forward Euler (BEFE), proposed by Mu and Zhu in [27] (and referred to as DBES therein). This method applies Backward Euler discretization for the subproblems and treats the coupling terms by explicit Forward Euler:

Algorithm 2 Backward Euler-Forward Euler (BEFE)

Given $(u_h^n, p_h^n, \phi_h^n) \in X_f^h \times Q_f^h \times X_p^h$, find $(u_h^{n+1}, p_h^{n+1}, \phi_h^{n+1}) \in X_f^h \times Q_f^h \times X_p^h$ such that for all $v_h \in X_f^h, q_h \in Q_f^h, \psi_h \in X_p^h$,

$$(\frac{u_h^{n+1} - u_h^n}{\Delta t}, v_h)_f + a_f(u_h^{n+1}, v_h) - (p_h^{n+1}, \nabla \cdot v_h)_f + c_I(v_h, \phi_h^n) = (\mathbf{f}_f^{n+1}, v_h)_f,$$
$$(q_h, \nabla \cdot u_h^{n+1})_f = 0,$$
$$gS_0(\frac{\phi_h^{n+1} - \phi_h^n}{\Delta t}, \psi_h)_p + a_p(\phi_h^{n+1}, \psi_h) - c_I(u_h^n, \psi_h) = g(f_p^{n+1}, \psi_h)_p.$$

A stability analysis for BEFE was given in [27]. These results only apply for bounded time intervals $[0, T]$ with $T < \infty$, as the estimates include e^{cT} multipliers and thus grow exponentially with T. The long-time stability of BEFE was established in [28]. An important feature of this proof, also of other long-time results coming next, is that no form of Gronwall's inequality was used. This result can be stated as follows.

Proposition 1 (Long-time stability of BEFE, [28]). *Consider the scheme BEFE. Assume the following time-step condition is satisfied*

$$\Delta t \lesssim \min\{\nu k_{\min}^2, S_0 \nu^2 k_{\min}\}.$$

Then, the following hold:

(i) *If* $\mathbf{f}_f \in L^\infty(0, \infty; L^2(\Omega_f))$, $f_p \in L^\infty(0, \infty; L^2(\Omega_p))$, *then*

$$\|u_h^n\|_f^2 + \|\phi_h^n\|_p^2 \le C, \quad \forall n \ge 0.$$

(ii) *If* $|||\mathbf{f}_f|||_{L^2(0,\infty;L^2(\Omega_f))}$ *and* $|||f_p|||_{L^2(0,\infty;L^2(\Omega_p))}$ *are uniformly bounded in* Δt, *then*

$$\|u_h^n\|_f^2 + \|\phi_h^n\|_p^2 + \Delta t \sum_{\ell=0}^{n} \left(\|\nabla u_h^\ell\|_f^2 + \|\nabla \phi_h^\ell\|_p^2\right) \le C, \quad \forall n \ge 0.$$

BEFE is first order in time. A convergence analysis of this scheme can be found in [27], with a very recent improvement in [43].

3.2. Backward Euler–Leap Frog

Backward Euler–Leap Frog (BELF) is another IMEX scheme, first proposed in [28]. This method is a combination of the three level implicit method with the coupling terms treated by the explicit Leap-Frog method. Approximations are needed at the first two time steps to begin. The stability region of the usual Leap-Frog time discretization for $y' = \lambda y$ is exactly the interval of the imaginary axis $-1 \leq Im(\Delta t\lambda) \leq +1$. Thus, LF is unstable for every problem except for ones that are exactly skew symmetric such as the coupling herein.

The Backward Euler–Leap Frog scheme can be formulated as follows:

Algorithm 3 Backward Euler–Leap Frog (BELF)

Given $(u_h^{n-1}, p_h^{n-1}, \phi_h^{n-1}), (u_h^n, p_h^n, \phi_h^n) \in X_f^h \times Q_f^h \times X_p^h$, find $(u_h^{n+1}, p_h^{n+1}, \phi_h^{n+1}) \in X_f^h \times Q_f^h \times X_p^h$ such that for all $v_h \in X_f^h, q_h \in Q_f^h, \psi_h \in X_p^h$,

$$
\begin{aligned}
&(\frac{u_h^{n+1} - u_h^{n-1}}{2\Delta t}, v_h)_f + a_f(u_h^{n+1}, v_h) - (p_h^{n+1}, \nabla \cdot v_h)_f + c_I(v_h, \phi_h^n) = (\mathbf{f}_f^{n+1}, v_h)_f, \\
&(q_h, \nabla \cdot u_h^{n+1})_f = 0, \\
&gS_0(\frac{\phi_h^{n+1} - \phi_h^{n-1}}{2\Delta t}, \psi_h)_p + a_p(\phi_h^{n+1}, \psi_h) - c_I(u_h^n, \psi_h) = g(f_p^{n+1}, \psi_h)_p.
\end{aligned}
$$

As with any explicit scheme, BELF inherits a time-step restriction for the stability. The following long-time stability result was established in [28].

Proposition 2 (Long-time stability of BELF, [28]). *Consider the scheme BELF. Assume that the following time-step condition is satisfied*

$$\Delta t \lesssim \min\{\sqrt{\nu k_{\min}}, S_0\sqrt{\nu k_{\min}}, \nu k_{\min}^2, S_0 \nu^2 k_{\min}\};$$

then, BELF possesses the same stability properties as those for BEFE in Proposition 1. More precisely,

(i) *If $\mathbf{f}_f \in L^\infty(0,\infty; L^2(\Omega_f))$, $f_p \in L^\infty(0,\infty; L^2(\Omega_p))$, then*

$$\|u_h^n\|_f^2 + \|\phi_h^n\|_p^2 \leq C, \quad \forall n \geq 0.$$

(ii) *If $|||\mathbf{f}_f|||_{L^2(0,\infty;L^2(\Omega_f))}$, $|||f_p|||_{L^2(0,\infty;L^2(\Omega_p))}$ are uniformly bounded in Δt, then*

$$\|u_h^n\|_f^2 + \|\phi_h^n\|_p^2 + \Delta t \sum_{\ell=0}^{n} \left(\|\nabla u_h^\ell\|_f^2 + \|\nabla \phi_h^\ell\|_p^2 \right) \leq C, \quad \forall n \geq 0.$$

It was also proved that BELF achieves the optimal convergence rate uniformly in time, as shown below.

Proposition 3 (Error estimate of BELF, [28]). *Consider the scheme BELF. Assume the following time-step condition is satisfied*

$$\Delta t \lesssim \min\{\sqrt{\nu k_{\min}}, S_0\sqrt{\nu k_{\min}}, \nu k_{\min}^2, S_0 \nu^2 k_{\min}\},$$

as in Proposition 2. If the solution of the Stokes–Darcy problem (1) *is long-time regular in the sense that*

$$u \in W^{2,\infty}(0,\infty; L^2(\Omega_f)) \cap W^{1,\infty}(0,\infty; H^{k+1}(\Omega_f)),$$
$$\phi \in W^{2,\infty}(0,\infty; L^2(\Omega_p)) \cap W^{1,\infty}(0,\infty; H^{k+1}(\Omega_p)),$$
$$p \in L^{\infty}(0,\infty; H^k(\Omega_f)),$$

then the solution of BELF satisfies the uniform in time error estimates

$$\|u(t_n) - u_h^n\|_f^2 + \|\phi(t_n) - \phi_h^n\|_p^2 \le C(\Delta t^2 + h^{2k}), \ \forall n \ge 0.$$

Numerical tests illustrating the theoretical stability and convergence properties of BEFE and BELF were presented in [28]. In particular, the stability of these two methods is compared to that of the fully implicit method in the case of small $k_{\min}$ for a Stokes–Darcy flow on $\Omega_f = (0,1) \times (1,2)$ and $\Omega_p = (0,1) \times (0,1)$ with the interface $I = (0,1) \times \{1\}$. Given the source terms $\mathbf{f}_f \equiv 0, f_p \equiv 0$, the initial condition

$$\begin{aligned} u(x,y,0) &= \left(x^2(y-1)^2 + y, \ -\frac{2}{3}x(y-1)^3 + 2 - \pi\sin(\pi x)\right), \\ p(x,y,0) &= (2 - \pi\sin(\pi x))\sin\left(\frac{\pi}{2}y\right), \ \phi(x,y,0) = (2 - \pi\sin(\pi x))(1 - y - \cos(\pi y)), \end{aligned} \tag{3}$$

set all the physical parameters (except for ν and $k_{\min}$) to 1. Letting $h = \frac{1}{10}$, $\nu = \frac{1}{10}$, the evolution of the energy $E^n = \|u_h^n\|_f^2 + \|\phi_h^n\|_p^2$ with $k_{\min} = 10^{-6}$ is shown in Figure 1. Since the true solution decays as $t \to \infty$, any growth in E^n indicates instability. The plot reveals that while not unconditionally stable like the fully implicit method, BEFE and BELF only require mild constraints on Δt for their stability. Indeed, BELF is already stable for $\Delta t \simeq \frac{1}{30}$, followed by BEFE at $\Delta t \simeq \frac{1}{50}$. These conditions are much weaker than those predicted by the theory.

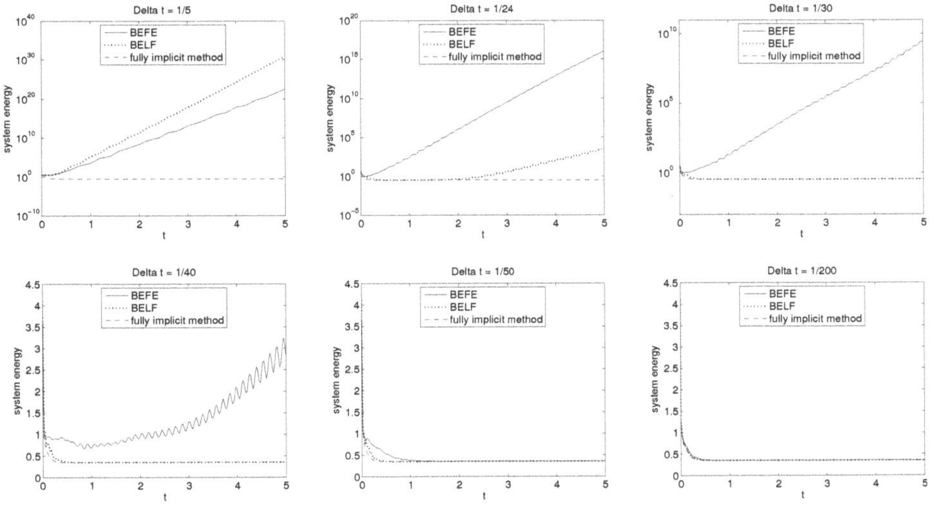

Figure 1. The evolution of system energy with $k_{\min} = 10^{-6}$ for different choices of time step [28].

In summary, the first-order IMEX methods BEFE and BELF allow a parallel, non-iterative uncoupling of the Stokes–Darcy system at each time step and, on the other hand, enjoy the desirable strong stability and convergence properties. One disadvantage, as shown in their time-step restrictions, is that they may become highly unstable when one of the parameters S_0 and $k_{\min}$ is small. In the next subsection, this constraint is relaxed with another type of first-order partitioned methods, splitting schemes.

3.3. First Order Sequential Splitting Schemes

In [34], several methods for non-iterative, sub-physics uncoupling the evolutionary Stokes–Darcy problem were proposed, using ideas from splitting methods. The estimates and tests therein suggest that these methods are stable for larger timesteps than the IMEX based partitioned methods BEFE and BELF, and in particular, a very good option when either $k_{\min}$ or S_0 (but not both) is small. Here, the Stokes and Darcy systems are uncoupled, but, unlike the aforementioned IMEX schemes, sequentially solved.

In the first Backward Euler time-split (BEsplit1) scheme, the coupling term in the ϕ equation is evaluated at the newly computed value u_h^{n+1} so we compute $\phi_h^n \to u_h^{n+1} \to \phi_h^{n+1}$.

Algorithm 4 Backward Euler time-split 1 (BEsplit1)

Given $(u_h^n, p_h^n, \phi_h^n) \in X_f^h \times Q_f^h \times X_p^h$, find $(u_h^{n+1}, p_h^{n+1}, \phi_h^{n+1}) \in X_f^h \times Q_f^h \times X_p^h$ such that for all $v_h \in X_f^h$, $q_h \in Q_f^h$, $\psi_h \in X_p^h$,

$$
\begin{aligned}
&(\frac{u_h^{n+1} - u_h^n}{\Delta t}, v_h)_f + a_f(u_h^{n+1}, v_h) - (p_h^{n+1}, \nabla \cdot v_h)_f + c_I(v_h, \phi_h^n) = (f_f^{n+1}, v_h)_f, \\
&(q_h, \nabla \cdot u_h^{n+1})_f = 0, \\
&gS_0(\frac{\phi_h^{n+1} - \phi_h^n}{\Delta t}, \psi_h)_p + a_p(\phi_h^{n+1}, \psi_h) - c_I(u_h^{n+1}, \psi_h) = g(f_p^{n+1}, \psi_h)_p.
\end{aligned}
$$

The long-time stability of this scheme can be stated as follows.

Proposition 4 (Long-time stability of BEsplit1, [34])**.** *Consider the scheme BEsplit1. Assume the following time-step condition is satisfied:*

$$\Delta t \lesssim \max\{S_0 \nu k_{\min}, S_0 h, S_0 \nu h, k_{\min}\}.$$

If $|||\mathbf{f}_f|||_{L^2(0,\infty;L^2(\Omega_f))}$ *and* $|||f_p|||_{L^2(0,\infty;L^2(\Omega_p))}$ *are uniformly bounded in* Δt*, then*

$$\|u_h^n\|_f^2 + \|\phi_h^n\|_p^2 \leq C, \ \forall n \geq 0.$$

The second Backward Euler time-split (BEsplit2) is the previous method in the opposite order, i.e., computing $u_h^n \to \phi_h^{n+1} \to u_h^{n+1}$. The analysis in [34] revealed that control was needed for a term $\|u_h^{n+1} - u_h^n\|_{\text{div}}$. This led to the insertion of the grad-div stabilization term $(\nabla \cdot (u_h^{n+1} - u_h^n)/\Delta t, \nabla \cdot v_h)$ acting on the time discretization of u_t. This term is exactly zero for the continuous problem so it does not increase the method's consistency error.

Algorithm 5 Backward Euler time-split 2 (BEsplit2)

Given $(u_h^n, p_h^n, \phi_h^n) \in X_f^h \times Q_f^h \times X_p^h$, find $(u_h^{n+1}, p_h^{n+1}, \phi_h^{n+1}) \in X_f^h \times Q_f^h \times X_p^h$ such that for all $v_h \in X_f^h$, $q_h \in Q_f^h$, $\psi_h \in X_p^h$,

$$gS_0(\frac{\phi_h^{n+1} - \phi_h^n}{\Delta t}, \psi_h)_p + a_p(\phi_h^{n+1}, \psi_h) - c_I(u_h^n, \psi_h) = g(f_p^{n+1}, \psi_h)_p,$$

$$(\frac{u_h^{n+1} - u_h^n}{\Delta t}, v_h)_f + \left(\nabla \cdot \frac{u_h^{n+1} - u_h^n}{\Delta t}, \nabla \cdot v_h\right)_f + a_f(u_h^{n+1}, v_h) - (p_h^{n+1}, \nabla \cdot v_h)_f + c_I(v_h, \phi_h^{n+1}) = (f_f^{n+1}, v_h)_f,$$

$$(q_h, \nabla \cdot u_h^{n+1})_f = 0.$$

The stability result of BEsplit2 is presented below.

Proposition 5 (Long-time stability of BEsplit2, [34]). *Consider the scheme BEsplit2. Assume the following time-step condition is satisfied*

$$\Delta t \lesssim \max\{S_0 \nu k_{\min}, S_0 h, k_{\min} h, k_{\min}\}.$$

If $|||\mathbf{f}_f|||_{L^2(0,\infty;L^2(\Omega_f))}$ *and* $|||f_p|||_{L^2(0,\infty;L^2(\Omega_p))}$ *are uniformly bounded in* Δt*, then*

$$\|u_h^n\|_f^2 + \|\phi_h^n\|_p^2 + \Delta t \sum_{\ell=0}^{n} \left(\|\nabla u_h^\ell\|_f^2 + \|\nabla \phi_h^\ell\|_p^2\right) \leq C, \quad \forall n \geq 0.$$

Propositions 4 and 5 impose two slightly different conditions on Δt, both of which are mild when *one* of S_0 and $k_{\min}$ is small. In those cases, BEsplit1 and BEsplit2 are preferable choices to first-order IMEX schemes, with the small price of solving the uncoupled subproblems sequentially, instead of in parallel. However, it is worth remarking that these methods may become unstable if *both* parameters are small. Finally, BEsplit1 and BEsplit2 can be shown to be optimally convergent under the same time-step conditions for stability. For a thorough analysis, we refer to [44].

Two other splitting methods were proposed in [34], whose details are omitted here for brevity. SDsplit is a first order scheme, long-time stable under the condition $\Delta t \lesssim \min\{S_0, k_{min}\}\, h$, and thus seems less favorable than BEsplit1 and BEsplit2 in theory. CNsplit, on the other hand, is stable with $\Delta t \lesssim \sqrt{S_0} h$ and a very good option in case of small $k_{\min}$. This scheme is second order.

Numerical tests checking and comparing the largest time step for which the four methods are stable over long-time intervals were also performed in [34]. Taking the initial condition as in (3), the body sources to be 0, the system parameters (except $k_{\min}$ and S_0) to be 1.0 and mesh size $h = \frac{1}{10}$, the authors computed the system energy E^N at final time $T = 10$ with different time-step sizes. Since the true solution decays as $t \to \infty$, large E^N indicates instability. The performance of presented splitting methods was plotted for three cases: (i) $O(1)$ $k_{\min}$ and small S_0, (ii) small $k_{\min}$ and $O(1)$ S_0, and (iii) small $k_{\min}$ and small S_0 (see Figures 2 and 3). These plots show that for small parameter $k_{\min}$ **or** S_0, BEsplit1 and BEsplit2 are stable for large time steps. The performance of SDsplit is close to those of BEsplit1 and BEsplit2, suggesting that its theoretical condition was not optimized. These three first order splitting methods display superior stability to IMEX methods in our previous tests. The second order CNsplit in general requires a much smaller time step, but still possesses strong stability in the case of small $k_{\min}$ and large S_0.

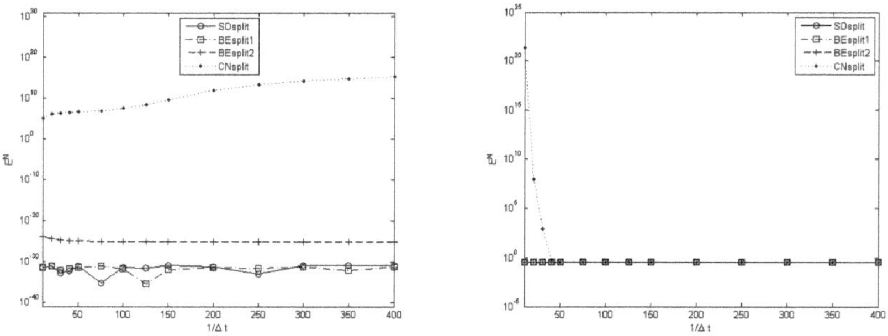

(**a**) (**b**)

Figure 2. A comparison of E^N computed using different choices of time step for different splitting methods with (**a**) $O(1)$ $k_{\min}$ and 'extremely small' S_0, i.e., $k_{\min} = 1$ and $S_0 = 10^{-12}$, and (**b**) with 'extremely small' $k_{\min}$ and $O(1)$ S_0, i.e., $k_{\min} = 10^{-12}$ and $S_0 = 1$, [34].

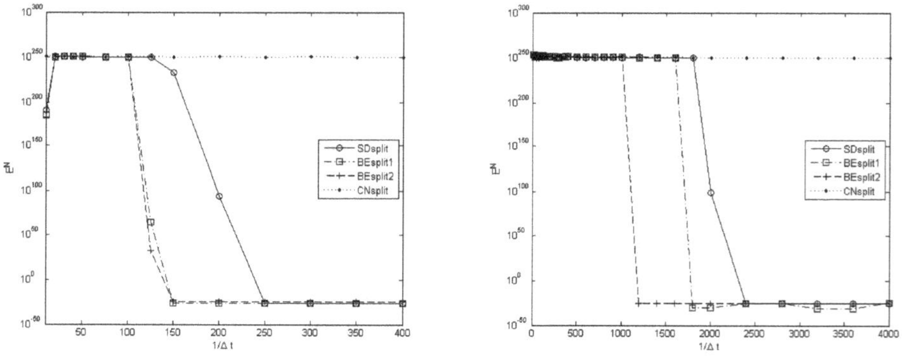

(**a**) (**b**)

Figure 3. A comparison of E^N computed using different choices of time step for different splitting methods with 'moderately small' $k_{\min}$ and 'moderately small' S_0. (**a**): $k_{\min} = 10^{-3}$ and $S_0 = 10^{-3}$. (**b**): $k_{\min} = 10^{-4}$ and $S_0 = 10^{-4}$, [34].

4. High Order Partitioned Schemes

In this section, we review recent developments in high-order partitioned schemes for uncoupling the Stokes–Darcy system. Thus far, most of the proposed methods are of IMEX type (except for CNsplit, which was mentioned above). The IMEX schemes we will discuss in detail here include Crank–Nicolson–Leap Frog [29,32,33], second-order backward-differentiation with Gear's extrapolation [30], and Adam-Moulton-Bashforth [30,31].

4.1. Crank–Nicolson–Leap Frog

The Crank–Nicolson–Leap Frog (CNLF) method is a second-order scheme that employs the implicit Crank–Nicolson discretization of subdomain terms and treats the interface terms explicitly with Leap Frog. CNLF was developed for uncoupling systems of evolutionary equations in [45–47]. This method was first applied to uncouple Stokes–Darcy system in [29].

CNLF is a three-level method. The first terms, (u_h^0, p_h^0, ϕ_h^0), arise from the initial conditions of the problem. To obtain (u_h^1, p_h^1, ϕ_h^1), one must use another numerical method. Note that approximations in this first step will affect the overall convergence rate of the method, and as usual, we assume them to be sufficiently accurate. This scheme can be stated as follows:

Algorithm 6 Crank–Nicolson–Leap Frog (CNLF)

Given $(u_h^{n-1}, p_h^{n-1}, \phi_h^{n-1})$, $(u_h^n, p_h^n, \phi_h^n) \in X_f^h \times Q_f^h \times X_p^h$, find $(u_h^{n+1}, p_h^{n+1}, \phi_h^{n+1}) \in X_f^h \times Q_f^h \times X_p^h$ such that for all $v_h \in X_f^h, q_h \in Q_f^h, \psi_h \in X_p^h$,

$$\left(\frac{u_h^{n+1} - u_h^{n-1}}{2\Delta t}, v_h\right)_f + a_f\left(\frac{u_h^{n+1} + u_h^{n-1}}{2}, v_h\right) - \left(\frac{p_h^{n+1} + p_h^{k-1}}{2}, \nabla \cdot v_h\right)_f + c_I(v_h, \phi_h^n) = (\mathbf{f}_f^n, v_h)_f,$$
$$\left(q_h, \nabla \cdot \left(\frac{u_h^{n+1} + u_h^{n-1}}{2}\right)\right)_f = 0,$$
$$gS_0\left(\frac{\phi_h^{n+1} - \phi_h^{n-1}}{2\Delta t}, \psi_h\right)_p + a_p\left(\frac{\phi_h^{n+1} + \phi_h^{n-1}}{2}, \psi_h\right) - c_I(u_h^n, \psi_h) = g(f_p^n, \psi_h)_p.$$

CNLF provably possesses a strong stability and convergence properties, as shown in [29] and presented below. Specifically, the time-step condition for CNLF does not depend on $k_{\min}$, making this method a very good choice for fluid-porous media systems with small $k_{\min}$.

Proposition 6 (Long-time stability of CNLF, [29])**.** *Consider the scheme CNLF. Assume the following time-step condition is satisfied:*

$$\Delta t \lesssim \max\{\min\{h^2, S_0\}, \min\{h, S_0 h\}\}.$$

If $|||\mathbf{f}_f|||_{L^2(0,\infty;L^2(\Omega_f))}$ and $|||f_p|||_{L^2(0,\infty;L^2(\Omega_p))}$ are uniformly bounded in Δt, then

$$\|u_h^n\|_f^2 + \|\phi_h^n\|_p^2 \leq C, \ \forall n \geq 0.$$

Proposition 7 (Error estimate of CNLF, [29])**.** *Consider the scheme CNLF. Assume the following time-step condition is satisfied:*

$$\Delta t \lesssim \max\{\min\{h^2, S_0\}, \min\{h, S_0 h\}\},$$

as in Proposition 6. If the solution of the Stokes–Darcy problem (1) *is regular in the sense that*

$$\begin{aligned}
&u \in H^1(0,\infty; H^{k+1}(\Omega_f)) \cap L^\infty(0,\infty; H^{k+1}(\Omega_f)) \cap H^3(0,\infty; H^1(\Omega_f)),\\
&\phi \in H^1(0,\infty; H^{k+1}(\Omega_p)) \cap L^\infty(0,\infty; H^{k+1}(\Omega_p)) \cap H^3(0,\infty; H^1(\Omega_p)),\\
&p \in L^2(0,\infty; H^k(\Omega_f)),
\end{aligned}$$

and the discrete norms of u, p and ϕ in $L^2(0,\infty; H^{k+1}(\Omega_f))$, $L^2(0,\infty; H^{s+1}(\Omega_f))$, and $L^2(0,\infty; H^{k+1}(\Omega_p))$ are uniformly bounded in Δt, then

$$\|u(t_n) - u_h^n\|_f^2 + \|\phi(t_n) - \phi_h^n\|_p^2 \leq C(\Delta t^4 + h^{2k}), \ \forall n \geq 0.$$

While independent of $k_{\min}$, the conditional stability of CNLF may still be restrictive when faced with small S_0. To tackle this difficulty, the authors of [32,33] proposed a strategy to improve the stability property of CNLF by adding appropriate stabilization terms to both the Stokes as well as the groundwater flow equation.

Algorithm 7 Stabilized Crank–Nicolson–Leap Frog (CNLFstab)

Given $(u_h^{n-1}, p_h^{n-1}, \phi_h^{n-1})$, $(u_h^n, p_h^n, \phi_h^n) \in X_f^h \times Q_f^h \times X_p^h$, find $(u_h^{n+1}, p_h^{n+1}, \phi_h^{n+1}) \in X_f^h \times Q_f^h \times X_p^h$ such that for all $v_h \in X_f^h, q_h \in Q_f^h, \psi_h \in X_p^h$,

$$\left(\frac{u_h^{n+1} - u_h^{n-1}}{2\Delta t}, v_h\right)_f + \left(\nabla \cdot \left(\frac{u_h^{n+1} - u_h^{n-1}}{2\Delta t}\right), \nabla \cdot v_h\right)_f + a_f\left(\frac{u_h^{n+1} + u_h^{n-1}}{2}, v_h\right)$$
$$- \left(\frac{p_h^{n+1} + p_h^{k-1}}{2}, \nabla \cdot v_h\right)_f + c_I(v_h, \phi_h^n) = (\mathbf{f}_f^n, v_h)_f,$$

$$\left(q_h, \nabla \cdot u_h^{n+1}\right)_f = 0,$$

$$gS_0\left(\frac{\phi_h^{n+1} - \phi_h^{n-1}}{2\Delta t}, \psi_h\right)_p + a_p\left(\frac{\phi_h^{n+1} + \phi_h^{n-1}}{2}, \psi_h\right) - c_I(u_h^n, \psi_h)$$
$$+ \Delta t g^2 C_{f,p}^2 \left\{(\phi_h^{k+1} - \phi_h^{k-1}, \psi_h)_p + (\nabla(\phi_h^{k+1} - \phi_h^{k-1}), \nabla\psi_h)_p\right\} = g(f_p^n, \psi_h)_p.$$

The resulting numerical scheme, denoted CNLFstab, is unconditionally, uniformly in time stable, as well as second-order convergent. More specifically, it was shown in [33] that the added stabilization terms,

$$\left(\nabla \cdot \left(\frac{u_h^{n+1} - u_h^{n-1}}{2\Delta t}\right), \nabla \cdot v_h\right)_f,$$
$$\text{and} \quad \Delta t g^2 C_{f,p}^2 \left\{(\phi_h^{k+1} - \phi_h^{k-1}, \psi_h)_p + (\nabla(\phi_h^{k+1} - \phi_h^{k-1}), \nabla\psi_h)_p\right\},$$

eliminate the time-step restriction without affecting the second-order accuracy of the method. Indeed, among all the methods we have discussed so far, CNLFstab exhibits the best stability and convergence properties. Here, $C_{f,p}$ is a constant satisfying

$$|c_I(u, \phi)| \le g C_{f,p} \|u\|_{\text{div},f} \|\phi\|_{1,p}.$$

In the special case of a flat interface I, with Ω_f and Ω_p being arbitrary domains, $C_{f,p}$ equals 1 (see [48, Lemmas 3.1 and 3.2]).

We make the above discussion rigorous with the following results.

Proposition 8 (Long-time, unconditional stability of CNLFstab, [33]). *Consider the scheme CNLFstab. If* $|||\mathbf{f}_f|||_{L^2(0,\infty;L^2(\Omega_f))}$ *and* $|||f_p|||_{L^2(0,\infty;L^2(\Omega_p))}$ *are uniformly bounded in* Δt, *then*

$$\|u_h^n\|_{\text{div},f}^2 + \|\phi_h^n\|_p^2 \le C, \quad \forall n \ge 0.$$

Proposition 9 (Error estimate of CNLFstab, [33]). *Consider the scheme CNLFstab. Assume u, p and ϕ satisfy the same regularity condition as in Proposition 7, then the solution of CNLFstab satisfies the unconditional, uniform in time error estimate*

$$\|u(t_n) - u_h^n\|_{\text{div},f}^2 + \|\phi(t_n) - \phi_h^n\|_p^2 \le C(\Delta t^4 + h^{2k}), \quad \forall n \ge 0.$$

The stability and convergence properties of CNLF and CNLFstab were illustrated and compared via numerical tests in [33,44]. Set the body sources to be 0, the system parameters (except $k_{\min}$

and S_0) to be 1.0 and mesh size $h = \frac{1}{10}$, the authors computed the system energy E^N at final time $T = 10$ with different time-step sizes. Since the true solution decays as $t \to \infty$, large E^N indicates instability. The performance of CNLF and CNLFstab was plotted for four cases: (i) small $k_{\min}$ and small S_0, (ii) small S_0, (iii) small $k_{\min}$, and (iv) $k_{\min} = S_0 = 1.0$. These plots show that, in all situations, CNLFstab is stable for large time steps, regardless of the size of $k_{\min}$ and S_0. This is a vast improvement over CNLF, which, for $\Delta t \leq 1/80$, was only stable when $k_{\min} = S_0 = 1.0$, and unstable in all other cases (see Figure 4).

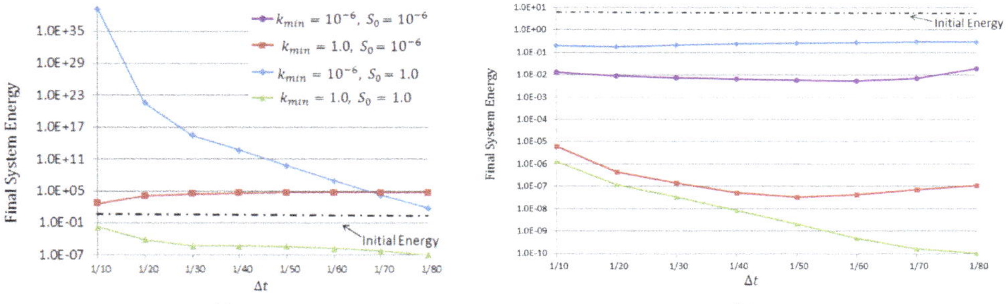

Figure 4. Final system energy E^N versus time-step size for (**a**) CNLF, and (**b**) CNLFstab, [44].

4.2. Second-Order Backward-Differentiation Formula with Gear's Extrapolation

In [30], the authors introduced a second-order scheme, which discretizes in time via a second-order backward-differentiation formula (BDF2), and treats the interface term via a second-order explicit Gear's extrapolation formula.

Algorithm 8 Second-order backward-differentiation (BDF2)

Given $(u_h^{n-1}, p_h^{n-1}, \phi_h^{n-1})$, $(u_h^n, p_h^n, \phi_h^n) \in X_f^h \times Q_f^h \times X_p^h$, find $(u_h^{n+1}, p_h^{n+1}, \phi_h^{n+1}) \in X_f^h \times Q_f^h \times X_p^h$ such that for all $v_h \in X_f^h, q_h \in Q_f^h, \psi_h \in X_p^h$,

$$\left(\frac{3u_h^{n+1} - 4u_h^n + u_h^{n-1}}{2\Delta t}, v_h\right)_f + a_f\left(u_h^{n+1}, v_h\right) - \left(p_h^{n+1}, \nabla \cdot v_h\right)_f + c_I(v_h, 2\phi_h^n - \phi_h^{n-1}) = (\mathbf{f}_f^{n+1}, v_h)_f,$$

$$\left(q_h, \nabla \cdot u_h^{n+1}\right)_f = 0,$$

$$gS_0\left(\frac{3\phi_h^{n+1} - 4\phi_h^n + \phi_h^{n-1}}{2\Delta t}, \psi_h\right)_p + a_p\left(\phi_h^{n+1}, \psi_h\right) - c_I(2u_h^n - u_h^{n-1}, \psi_h)$$
$$+ \gamma_p(\phi_h^{n+1} - 2\phi_h^n + \phi_h^{n-1}, \psi_h)_I = g(f_p^{n+1}, \psi_h)_p.$$

BDF2 provably possesses a strong stability and convergence properties, as stated below. Specifically, an inspection of the analysis argument in [30] shows that the time-step condition for BDF2 does not depend on S_0, making this method a very good choice for fluid-porous media systems with small S_0 (and moderate $k_{\min}$).

Proposition 10 (Long-time stability of BDF2, [30]). *Consider the scheme BDF2. Assume the following time-step restriction is satisfied:*

$$\Delta t \lesssim \min\{\nu^3, k_{\min}^3\}.$$

If $\mathbf{f}_f \in L^\infty(0,\infty; L^2(\Omega_f))$, $f_p \in L^\infty(0,\infty; L^2(\Omega_p))$, *then*

$$\|u_h^n\|_f^2 + \|\phi_h^n\|_p^2 \le C, \ \forall n \ge 0.$$

We note that the solution of BDF2 was also proved to be long-time stable in H^1 norm, i.e., $\|\nabla u_h^n\|_f^2 + \|\nabla \phi_h^n\|_p^2 \le C$, $\forall n \ge 0$, in [30], with a more restrictive time-step condition. A strong convergence property of BDF2 is stated below.

Proposition 11 (Error estimates of BDF2, [30]). *Consider the scheme BDF2. Assume Δt is sufficiently small (independent of mesh size and final time). If the solution of the Stokes–Darcy problem* (1) *is long-time regular in the sense that*

$$u \in W^{3,\infty}(0,\infty; H^1(\Omega_f)) \cap W^{2,\infty}(0,\infty; H^{k+1}(\Omega_f)),$$
$$\phi \in W^{3,\infty}(0,\infty; H^1(\Omega_p)) \cap W^{2,\infty}(0,\infty; H^{k+1}(\Omega_p)),$$

then the solution of BDF2 satisfies the uniform in time error estimates

$$\|u(t_n) - u_h^n\|_f^2 + \|\phi(t_n) - \phi_h^n\|_p^2 \le C(\Delta t^4 + h^{2(k+1)}), \qquad \forall n \ge 0,$$
$$\|\nabla(u(t_n) - u_h^n)\|_f^2 + \|\nabla(\phi(t_n) - \phi_h^n)\|_p^2 + \|p(t_n) - p_h^n\|_f^2 \le C(\Delta t^2 + h^{2k}), \qquad \forall n \ge 0.$$

This result showed not only the error estimate of the velocity with respect to the L^2 norm, but also with respect to the H^1 norm, as well as the error estimate of the pressure. The two latter estimates are not second order in time; however, these were also observed in the numerical experiments [30]. Finally, the authors suggested that the stabilization terms

$$\gamma_f((u_h^{n+1} - 2u_h^n + u_h^{n-1}) \cdot \widehat{n}_f, v_h \cdot \widehat{n}_f)_I \quad \text{and} \quad \gamma_p(\phi_h^{n+1} - 2\phi_h^n + \phi_h^{n-1}, \psi_h)_I$$

may be added to the Stokes and Darcy solves correspondingly, with parameters $\gamma_f, \gamma_p \ge 0$. While the analysis does not take advantage of the stabilization term, the numerical experiments demonstrate the benefit of this strategy in the sense that the presence of the stabilization term relaxes the time-step restriction.

4.3. Adam–Moulton–Bashforth

The second-order Adam–Moulton–Bashforth method (AMB2), studied in [30], combines the second-order implicit Adams–Moulton treatment of the symmetric terms and the second-order explicit Adams–Bashforth treatment of the interface term.

Algorithm 9 Second-order Adam–Moulton–Bashforth method (AMB2)

Given $(u_h^{n-1}, p_h^{n-1}, \phi_h^{n-1})$, $(u_h^n, p_h^n, \phi_h^n) \in X_f^h \times Q_f^h \times X_p^h$, find $(u_h^{n+1}, p_h^{n+1}, \phi_h^{n+1}) \in X_f^h \times Q_f^h \times X_p^h$ such that for all $v_h \in X_f^h, q_h \in Q_f^h, \psi_h \in X_p^h$,

$$\left(\frac{u_h^{n+1}-u_h^n}{\Delta t}, v_h\right)_f + a_f\left(\alpha u_h^{n+1} + (\frac{3}{2}-2\alpha)u_h^n + (\alpha-\frac{1}{2})u_h^{n-1}, v_h\right)$$
$$-\left(\alpha p_h^{n+1} + (\frac{3}{2}-2\alpha)p_h^n + (\alpha-\frac{1}{2})p_h^{n-1}, \nabla\cdot v_h\right)_f + c_I\left(v_h, \frac{3}{2}\phi_h^n - \frac{1}{2}\phi_h^{n-1}\right) = (\mathbf{f}_f^{n+1/2}, v_h)_f,$$
$$\left(q_h, \nabla\cdot\left(\alpha u_h^{n+1} + (\frac{3}{2}-2\alpha)u_h^n + (\alpha-\frac{1}{2})u_h^{n-1}\right)\right)_f = 0,$$
$$gS_0\left(\frac{\phi_h^{n+1}-\phi_h^n}{\Delta t}, \psi_h\right)_p + a_p\left(\alpha\phi_h^{n+1} + (\frac{3}{2}-2\alpha)\phi_h^n + (\alpha-\frac{1}{2})\phi_h^{n-1}, \psi_h\right)$$
$$-c_I\left(\frac{3}{2}u_h^n - \frac{1}{2}u_h^{n-1}, \psi_h\right) = g(f_p^{n+1/2}, \psi_h)_p.$$

The third-order Adams–Moulton–Bashforth method (AMB3), studied in [31], is a combination of the third-order explicit Adams–Bashforth treatment for the coupling term and the third-order Adams–Moulton method for the remaining terms. To the best of our knowledge, this has been the only third order IMEX scheme that was applied to uncouple the Stokes–Darcy equation so far.

Algorithm 10 Third-order Adam–Moulton–Bashforth method (AMB3)

Given $(u_h^{n-3}, p_h^{n-3}, \phi_h^{n-3}), \ldots, (u_h^n, p_h^n, \phi_h^n) \in X_f^h \times Q_f^h \times X_p^h$, find $(u_h^{n+1}, p_h^{n+1}, \phi_h^{n+1}) \in X_f^h \times Q_f^h \times X_p^h$ satisfying for all $v_h \in X_f^h, q_h \in Q_f^h, \psi_h \in X_p^h$:

$$\left(\frac{u_h^{n+1}-u_h^n}{\Delta t}, v_h\right)_f + a_f\left(\frac{2}{3}u_h^{n+1} + \frac{5}{12}u_h^{n-1} - \frac{1}{12}u_h^{n-3}, v_h\right) - \left(\frac{2}{3}p_h^{n+1} + \frac{5}{12}p_h^{n-1} - \frac{1}{12}p_h^{n-3}, \nabla\cdot v_h\right)_f$$
$$+ c_I\left(v_h, \frac{23}{12}\phi_h^n - \frac{4}{3}\phi_h^{n-1} + \frac{5}{12}\phi_h^{n-2}\right) = \left(\frac{2}{3}\mathbf{f}_f^{n+1} + \frac{5}{12}\mathbf{f}_f^{n-1} - \frac{1}{12}\mathbf{f}_f^{n-3}, v_h\right)_f,$$
$$\left(q_h, \nabla\cdot\left(\frac{2}{3}u_h^{n+1} + \frac{5}{12}u_h^{n-1} - \frac{1}{12}u_h^{n-3}\right)\right)_f = 0,$$
$$gS_0\left(\frac{\phi_h^{n+1}-\phi_h^n}{\Delta t}, \psi_h\right)_p + a_p\left(\frac{2}{3}\phi_h^{n+1} + \frac{5}{12}\phi_h^{n-1} - \frac{1}{12}\phi_h^{n-3}, \psi_h\right)$$
$$-c_I\left(\frac{23}{12}u_h^n - \frac{4}{3}u_h^{n-1} + \frac{5}{12}u_h^{n-2}, \psi_h\right) = g\left(\frac{2}{3}f_p^{n+1} + \frac{5}{12}f_p^{n-1} - \frac{1}{12}f_p^{n-3}, \psi_h\right)_p.$$

These methods proved to be long-time stable under small time-step restrictions. The explicit dependence of these conditions on the system parameters can be elaborated from [30,31] as follows.

Proposition 12 (Long-time stability of AMB2 and AMB3, [30,31]). *Consider the schemes AMB2 and AMB3. Assume the time-step restrictions*

$$\Delta t \lesssim \min\{\nu^3, k_{\min}^3\} \text{ (in case of AMB2), and } \Delta t \lesssim \min\{\nu, k_{\min}\} \text{ (in case of AMB3).}$$

If $\mathbf{f}_f \in L^\infty(0,\infty;L^2(\Omega_f)), f_p \in L^\infty(0,\infty;L^2(\Omega_p))$, *then*

$$\|u_h^n\|_f^2 + \|\phi_h^n\|_p^2 \le C, \ \forall n \ge 0.$$

Similar to CNLF and BDF2, appropriate stabilization terms can be added for both algorithms AMB2 and AMB3, which were numerically shown to significantly relax the time-step constraint (see [30,31]).

5. Asynchronous Schemes

In surface water–groundwater models, the flow in fluid regions is often associated with higher velocities, compared to flow in porous media region. In such cases, it may be desirable to apply an asynchronous scheme (aka, multiple-time-step scheme, multi-rate scheme), which computes fast solutions using a small time step and consider a larger time step for slow solutions. The first partitioned scheme that allows different time steps in the fluid and porous region for the nonstationary Stokes–Darcy problem was probably proposed and analyzed in [35]. In that work, the decoupling is based on lagging the interfacial coupling terms following the BEFE method; thus, we will refer to the scheme as asynchronous BEFE or BEFE-as1. Let Δs be the (small) time-step size in the fluid region Ω_f and Δt be the (large) time-step size in the porous region Ω_p such that $\Delta t = r\Delta s$. In addition, define $n_k := kr$ and let $N = T/\Delta s$, the number of small time step, and $M = T/\Delta t = N/r$, the number of large time step. The algorithm in [35] reads as follows.

Algorithm 11 Asynchronous Backward Euler-Forward Euler 1 (BEFE-as1)

For $k = 0$ to $\frac{T}{\Delta t} - 1$, do the following:

1. Find $(u_h^{n+1}, p_h^{n+1}) \in X_f^h \times Q_f^h$ with $n = n_k, \dots, n_{k+1} - 1$ satisfying for all $v_h \in X_f^h, q_h \in Q_f^h$:

$$(\frac{u_h^{n+1} - u_h^n}{\Delta s}, v_h)_f + a_f(u_h^{n+1}, v_h) - (p_h^{n+1}, \nabla \cdot v_h)_f + c_I(v_h, \phi_h^{n_k}) = (\mathbf{f}_f^{n+1}, v_h)_f,$$
$$(q_h, \nabla \cdot u_h^{n+1})_f = 0.$$

2. Set $U^{n_k} := \frac{1}{r}\sum_{i=n_k}^{n_{k+1}-1} u_h^i$.
3. Find $\phi_{n_{k+1}} \in X_p^h$ such that for all $\psi_h \in X_p^h$:

$$gS_0(\frac{\phi_h^{n_{k+1}} - \phi_h^{n_k}}{\Delta t}, \psi_h)_p + a_p(\phi_h^{n_{k+1}}, \psi_h) - c_I(U^{n_k}, \psi_h) = g(f_p^{n_{k+1}}, \psi_h)_p.$$

4. Set $k := k + 1$.

The stability of BEFE-as1 was proved in [35] over a bounded interval. In particular, one has the following:

Proposition 13 (Stability of BEFE-as1). *Consider the scheme BEFE-as1. Let $T > 0$ be any fixed time. Assume the following condition on the small time step*

$$\Delta s \lesssim \sqrt{S_0 \nu k_{\min}}.$$

If $|||\mathbf{f}_f|||_{L^2(0,T;L^2(\Omega_f))}$ *and* $|||f_p|||_{L^2(0,T;L^2(\Omega_p))}$ *are uniformly bounded in* Δt*, then, for all* $1 \le k \le M - 1$, $0 \le i \le r$*, there holds*

$$\|u_h^{n_k+i}\|_f^2 + \|\phi_h^{n_k}\|_p^2 + \Delta s \sum_{\ell=0}^{n_k+i} \|\nabla u_h^\ell\|_f + \Delta t \sum_{j=0}^{k} \|\nabla \phi_h^{n_j}\|_p \le Ce^{cT}$$

The study on multi-rate schemes continues with [36], in which the second asynchronous strategy based on BEFE was proposed. This method in computing the hydraulic head ϕ, instead of using free flow velocity averaged over multiple previous steps as in BEFE-as1, only uses the free flow velocity value at the immediately previous time level. As such, the long-time stability was acquired, with the time-step restriction depending not only on the model parameters, but also including the ratio between the time steps applied in the free flow and porous medium domains. A remarkable property of this method is that it conserves mass across the interface, which does not seem possible with BEFE-as1. To be precise, we state the method as follows.

Algorithm 12 Asynchronous Backward Euler-Forward Euler 2 (BEFE-as2)

For $k = 0$ to $\frac{T}{\Delta t} - 1$, follow the same procedure as in BEFE-as1, except for Step 2, where it is replaced by:

$$2. \text{ set } U^{n_k} := u_h^{n_{k+1}}.$$

The stability results of BEFE-as2 can be established as follows.

Proposition 14 (Long-time stability of BEFE-as2, [36])**.** *Consider the scheme BEFE-as2. Assume following condition on the time-step condition is satisfied:*

$$\Delta s \lesssim \min\left\{\frac{k_{\min}}{\nu(r-1)^2}, \frac{S_0 \nu k_{\min}}{r}\right\}.$$

If $|||\mathbf{f}_f|||_{L^2(0,\infty;L^2(\Omega_f))}$ *and* $|||f_p|||_{L^2(0,\infty;L^2(\Omega_p))}$ *are uniformly bounded in* Δt*, then*

$$\|u_h^{n_k}\|_f^2 + \|\phi_h^{n_k}\|_p^2 \leq C, \quad \forall k \geq 0.$$

For the error analysis and numerical experiments illustrating the convergence rate and mass conservation, we refer to [36].

6. Conclusions

In solving the coupled Stokes–Darcy equations, the non-iterative partitioned approach is an attractive alternative to fully implicit, monolithic discretization (combining with either direct coupled problem solve or iterative domain decomposition methods). First, these uncoupling schemes allow the use of legacy sub-problems' codes, in which the spatial mesh, time step and numerical method may be optimized according to each subprocess. Second, this approach, by exploiting the interface information obtained in previous time steps, only needs a single Stokes solve and a single Darcy solve per time level (some splitting methods may require a couple solves), and is therefore very cost effective. Since the coupling terms are treated in an explicit manner, obtaining optimal accuracy and good stability properties is a major concern with these methods. In recent years, many proposed partitioned schemes have surpassed this challenge, with proven long-time stability and optimal convergence properties. Further improvements in efficiency include high-order discretizations, stabilization strategies, and asynchronous schemes. Table 3 summarizes and compares the stability and convergence properties for all numerical schemes surveyed herein. There are, however, several important questions that remain open, in our point of view.

1. For the long-time stability, most of the current methods require time-step conditions sensitive to the sizes of system parameters. These conditions may become restrictive for the fluid-porous media coupling with small parameters, particularly S_0 and k_{min}. See Table 3 for the stability dependence on problem parameters of all methods discussed here. While there are a few methods achieving some successes in this case, e.g., CNLFstab, in our opinion, long-time stable

and accurate schemes in case of small parameters, in particular, when both S_0 and $k_{\min}$ are small, are worth further study.

2. Most of the existing methods have not accounted for the dependence of the time step on the domain size. This is an important problem, especially for domains with large aspect ratios.
3. To our knowledge, there are no adaptations of the asynchronous approach beyond first-order schemes. High order asynchronous methods are desirable and the next logical step.
4. The primary motivation for modeling the fully evolutionary Stokes–Darcy flow is transport contaminant tracking, a major concern in several modern environmental problems. However, the problem of coupling numerical methods for the time-dependent Stokes–Darcy equation, in particular non-iterative partitioned methods discussed herein, with the transport equation to simulate the path of chemicals is largely open.

Table 3. A compilation of the surveyed partitioned methods, grouped by temporal convergence, with stability restrictions presented to highlight potential sensitivity to key problem parameters.

	First-Order in Time	
Method	**Type**	**Stability Condition**
BEFE	parallel	$\Delta t \lesssim \min\{\nu k_{\min}^2, S_0\nu^2 k_{\min}\}$
BELF	parallel	$\Delta t \lesssim \min\{\sqrt{\nu k_{\min}}, S_0\sqrt{\nu k_{\min}}, \nu k_{\min}^2, S_0\nu^2 k_{\min}\}$
BEsplit1	sequential	$\Delta t \lesssim \max\{S_0\nu k_{\min}, S_0 h, S_0\nu h, k_{\min}\}$
BEsplit2	sequential	$\Delta t \lesssim \max\{S_0\nu k_{\min}, S_0 h, k_{\min} h, k_{\min}\}$
SDsplit	sequential	$\Delta t \lesssim \min\{S_0\nu, k_{\min}\} h$
BEFE-as1	asynchronous	$\Delta s \lesssim \sqrt{S_0\nu k_{\min}}, \Delta t = r\Delta s$
BEFE-as2	asynchronous	$\Delta s \lesssim \min\left\{\frac{k_{\min}}{\nu(r-1)^2}, \frac{S_0\nu k_{\min}}{r}\right\}, \Delta t = r\Delta s$
	Second-Order in Time	
Method	**Type**	**Stability Condition**
CNLF	parallel	$\Delta t \lesssim \max\{\min\{h^2, S_0\}, \min\{h, S_0 h\}\}$
CNLFstab	parallel	*none*
CNsplit	sequential	$\Delta t \lesssim \sqrt{S_0} h$
BDF2	parallel	$\Delta t \lesssim \min\{\nu^3, k_{\min}^3\}$
AMB2	parallel	$\Delta t \lesssim \min\{\nu^3, k_{\min}^3\}$
	Third-Order in Time	
Method	**Type**	**Stability Condition**
AMB3	parallel	$\Delta t \lesssim \min\{\nu, k_{\min}\}$

Acknowledgments: The second author acknowledges support by the U.S. Defense Advanced Research Projects Agency, Defense Sciences Office under Grant HR0011619523.

Author Contributions: MK and HT did an equal amount of work on this manuscript, including the analysis, experiments, and writeup.

Conflicts of Interest: The authors declare no conflict of interest.

References

1. Pinder, G.; Celia, M. *Subsurface Hydrology*; John Wiley and Sons: Hoboken, NJ, USA, 2006.
2. Watson, I.; Burnett, A. *Hydrology: An Environmental Approach*; CRC Press: Boca Raton, FL, USA, 1995.
3. Bear, J. *Dynamics of Fluids in Porous Media*; Courier Corporation: North Chelmsford, MA, USA, 1988.
4. Kundu, P.; Cohen, I.; Dowling, D. *Fluid Mechanics*, 5th ed.; Academic Press: Cambridge, MA, USA, 2012.
5. Beavers, G.; Joseph, D. Boundary conditions at a naturally permeable wall. *J. Fluid Mech.* **1967**, *30*, 197–207.
6. Saffman, P. On the boundary condition at the interface of a porous medium. *Stud. Appl. Math.* **1971**, *1*, 93–101.

7. Payne, L.E.; Song, J.C.; Straughan, B. Continuous dependence and convergence results for Brinkman and Forchheimer models with variable viscosity. *Proc. Math. Phys. Eng. Sci.* **1999**, *455*, 2173–2190.
8. Payne, L.E.; Straughan, B. Analysis of the boundary condition at the interface between a viscous fluid and a porous medium and related modelling questions. *J. Math. Pures Appl.* **1998**, *77*, 317–354.
9. Discacciati, M.; Miglio, E.; Quarteroni, A. Mathematical and numerical models for coupling surface and groundwater flows. *Appl. Numer. Math.* **2002**, *43*, 57–74. 19th Dundee Biennial Conference on Numerical Analysis (2001).
10. Layton, W.J.; Schieweck, F.; Yotov, I. Coupling fluid flow with porous media flow. *SIAM J. Numer. Anal.* **2002**, *40*, 2195–2218.
11. Rivière, B. Analysis of a discontinuous finite element method for the coupled Stokes and Darcy problems. *J. Sci. Comp.* **2005**, *22*, 479–500.
12. Rivière, B.; Yotov, I. Locally conservative coupling of Stokes and Darcy flows. *SIAM J. Numer. Anal.* **2005**, *42*, 1959–1977.
13. Burman, E.; Hansbo, P. A unified stabilized method for Stoke's and Darcy's equations. *J. Comput. Appl. Math.* **2007**, *198*, 35–51.
14. Discacciati, M.; Quarteroni, A. Analysis of a domain decomposition method for the coupling of Stokes and Darcy equations. In *Numerical Mathematics and Advanced Applications*; Springer: Milan, Italy, 2003; pp. 3–20.
15. Miglio, E.; Quarteroni, A.; Saleri, F. Coupling of free surface flow and groundwater flows. *Comput. Fluids* **2003**, *23*, 73–83.
16. Discacciati, M. Domain Decomposition Methods for the Coupling of Surface and Groundwater Flows. Ph.D. Thesis, Ecole Polytechnique Federale de Lausanne, Lausanne, Switzerland, 2004.
17. Discacciati, M.; Quarteroni, A.; Valli, A. Robin-Robin domain decomposition methods for the Stokes-Darcy coupling. *SIAM J. Numer. Anal.* **2007**, *45*, 1246–1268.
18. Hoppe, R.; Porta, P.; Vassilevski, Y. Computational issues related to iterative coupling of subsurface and channel flows. *Calcolo* **2007**, *44*, 1–20.
19. Mu, M.; Xu, J. A two-grid method of a mixed Stokes-Darcy model for coupling fluid flow with porous media flow. *SIAM J. Numer. Anal.* **2007**, *45*, 1801–1813.
20. Cai, M.; Mu, M.; Xu, J. Preconditioning techniques for a mixed Stokes/Darcy model in porous media applications. *J. Comput. Appl. Math.* **2009**, *233*, 346–355.
21. Cao, Y.; Gunzburger, M.; He, X.M.; Wang, X. Robin-Robin domain decomposition methods for the steady-state Stokes-Darcy system with the Beavers-Joseph interface condition. *Numer. Math.* **2010**, *117*, 601–629.
22. Jiang, B. A parallel domain decomposition method for coupling of surface and groundwater flows. *Comput. Methods Appl. Mech. Eng.* **2009**, *198*, 947–957.
23. Vassilev, D.; Wang, C.; Yotov, I. Domain decomposition for coupled Stokes and Darcy flows. *Comput. Methods Appl. Mech. Eng.* **2014**, *268*, 264–283.
24. Chen, W.; Gunzburger, M.; Hua, F.; Wang, X. A parallel Robin-Robin domain decomposition method for the Stokes–Darcy system. *SIAM J. Numer. Anal.* **2011**, *49*, 1064–1084.
25. Cao, Y.; Gunzburger, M.; Hua, F.; Wang, X. Coupled Stokes-Darcy model with Beavers-Joseph interface boundary condition. *Commun. Math. Sci.* **2010**, *8*, 1–25.
26. Cao, Y.; Gunzburger, M.; Hu, X.; Hua, F.; Wang, X.; Zhao, W. Finite element approximations for Stokes-Darcy flow with Beavers-Joseph interface conditions. *SIAM J. Numer. Anal.* **2010**, *47*, 4239–4256.
27. Mu, M.; Zhu, X. Decoupled schemes for a non-stationary mixed Stokes-Darcy model. *Math. Comput.* **2010**, *79*, 707–731.
28. Layton, W.; Tran, H.; Trenchea, C. Analysis of long-time stability and Errors of Two Partitioned Methods for Uncoupling Evolutionary groundwater–surface Water Flows. *SIAM J. Numer. Anal.* **2013**, *51*, 248–272.
29. Kubacki, M. Uncoupling Evolutionary groundwater–surface Water Flows Using the Crank–Nicolson LeapFrog Method. *Numer. Methods Part. Differ. Equ.* **2013**, *29*, 1192–1216.
30. Chen, W.; Gunzburger, M.; Sun, D.; Wang, X. Efficient and long-time accurate second-order methods for Stokes–Darcy system. *SIAM J. Numer. Anal.* **2013**, *51*, 2563–2584.
31. Chen, W.; Gunzburger, M.; Sun, D.; Wang, X. An efficient and long-time accurate third-order algorithm for the Stokes–Darcy system. *Numer. Math.* **2016**, *134*, 857–879.

32. Jiang, N.; Kubacki, M.; Layton, W.; Moraiti, M.; Tran, H. A Crank–Nicolson Leapfrog stabilization: Unconditional stability and two applications. *J. Comput. Appl. Math.* **2015**, *281*, 263–276.
33. Kubacki, M.; Moraiti, M. Analysis of a second-order, unconditionally stable, partitioned method for the evolutionary Stokes–Darcy model. *Int. J. Numer. Anal. Model.* **2015**, *12*, 704–730.
34. Layton, W.; Tran, H.; Xiong, X. Long-time stability of Four Methods for Splitting the Evolutionary Stokes-Darcy Problem into Stokes and Darcy Sub-problems. *J. Comput. Appl. Math.* **2012**, *236*, 3198–3217.
35. Shan, L.; Zheng, H.; Layton, W. A Decoupling Method with Different Subdomain Time Steps for the Nonstationary Stokes–Darcy Model. *Numer. Methods Part. Differ. Equ.* **2013**, *29*, 549–583.
36. Rybak, I.; Magiera, J. A multiple-time-step technique for coupled free flow and porous medium systems. *J. Comput. Phys.* **2014**, *272*, 327–342.
37. Domenico, P.; Mifflin, M. Water from low-permeability sediments and land subsidence. *Water Resour. Res.* **1965**, *1*, 563–576.
38. Johnson, A. Specific yield—Compilation of specific yields for various materials. In *USGS Water Supply Paper*; U.S. Government Printing Office: Washington, DC, USA, 1967.
39. Jäger, W.; Mikelić, A. On the interface boundary condition of Beavers, Joseph, and Saffman. *SIAM J. Appl. Math.* **2000**, *60*, 1111–1127.
40. Shan, L.; Zheng, H. Partitioned Time Stepping Method for Fully Evolutionary Stokes–Darcy Flow with Beavers–Joseph Interface Conditions. *SIAM J. Numer. Anal.* **2013**, *51*, 813–839.
41. Girault, V.; Raviart, P.A. *Finite Element Methods for Navier-Stokes Equations:Theory and Algorithms*; Springer Series in Computational Mathematics; Springer: Berlin, Germany, 1986; Volume 5, 376p.
42. Cao, Y.; Gunzburger, M.; He, X.; Wang, X. Parallel, non-iterative, multi-physics domain decomposition methods for time-dependent Stokes-Darcy systems. *Math. Comput.* **2014**, *83*, 1617–1644.
43. Shan, L.; Zhang, Y. Error estimates of the partitioned time stepping method for the evolutionary Stokes–Darcy flows. *Comput. Math. Appl.* **2017**, *73*, 713–726.
44. Kubacki, M. Higher-Order, Strongly Stable Methods for Uncoupling groundwater–surface Water Flow. Ph.D. Thesis, University of Pittsburgh, Pittsburgh, PA, USA, 2014.
45. Layton, W.; Trenchea, C. Stability of two IMEX methods, CNLF and BDF2-AB2, for uncoupling systems of evolution equations. *Appl. Numer. Math.* **2011**, *62*, 112–120.
46. Layton, W.J.; Takhirov, A.; Sussman, M. Instability of Crank-Nicolson Leap-Frog for nonautonomous systems. *Int. J. Numer. Anal. Mod. Ser. B* **2014**, *5*, 289–298.
47. Hurl, N.; Layton, W.; Li, Y.; Moraiti, M. The unstable mode in the Crank-Nicolson Leap-Frog method is stable. *Int. J. Numer. Anal. Model.* **2016**, *13*, 753–762.
48. Moraiti, M. On the quasistatic approximation in the Stokes–Darcy model of groundwater–surface water flows. *J. Math. Anal. Appl.* **2012**, *394*, 796–808.

fluids

MDPI

Article

Lagrangian Modeling of Turbulent Dispersion from Instantaneous Point Sources at the Center of a Turbulent Flow Channel

Quoc Nguyen , Samuel E. Feher and Dimitrios V. Papavassiliou *

School of Chemical, Biological and Materials Engineering, The University of Oklahoma, Norman, OK 73019, USA; quocnguyen@ou.edu (Q.N.); Samuel.E.Feher-1@ou.edu (S.E.F.)

* Correspondence: dvpapava@ou.edu; Tel.: +1-405-325-5811

Received: 5 August 2017; Accepted: 6 September 2017; Published: 8 September 2017

Abstract: The paper is focused on the simulation and modeling of the dispersion from an instantaneous source of heat or mass located at the center of a turbulent flow channel. The flow is modeled with a direct numerical simulation, and the dispersion is modeled with Lagrangian methods based on Lagrangian scalar tracking (LST). The LST technique allows the simulation of scalar sources that span a range of Prandtl or Schmidt numbers that cover orders of magnitude. The trajectories of individual heat or mass markers are tracked, generating a probability distribution function that describes the behavior of instantaneous point sources of a scalar in the turbulent field. The effect of the Prandtl or Schmidt number on turbulent dispersion is examined, with emphasis on the dispersion pattern. Results for Prandtl or Schmidt numbers between 0.1 and 15,000 are presented. For an instantaneous source at the channel center, it is found that there are two zones of cloud development: one where molecular diffusion plays a role at very small times (early stage of the dispersion), and one where turbulent convection dominates. The asphericity of the scalar marker cloud is found to increase monotonically, in contrast to published results for isotropic, homogenous turbulence, where the asphericity goes through a maximum.

Keywords: turbulent transport; Lagrangian modeling; turbulent dispersion; direct numerical simulation

---

1. Introduction

As turbulence is the rule rather than the exception in fluid flows, in industry as well as in the environment, a strong and ongoing effort by scientists and engineers has focused on the modeling and simulation of turbulent flows. The development of both experimental [1–3] and simulation techniques [4–6] has improved our understanding of the mechanisms of turbulence generation and dissipation [7–9] and as a consequence has enabled the use of computational fluid dynamics (CFD) techniques to reliably design processes and equipment where turbulence dominates. In addition, the case when a scalar is dispersed within a turbulent flow field is important in a host of applications where the flow is coupled with transport of heat or mass. Typical examples from everyday life are the dispersion of smoke and chemicals from industrial smoke stacks, agitated mixing, etc. Other examples with industrial applications are heat exchangers, industrial mixing, flow in chemical reactors and in catalyst regeneration units, heat transfer over moving blades, and the dispersion of pollutants in rivers and oceans.

While a much broader research effort has been devoted to the modeling of turbulent flow [4,5], the effort to simulate and develop models for turbulent heat transfer has been more limited [10–17]. The simulations are mainly divided into two categories: those that are based on the Eulerian approach to describing transport phenomena (and where the system of reference is not moving), and those that are based on the Lagrangian approach to transport phenomena (where the system of reference moves with the scalar being dispersed).

Eulerian techniques are employed by typical CFD software packages. In these, models that are by-and-large based on the Reynolds analogy for heat or mass transport in turbulent flows are utilized. These models typically are based on the concepts of eddy diffusivity, the mixing length, the turbulent kinetic energy, and the rate of turbulence dissipation. Models based on the Reynolds analogy make the assumption that the eddy diffusivity of a scalar is related to the eddy viscosity for the flow through a form of the turbulent Prandtl number (Pr_t). Such models are empirical and require the calibration of several variables with experiments. They often fail to produce accurate results, for example in asymmetric flows, like the flow between a rough and a smooth plate, or transfer in annular flow [18]. Other models include those that are based on heat transfer equations that are analogous to the k-ε models suffer from the same shortcomings as the k-ε models for flow [17]. Second moment closure models, such as the analogues of Reynolds Stress Models for heat transfer, require the modeling of a large number of terms in the scalar flux transport equations and in the turbulent stress equations, which are difficult to measure experimentally (i.e., the terms of the pressure-strain term, and the turbulent dissipation term) and are known to be inaccurate close to solid boundaries [19–21]. In wall turbulence, where a solid boundary is present, the Reynolds analogy–based models fail because of a fundamental reason: the scales of turbulence that contribute to scalar transport are not the same as the scales that contribute to heat or mass transport, and the later depend on the Prandtl number (Pr) of the fluid [22–26]. (Note that from now on, we use terminology that can be applied to heat transfer and the Prandtl number, while the results and analysis apply to mass transfer and the Schmidt number.).

Lagrangian techniques have also been employed for the study of turbulent transport, but mostly for developing theoretical understanding. In isotropic, homogeneous turbulence, such simulation techniques have been very successful in exploring the stochastic modeling of turbulence over a range of Pr between 0.04 and 1024 [27–33], while in anisotropic wall turbulence, they have been used to describe transport over a range of Pr that covers six orders of magnitude (from $Pr = 0.01$ to $Pr = 50{,}000$) and different modes of heat transfer [34–36]. Hybrid Eulerian-Lagrangian methods have also been recently utilized [37]. An advantage in terms of computations for Lagrangian methods is the ability to simulate heat transfer cases with practically no limits on the Pr. Eulerian direct numerical simulation methods are limited not only to relatively low Reynolds numbers (Re), but also to a narrow range of small and intermediate Pr. This limitation arises because, in order to resolve all the scales of motion and temperature [38], the number of computational mesh points has to be proportional to $Pr^{3/2}Re^{9/4}$. An increase of Pr by one order of magnitude means an increase of the number of grid points by about thirty times. The main disadvantage of Lagrangian methods is the need to use a large number of scalar markers to simulate turbulent transfer and the slow convergence of particle-based methods [39].

In the present work, Lagrangian scalar tracking (LST) is used to investigate turbulent dispersion from instantaneous point sources of a scalar at the middle of a turbulent flow channel simulated with a direct numerical simulation (DNS). (The cloud of scalar markers resulting from an instantaneous point release is usually called a *puff*). The DNS method has been employed and validated with experiment results in Poiseuille and Poiseuille-Couette flow [40], while LST simulation results have been found to agree well with different experimental results for heat or mass transfer [41]. The significant advantage of the LST, in terms of physics, is the ability to release such particle puffs that form the most elementary unit for heat transfer. The heat markers in the puff can be followed individually and can be correlated with actual flow structures [25,26]. The contributions of this work are (a) to describe the computational method utilized to develop the LST results, (b) to investigate the characteristics of turbulent transport in a channel flow and its comparison to turbulent dispersion in an isotropic turbulent field, and (c) to explore possible Pr effects on turbulent dispersion, with the use of results for $Pr = 0.1$, 6, 100, 2400, and 15,000, which correspond to liquid metals, gases, liquids, and heavy oils.

2. Results

2.1. Statistics of the Marker Location and Prandtl number (Pr) Effects

The friction *Re* for the presented simulation results is $Re_\tau = 300$. Figure 1 is a plot of the mean streamwise position of the puffs that result from the release of 10,000 particles at an instant at a point source located at the middle of the turbulent flow channel at $x_0 = 0$ and $y_0 = 0$. The time of particle release is $t_0 = 0$. All space and time quantities presented in this study are in viscous wall units. The bottom wall of the channel is considered to be at $y = -300$ and the top wall at $y = 300$. The mean puff position does not change significantly with the *Pr*, as expected.

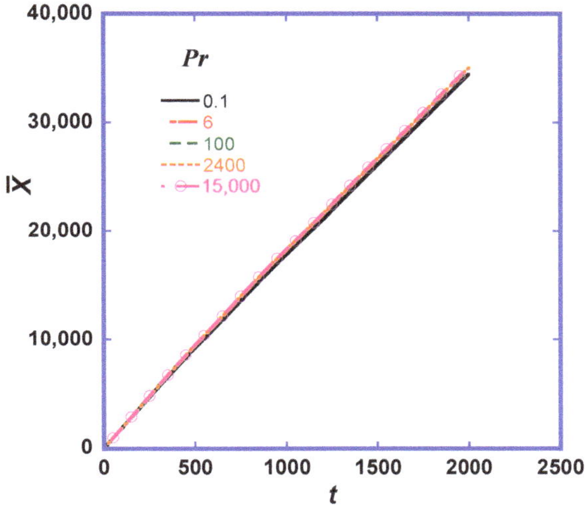

Figure 1. Mean streamwise position (in viscous wall units) of particles released instantaneously from a point source at the center of a turbulent flow channel. The slope of the lines changes from representing the maximum mean fluid velocity at small times to representing the bulk fluid velocity in the channel at longer times.

While the first order statistics of the puff location are not expected to change with *Pr*, the second order statistics can be a better measure to show the *Pr* effects on the puff motion, especially at small times since the scalar marker release. In Figure 2 we present the second order statistics of the puff motion in the three spatial directions as a function of the time elapsed from the release of the puff. The variance of the distribution of the location of the markers in each cloud and at each time is calculated as follows:

$$\sigma_X = \overline{(\overline{X} - X)^2}^{1/2}, \sigma_Y = \overline{(\overline{Y} - Y)^2}^{1/2} \text{ and } \sigma_Z = \overline{(\overline{Z} - Z)^2}^{1/2} \quad (1)$$

where the overbar designates average over all the markers in the flow field, and X, Y, and Z are the locations of markers in the x, y, and z directions.

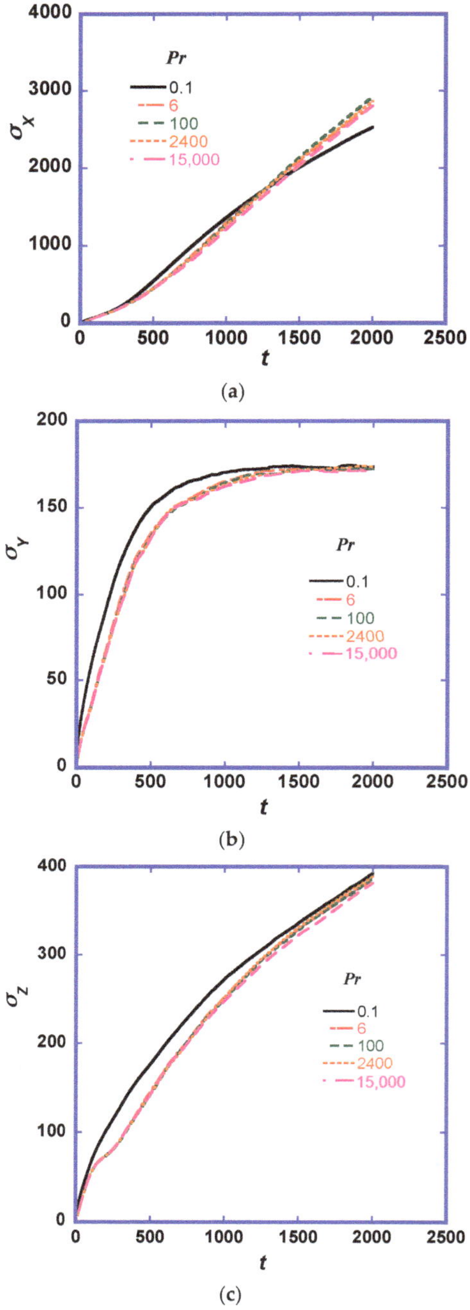

Figure 2. Standard deviation of the scalar marker location in the (**a**) streamwise direction, x; (**b**) wall normal direction, y; and (**c**) the spanwise direction z.

2.2. *Shape of Puff and Differences from Puffs Released in Isotropic Turbulence*

The shape of the puffs in anisotropic turbulence is expected to be changing in the three dimensions, starting from a spherical shape and changing to a spheroid as a function of time and *Pr*. In isotropic turbulence, the flow and molecular effects would result in spherical puffs irrespective of *Pr*. Figure 3 is a depiction of the marker puff at time t = 2000 after the puff release. It is seen that the puff extends longer in the x direction at higher *Pr*, in agreement with Figure 2a.

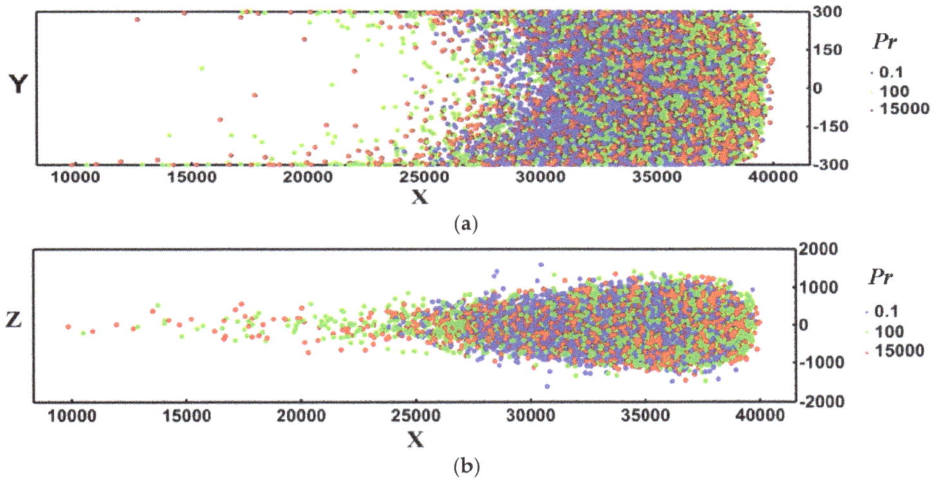

Figure 3. A view of the puff shape at t = 2000. (**a**) side view in the x-y plane; (**b**) top view in the x-z plane. Viscous wall units are used for space and time.

A measure of the departure of the puff shape from the spherical shape is the *asphericity* (A_s), which can be used to quantify the deformation of a puff from a spherically symmetric geometry [42,43]. Asphericity of a body varies from 0 to 1 (A_s = 0 for a perfectly spherical cloud; A_s = 0.25 for a two-dimensional circle without width, a disc; and A_s = 1 for an infinite cylinder) [44]. The values of A_s were calculated from the moment of inertia tensor (T) of each puff [44],

$$A_s = \frac{\left(R_1^2 - R_2^2\right)^2 + \left(R_2^2 - R_3^2\right)^2 + \left(R_3^2 - R_1^2\right)^2}{2R_g^4} \tag{2}$$

$$T_{ij} = \frac{\sum_{m=1}^{N}\left(S_{im} - S_i^{CM}\right)\left(S_{jm} - S_j^{CM}\right)}{N} \tag{3}$$

where N is the total number of markers in the puff, R_1^2, R_2^2, and R_3^2 are the three eigenvalues of the tensor T (i.e., the three principal radii of gyration squared for all Nmarkers), S_{im} is the position of marker m in the i-th Cartesian dimension (i denotes x, y, or z), and S_i^{CM} is the center of mass of the N markers in coordinate i. The puff asphericity can be seen in Figure 4 for the puff cases simulated herein.

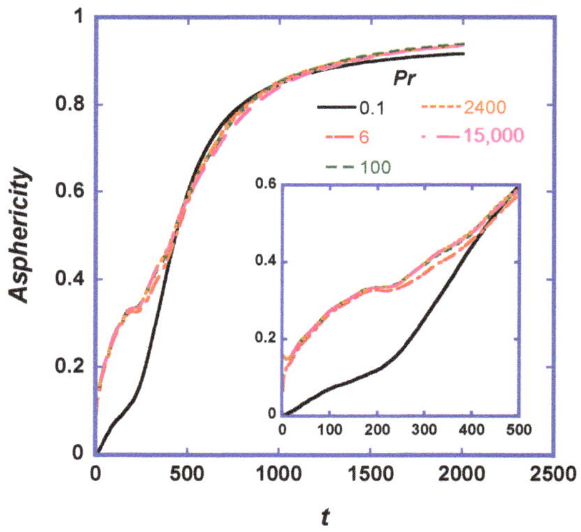

Figure 4. Asphericity of puff shapes as a function of time and *Pr*. At times less than 50, the value of asphericity is less than 0.2, but increases at larger times.

3. Discussion

For a puff released at the channel center, one would expect that the mean puff velocity (the slope of the lines in Figure 1) would be very close to the mean flow velocity at the channel center (which is 19.15 in the simulation) at small times after the particle release. As the time elapses, the puff of markers should disperse in the vertical and spanwise directions and would eventually extend to cover the channel width. From that point and on, the mean puff velocity would be very close to the bulk velocity of the fluid in the channel. In Figure 1, it is seen that the slope of the mean streamwise position of the puff is $\frac{d\overline{X}}{dt} = 19.1$ for times less than $t = 500$, and the slope of the line changes to $\frac{d\overline{X}}{dt} = 16.7$ as times $t > 1500$. The change in slope, occurring between time $t = 500$ and $t = 1000$, is the same for all cases of *Pr*. The average *Y* and *Z* positions (not shown here) remain $\overline{Y} = 0$ and $\overline{Z} = 0$, since the turbulent channel flow is symmetric around the center plane regarding the statistics of the flow. There is no reason for the markers in the puff to move on average toward the top or toward the bottom of the channel.

Scalar dispersion is dominated by molecular effects at very small times and then by turbulent effects at longer times. In the particular case of flow in a channel examined here, the cloud of markers that comprise a puff will eventually distribute uniformly across the channel. It is expected that at long times, the standard deviation of the particle location in the *y* direction will be that of a uniform distribution in the *y* direction. The value of σ_Y increases and it is expected to tend to the constant value of $(600^2/12)^{1/2} = 173.2$ for all *Pr*, (which is the standard deviation for a uniform distribution between the values -300 and 300). It takes somewhat longer for higher *Pr* markers to get to this predicted value because at small times, there is a molecular diffusion effect. While the marker cloud extends in the normal direction, markers start to enter the log layer and then the viscous wall region closer to the channel walls. When this occurs, the effects of the *Pr* become more important and manifest themselves in the values of σ_X. When high *Pr* markers enter the near wall region, the molecular motion is rather limited, and it is more difficult for them to jump out of that region with molecular means (contrary to low *Pr* markers that can jump out of the near wall region because of turbulent velocity fluctuations and also because of molecular motion). In this respect, higher *Pr* markers stay longer in the viscous wall region, where the mean streamwise velocity is smaller than the bulk fluid velocity, and the high *Pr* clouds tend to extend longer in the *x* direction. This is seen in Figure 2a by observing the slope of the σ_X line that increases for high *Pr* puffs. The existence of slower moving markers is also seen in

Figure 3, where the individual markers trailing at the edge of the puffs for $Pr = 100$ and $Pr = 15{,}000$ are seen.

Regarding the puff shape, all of the puffs started as sphere-like at very small times, but changed their shape at larger times (see Figure 4). Intermediate and high Pr puffs ($Pr \geq 6$) have almost the same asphericity as time elapses, but the low Pr puff exhibits differences at times up to $t = 400$. The reason for these differences, and the reason that the $Pr = 0.1$ puff is shaped almost like a sphere for a longer time, is the high molecular diffusion of this puff relative to the others. Due to the random Brownian motion, the markers are initially like particles within the volume of a balloon that expands with time. At larger times, the markers start to enter the logarithmic region and the viscous wall region, and the puff starts to elongate, departing from the spherical, balloon-like shape. Recent results about the asphericity and the puff shape of passive particles released in isotropic, homogeneous turbulence [45], indicate that the marker puffs start as spherical and quickly deform into elongated structures, showing a maximum value of asphericity of about 0.7 within 13 Kolmogorov time scales. Beyond that point, the puffs tend to slowly return to a more spherical shape with asphericity values close to 0.2 at longer dispersion times, longer than the eddy turnover time scale. This is a notable difference with the present case, where the deformation of the shape of the puff is not reversible.

In prior work about the dispersion of a puff released at the wall of a turbulent flow channel, it has been found that there are three zones of puff development, depending on the physical mechanism that dominates dispersion [46]. Zone I is dominated by molecular diffusion effects, Zone II is a transition zone where both molecular and convection effects are important, and Zone III is the zone where turbulent convection is dominant. The time that it takes for a puff to transition between these zones depends on the Pr. For example, $Pr = 0.1$ puffs were found to exit Zone I at about $t = 20$, while $Pr = 2400$ puffs would exit this zone at $t = 200$. In the present case, where the puffs are released in the center of the channel, there appear to be only two Zones of puff development. Zones I and II (where molecular effects can be important) are very short for $Pr \geq 6$. When the beginning of Zone III is identified as the time period in which the asphericity becomes larger than 0.1, the transition to turbulent convection dominance does not take more than $t = 5$. Only for $Pr = 0.1$, the asphericity stays below 0.1 for times up to $t = 150$. A prior study for dispersion from the center of a channel flow also indicated that Pr effects are more important for low Pr dispersion (Pr on the order of one and lower) [47]. However, that study was for a $Re_\tau = 150$ channel, where the convective effects are less prominent than in the current study, and it was limited to dispersion up to time $t = 125$.

4. Materials and Methods

The velocity field for a Newtonian and incompressible fluid is calculated using a DNS of fully developed turbulent flow in a plane channel. The dimensions of the computational box in the x (streamwise), y (wall-normal) and z (spanwise) directions are $16\ \pi h \times 2\ h \times 2\ \pi h$ with a half channel height (h) of 300. The flow is driven by a constant mean pressure gradient. For the present problem, the simulation is conducted on a $1024 \times 129 \times 256$ mesh in the x, y, z directions, respectively, with uniform spacing in the x and z, while Chebyshev polynomial collocation points are used in the wall normal direction. The DNS algorithm is based on the pseudospectral method published in [10,48]. The integrity of this method has been verified with comparisons to experiments at an equal Re [49]. The Reynolds number, Re, defined with the centerline mean velocity and half channel height, h, is 5760, corresponding to a Re defined based on the hydraulic radius of ~23,000 (this is the equivalent Re for flow in a pipe). The friction Reynolds number is $Re_\tau = h = 300$. The variables presented are dimensionless with the use of the friction velocity, u^*, and the kinematic viscosity of the fluid, ν, namely, the viscous wall units. The friction velocity is given as $u^* = (\tau_w/\rho)^{1/2}$, where τ_w is the shear stress at the wall and ρ is the fluid density. The assumptions of no slip and no penetration are used as boundary conditions on the wall, and the heat production by viscous dissipation is assumed to be negligible. The time step was 0.1 in viscous wall units, and the iterations were carried out for 20,000 time steps for stationary channel flow in order to simulate 2000 viscous time units of flow after a puff is released. Briefly, the DNS solves the rotational form of the dimensionless Navier-Stokes equations,

$$\frac{\partial \vec{U}}{\partial t} = \vec{U} \times \vec{\Omega} - \nabla \zeta + \frac{1}{h}\vec{i}_x + \nabla^2 \vec{U} \tag{4}$$

where $\vec{U}$ is the velocity vector, $\vec{\Omega}$ is the vorticity vector, and $\vec{i_x}$ is the unit vector in x direction. The continuity equation also applies,

$$\nabla \cdot \vec{U} = 0 \tag{5}$$

and

$$\zeta = p' + \frac{1}{2}\vec{U} \cdot \vec{U} \tag{6}$$

where the term $p\prime$ is the fluctuating component of the pressure in viscous wall units. The velocity is expanded in truncated Fourier series in the x and z directions, and a truncated Chebyshev polynomial series in the y direction. The boundary conditions in the x and z directions are periodic with periodicity length the size of the computational channel in each one of these directions. The Navier-Stokes equations are integrated in time using the pseudospectral fractional step method originally developed by Orszag and Kells [50] and the added correction suggested by Marcus [51] to correct with the pressure at the channel walls.

The trajectories of heat or mass markers released from a line parallel to the z axis at the center of the channel are calculated in the flow field created by the DNS. The total number of markers is 10,000 for each case of Pr, for a total of 50,000 markers. Since the flow is homogeneous in the z direction, it is not statistically important where on the line the markers are released, so in order to reduce the possible effects of releasing the markers in an idiosyncratic flow structure, the markers were released uniformly spaced on the line source. However, in order to obtain the statistics of the cloud location in the z direction, the z coordinate of each particle was determined after subtracting the z coordinate at the marker point of release. The flow field used for the Lagrangian scalar tracking of the markers is the same for all Pr cases, so that the effects of the Pr can be observed rather than the effects of the flow. The tracking of the passive scalar markers was based on the Kontomaris et al. [52] algorithm. The motion of the scalar markers is decomposed into a convection part and a molecular diffusion part. The convective part can be calculated from the fluid velocity at the particle position, so that the Lagrangian velocity at time t of a marker released at location $\vec{X_o}$ is assumed to be the same as the Eulerian velocity at that particle's location at the beginning of the convective step, i.e., $\vec{V}(\vec{X_o}, t) = \vec{U}[\vec{X}(\vec{X_o}, t), t]$. The equation of particle motion then is

$$\vec{V}(\vec{X_o}, t) = \frac{\partial \vec{V}(\vec{X_o}, t)}{\partial t} \tag{7}$$

The effect of molecular diffusion is simulated by adding a random movement on the particle motion at the end of each convective step. A similar method was used in simulating the random diffusion part of Lagrangian trajectories in [52]. This random motion is a random jump that takes values from a Gaussian distribution with zero mean and standard deviation $\sigma = \sqrt{2\Delta t / Pr}$, where Δt is the time step of the simulation in viscous wall units, ($\Delta t = 0.1$). The particle velocity is found by using a mixed Lagrangian-Chebyshev interpolation between the Eulerian velocity field values at the surrounding mesh points. The particle position integration in time is approximated with a second order Adams-Bashforth scheme [53].

As previously mentioned, Lagrangian methods require a large number of scalar markers to simulate turbulent transport. In this study, 10,000 scalar markers were released on a single line source. The question of whether one needs to increase the number of markers for higher accuracy is explored by comparing results in this study with second order statistics from another numerical experiment, which was conducted with 100,000 markers and $Pr = 6$ being released at several line sources along the streamwise direction. While the mean marker position did not show any differences in the streamwise direction (as expected), the second order statistics exhibit some differences. The standard deviation of the scalar marker location in the streamwise and wall normal direction of $Pr = 6$, with 10,000 and 100,000 markers, is plotted in Figure 5. It is seen that using 100,000 markers results in about 3.5% difference in σ_x at $t = 2000$ and about 4.4% difference for σ_y at $t = 500$ (the differences disappear at long times for σ_y). However, these small differences in the second order statistics are not large enough to change the qualitative findings of the present study or to change significantly the quantitative findings about the asphericity of the particle clouds. Other simulations with a higher number of particles could benefit the accuracy of the quantitative findings herein.

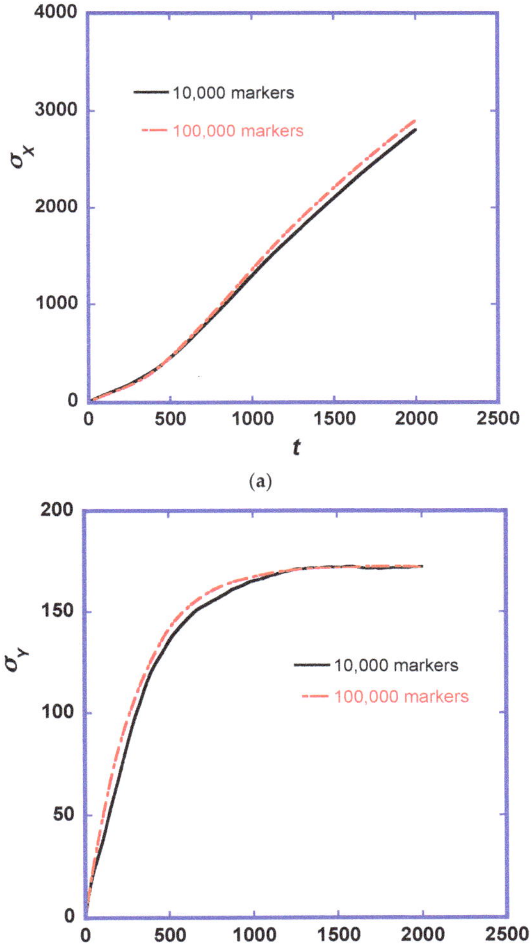

Figure 5. Standard deviation of the scalar marker location in the (**a**) streamwise direction, x, and (**b**) wall normal direction, y. Two experiments with 10,000 and 100,000 markers are presented, for $Pr = 6$.

5. Conclusions

A Lagrangian method to explore the dispersion of passive scalars in turbulent flow is described. The advantages of using this numerical method to model heat or mass transfer is that one can use a single velocity field obtained by running a DNS once to model the dispersion of scalars of several different *Pr*. In addition, very high and very low *Pr* cases can be modeled. Another advantage of the Lagrangian numerical approach is that the physical mechanism of turbulent dispersion can be revealed in a rather natural way. The dispersion of scalars of different *Pr* from the channel center showed that the resulting cloud of markers moves differently than puffs released from the channel wall and differently than puffs that result from an instantaneous release in an isotropic turbulent velocity field because of the relative importance of molecular diffusion in the marker dispersion in each one of these cases. Finally, puffs for *Pr* larger than 6 appear to behave the same way, while the $Pr = 0.1$ puff, where the effects of molecular diffusion are relatively stronger, exhibits different behavior.

Acknowledgments: The use of computing facilities at the University of Oklahoma Supercomputing Center for Education and Research (OSCER) and at the Extreme Science and Engineering Development Environment (XSEDE) (under allocation CTS-090025) is gratefully acknowledged.

Author Contributions: Dimitrios V. Papavassiliou conceived and designed the experiments; Quoc Nguyen and Samuel E. Feher performed the simulations and Dimitrios V. Papavassiliou, Quoc Nguyen, and Samuel E. Feher analyzed the data and wrote the paper.

Conflicts of Interest: The authors declare no conflict of interest.

References

1. Tao, B.; Katz, J.; Meneveau, C. Geometry and scale relationships in high Reynolds number turbulence determined from three-dimensional holographic velocimetry. *Phys. Fluids* **2000**, *12*, 941–944. [CrossRef]
2. Scarano, F. Tomographic PIV: Principles and practice. *Meas. Sci. Technol.* **2013**, *24*. [CrossRef]
3. Hu, H. Stereo particle imaging velocimetry techniques: Technical basis, system setup, and application. In *Handbook of 3D Machine Vision: Optical Metrology and Imaging*; Zhang, S., Ed.; CRC Press: Boca Raton, FL, USA, 2013; Chapter 4; pp. 71–100.
4. Moin, P.; Mahesh, K. Direct Numerical Simulation: A tool in turbulence research. *Annu. Rev. Fluid Mech.* **1998**, *30*, 539–578. [CrossRef]
5. Alfonsi, G. On direct numerical simulation of turbulent flows. *Appl. Mech. Rev.* **2011**, *64*. [CrossRef]
6. Lee, M.; Moser, R.D. Direct numerical simulation of turbulent channel flow up to $Re_\tau = 5200$. *J. Fluid Mech.* **2015**, *774*, 395–415. [CrossRef]
7. Marusic, I.; Mathis, R.; Hutchins, N. Predictive model for wall-bounded turbulent flow. *Science* **2010**, *329*, 193–196. [CrossRef] [PubMed]
8. Adrian, R.J. Closing in on models of wall turbulence. *Science* **2010**, *329*, 155–156. [CrossRef] [PubMed]
9. Smits, A.J.; McKeon, B.J.; Marusic, I. High-Reynolds number wall turbulence. *Annu. Rev. Fluid Mech.* **2011**, *43*, 353–375. [CrossRef]
10. Chakrabarti, M.; Kerr, R.M.; Hill, J.C. Direct Numerical simulation of chemical selectivity in homogeneous turbulence. *AIChE J.* **1995**, *41*, 2356–2370. [CrossRef]
11. Churchill, S.W. Progress in the thermal sciences: AIChE Institute Lecture. *AIChE J.* **2000**, *46*, 1704–1722. [CrossRef]
12. Lyons, S.L.; Hanratty, T.J.; McLaughlin, J.B. Direct numerical simulation of passive heat transfer in a turbulent channel flow. *Int. J. Heat Mass Transf.* **1991**, *34*, 1149–1161. [CrossRef]
13. Kasagi, N.; Tomita, Y.; Kuroda, A. Direct numerical simulation of passive scalar field in a turbulent channel flow. *J. Heat Transf.* **1992**, *114*, 598–606. [CrossRef]
14. Teitel, M.; Antonia, R.A. A step change in wall heat flux in turbulent channel flow. *Int. J. Heat Mass Transf.* **1993**, *36*, 1707–1709. [CrossRef]
15. Kawamura, H.; Ohsaka, K. DNS of turbulent heat transfer in channel flow with low to medium-high Prandtl number fluid. *Int. J. Heat Fluid Flow* **1998**, *19*, 482–491. [CrossRef]
16. Na, Y.; Papavassiliou, D.V.; Hanratty, T.J. Use of Direct Numerical Simulation to study the effect of Prandtl number on temperature fields. *Int. J. Heat Fluid Flow* **1999**, *20*, 187–195. [CrossRef]

17. Bradshaw, P. Understanding and prediction of turbulent flow-1996. *Int. J. Heat Fluid Flow* **1997**, *18*, 45–54. [CrossRef]
18. Churchill, S.W. A Critique of Predictive and Correlative Models for Turbulent Flow and Convection. *Ind. Eng. Chem. Res.* **1996**, *35*, 3122–3140. [CrossRef]
19. Speziale, C.G.; Xu, X.H. Towards the development of second-order closure models for nonequilibrium turbulent flows. *Int. J. Heat Fluid Flow* **1996**, *17*, 238–244. [CrossRef]
20. Speziale, C.G. Modeling of Turbulent Transport Equations. In *Simulation and Modeling of Turbulent Flows;* Lumley, J., Ed.; Oxford University Press: New York, NY, USA, 1996; pp. 185–242.
21. Liu, X.; Moreto, J.R.; Mitchell, S.S. Instantaneous Pressure Reconstruction from Measured Pressure Gradient using Rotating Parallel Ray Method. In Proceedings of the 54th AIAA Aerospace Sciences Meeting, San Diego, CA, USA, 4–8 January 2016.
22. Srinivasan, C.; Papavassiliou, D.V. Heat Transfer Scaling Close to the Wall for Turbulent Channel Flows. *Appl. Mech. Rev.* **2013**, *65*. [CrossRef]
23. Hasegawa, Y.; Kasagi, N. Low-pass filtering effects of viscous sublayer on high Schmidt number mass transfer close to a solid wall. *Int. J. Heat Fluid Flow* **2009**, *30*, 525–533. [CrossRef]
24. Na, Y.; Hanratty, T.J. Limiting behavior of turbulent scalar transport close to a wall. *Int. J. Heat Mass Trans.* **2000**, *43*, 1749–1758. [CrossRef]
25. Le, P.M.; Papavassiliou, D.V. A physical picture of the mechanism of turbulent heat transfer from the wall. *Int. J. Heat Mass Transf.* **2009**, *52*, 4873–4882. [CrossRef]
26. Karna, A.K.; Papavassiliou, D.V. Near-wall velocity structures that drive turbulent transport from a line source at the wall. *Phys. Fluids* **2012**, *24*. [CrossRef]
27. Antonia, R.A.; Orlandi, P. Effect of Schmidt number on small-scale passive scalar turbulence. *Appl. Mech. Rev.* **2003**, *56*, 615–632. [CrossRef]
28. Brethouwer, G.; Nieuwstadt, F.T.M. DNS of Mixing and Reaction of Two Species in a Turbulent Channel Flow: A Validation of the Conditional Moment Closure. *Flow Turbul. Combust.* **2001**, *66*, 209–239. [CrossRef]
29. Brethouwer, G.; Hunt, J.C.R.; Nieuwstadt, F.T.M. Micro-structure and Lagrangian statistics of the scalar field with a mean gradient in isotropic turbulence. *J. Fluid Mech.* **2003**, *474*, 193–225. [CrossRef]
30. Yeung, P.K.; Xu, S.; Sreenivasan, K.R. Schmidt number effects on turbulent transport with uniform mean scalar gradient. *Phys. Fluids* **2002**, *14*, 4178–4191. [CrossRef]
31. Yeung, P.K.; Xu, S.; Donzis, D.A.; Sreenivasan, K.R. Simulations of three-dimensional turbulent mixing for Schmidt numbers of the order 1000. *Flow Turbul. Combust.* **2004**, *72*, 333–347. [CrossRef]
32. Borgas, M.S.; Sawford, B.L.; Xu, S.; Donzis, D.A.; Yeung, P.K. High Schmidt number scalars in turbulence: Structure functions and Lagrangian theory. *Phys. Fluids* **2004**, *16*, 3888–3899. [CrossRef]
33. Buaria, D.; Yeung, P.K.; Sawford, B.L. A Lagrangian study of turbulent mixing: forward and backward dispersion of molecular trajectories in isotropic turbulence. *J. Fluid Mech.* **2016**, *799*, 352–382. [CrossRef]
34. Papavassiliou, D.V.; Hanratty, T.J. The use of Lagrangian methods to describe turbulent transport of heat from the wall. *Ind. Eng. Chem. Res.* **1995**, *34*, 3359–3367. [CrossRef]
35. Papavassiliou, D.V.; Hanratty, T.J. Transport of a passive scalar in a turbulent channel flow. *Int. J. Heat Mass Transf.* **1997**, *40*, 1303–1311. [CrossRef]
36. Mitrovic, B.M.; Le, P.M.; Papavassiliou, D.V. On the Prandtl or Schmidt number dependence of the turbulence heat or mass transfer coefficient. *Chem. Eng. Sci.* **2004**, *59*, 543–555. [CrossRef]
37. Lagaert, J.B.; Balarac, G.; Cottet, G.H. Hybrid spectral-particle method for the turbulent transport of a passive scalar. *J. Comput. Phys.* **2014**, *260*, 127–142. [CrossRef]
38. Tennekes, H.; Lumley, J.L. *A First Course In Turbulence;* MIT Press: Boston, NA, USA, 1972; p. 96.
39. Koumoutsakos, P. Multiscale simulations using particles. *Annu. Rev. Fluid Mech.* **2005**, *37*, 457–487. [CrossRef]
40. Nguyen, Q.; Papavassiliou, D.V. Turbulent plane Poiseuille-Couette flow as a model for fluid slip over superhydrophobic surfaces. *Phys. Rev. E* **2013**, *88*. [CrossRef] [PubMed]
41. Mitrovic, B.M.; Papavassiliou, D.V. Transport properties for turbulent dispersion from wall sources. *AIChE J.* **2003**, *49*, 1095–1108. [CrossRef]
42. Rudnick, J.; Gaspari, G. The asphericity of random walks. *Phys. A Math. Gen.* **1986**, *19*, 191–193. [CrossRef]
43. Vo, M.D.; Shiau, B.; Harwell, J.H.; Papavassiliou, D.V. Adsorption of anionic and non-ionic surfactants on carbon nanotubes in water with dissipative particle dynamics simulation. *J. Chem. Phys.* **2016**, *144*. [CrossRef] [PubMed]

44. Noguchi, H.; Yoshikawa, K. Morphological variation in a collapsed single homopolymer chain. *J. Chem. Phys.* **1998**, *109*, 5070–5077. [CrossRef]
45. Bianchi, S.; Biferale, L.; Celani, A.; Cencini, M. On the evolution of particle puffs in turbulence. *Eur. J. Mech. B/Fluids* **2016**, *55*, 324–329. [CrossRef]
46. Papavassiliou, D.V. Scalar dispersion from an instantaneous line source at the wall of a turbulent channel for medium and high Prandtl number fluids. *Int. J. Heat Fluid Flow* **2002**, *23*, 161–172. [CrossRef]
47. Kontomaris, K.; Hanratty, T.J. Effect of molecular diffusivity on point-source diffusion in the center of a numerically simulated turbulent channel flow. *Int. J. Heat Mass Transf.* **1994**, *37*, 1817–1828. [CrossRef]
48. Lyons, S.L.; Hanratty, T.J.; McLaughlin, J.B. Large-scale computer-simulation of fully-developed turbulent channel flow with heat-transfer. *Int. J. Numer. Methods Fluids* **1991**, *13*, 999–1028. [CrossRef]
49. Gunther, A.; Papavassiliou, D.V.; Warholic, M.D.; Hanratty, T.J. Turbulent flow in a channel at a low Reynolds number. *Exp. Fluids* **1998**, *25*, 503–511. [CrossRef]
50. Orszag, S.A.; Kells, L.C. Transition to turbulence in plane Poiseuille and plane Couette flow. *J. Fluid Mech.* **1980**, *96*, 159–205. [CrossRef]
51. Marcus, P.S. Simulation of Taylor-Couette flow. *J. Fluid Mech.* **1984**, *146*, 45–64. [CrossRef]
52. Abascal, A.J.; Castanedo, S.; Minguez, R.; Medina, R.; Liu, Y.; Weisberg, R.H. Stochastic Lagrangian trajectory modeling of surface drifters deployed during the deepwater horizon oil spill. In *Proceedings of the Thirty-Eighth AMOP Technical Seminar*; Environment Canada: Ottawa, ON, Canada, 2015; pp. 77–91.
53. Kontomaris, K.; Hanratty, T.J.; McLaughlin, J.B. An algorithm for tracking fluid particles in a spectral simulation of turbulent channel flow. *J. Comput. Phys.* **1992**, *103*, 231–242. [CrossRef]

fluids MDPI

Article

The Reduced NS-α Model for Incompressible Flow: A Review of Recent Progress

Abigail L. Bowers [1] and Leo G. Rebholz [2,*]

[1] Department of Mathematics, Florida Polytechnic University, Lakeland, FL 33805, USA; abowers@flpoly.org
[2] Department of Mathematical Sciences, Clemson University, Clemson, SC 29634, USA
* Correspondence: rebholz@clemson.edu; Tel.: +1-864-656-1840

Academic Editor: William Layton
Received: 3 June 2017; Accepted: 30 June 2017; Published: 6 July 2017

Abstract: This paper gives a review of recent results for the reduced Navier–Stokes-α (rNS-α) model of incompressible flow. The model was recently developed as a numerical approximation to the well known Navier–Stokes-α model, for the purpose of more efficiently computations in the C^0 finite element setting. Its performance in initial numerical tests was remarkable, which led to analytical studies and further numerical tests, all of which provided excellent results. This paper reviews the main results established thus far for rNS-α, and presents some open problems for future work.

Keywords: incompressible flow; regularization models; α models; turbulence

1. Introduction

This paper presents a review of results for the recently introduced reduced Navier–Stokes-α (rNS-α) model of incompressible, viscous flow, given by

$$-\alpha^2\Delta w_t + w_t + (\nabla \times Dw) \times w + \nabla p - \nu\Delta Dw = f, \quad (1)$$

$$\nabla \cdot w = 0, \quad (2)$$

$$w(0) = w_0, \quad (3)$$

where w represents velocity, p a Bernoulli-type pressure, f a body force, and D a deconvolution operator intended to approximate the inverse of the Helmholtz (also called α) differential filter $F = (-\alpha^2\Delta + I)^{-1}$. The parameter $\alpha > 0$ represents the filtering radius.

For simplicity, only van Cittert approximate deconvolution,

$$D := D_N = \sum_{n=0}^{N} (I - F)^n,$$

is considered herein. This is the most common type used in fluid flow modeling [1–3]. Other types of approximate deconvolution operators, such as Tikhonov–Lavrentiev, may yield similar results, but all numerical tests done to date for rNS-α have used van Cittert. Some advantages of van Cittert, which likely make it the most common, are that it is simple to use, easy to analyze, is formally very high accuracy, and has performed very well in many computational tests for several models [1–3]. Note that the groundbreaking idea of using approximate deconvolution in fluid models is due to Stolz, Adams and Kleiser [4–7]. Typically N is chosen small, and for all numerical simulations run so far with rNS-α, $N \leq 2$.

The rNS-α model was proposed in [8] as a numerical approximation to the well-known NS-α model (see e.g., [9–12]) that is more efficiently computable in a C^0 finite element framework. NS-α is given by

$$v_t + (\nabla \times v) \times \overline{v} + \nabla p - \nu \Delta v = f, \tag{4}$$

$$\nabla \cdot v = \nabla \cdot \overline{v} = 0, \tag{5}$$

$$-\alpha^2 \Delta \overline{v} + \overline{v} - v = 0. \tag{6}$$

This model (also known as Camassa–Holm equations [11–13]) has extensive and attractive theoretical properties (see [9,10] and references therein), including well-posedness [10], frame invariance [14], adherence to Kelvin's circulation theorem [9], conserving a model energy and helicity [9], requiring significantly fewer degrees of freedom than direct numerical simulation (DNS) of Navier–Stokes [9], and accurately predicting scalings of the turbulent boundary layer [15]. Most other fluid models do not have all (or even most) of these properties, and thus NS-α is widely considered to be a very 'physically accurate' model. However, how to construct stable, efficient algorithms for it in the C^0 finite element setting is an open problem. The essential issue is that the filtering seemingly cannot be decoupled from the momentum-mass system in a stable way, leaving the practitioner to solve the nonlinear problem at each time step. Not only does this create the need for many system solves at each time step, but if one wishes to use a Newton iteration, then the linear systems must solve simultaneously for the velocity, filtered velocity, and pressure. However, rNS-α can be computed efficiently in this setting, in the sense that it can stably decouple the filtering from the mass/momentum system, solve only one mass/momentum system at each time step, and obtain optimal accuracy. Initial testing of rNS-α in [8,16,17] revealed outstanding results for turbulent channel flow simulation and flow past a cylinder.

The purpose of this paper is to review the recent analytical and numerical results for rNS-α. In Section 2, the derivation of the model is shown and how it is related to other models and stabilizations. Section 3 reviews the analytical results established for the model, including a priori bounds, energy trapping, global well-posedness, energy dissipation rate, and energy spectra. Results for sensitivity of the model with respect to the filtering radius are discussed in Section 4. Numerical schemes and their analysis are given in Section 5, both for rNS-α and its associated sensitivity equations from Section 4. Finally, in Section 6, results of numerical tests for channel flow past a cylinder and turbulent channel flow are discussed. Section 7 presents some open problems for possible future work.

2. Derivation of the Model and Connections to Other Models

It will be helpful for the later discussion to present the derivation found in [8] of the rNS-α model, which starts from the NS-α model. Recall from above that the NS-α model is defined as follows, after writing $Fv = \overline{v}$:

$$v_t + (\nabla \times v) \times Fv + \nabla q - \nu \Delta v = f,$$

$$\nabla \cdot Fv = 0.$$

It is discussed in detail in [8] why this model is difficult to efficiently compute with in a C^0 finite element framework.

The rNS-α model is created through a series of transformations to NS-α, as follows. By using $I = F^{-1}F = (-\alpha^2 \Delta + I)F$ in NS-α, one gets

$$-\alpha^2 \Delta F v_t + F v_t + (\nabla \times F^{-1} F v) \times F v + \nabla q - \nu \Delta F^{-1} F v = f,$$

$$\nabla \cdot F v = 0.$$

Denote $\tilde{w} = Fv$, which provides

$$\begin{aligned} -\alpha^2\Delta\tilde{w}_t + w_t + (\nabla \times F^{-1}\tilde{w}) \times \tilde{w} + \nabla q - \nu\Delta F^{-1}\tilde{w} &= f, \\ \nabla \cdot \tilde{w} &= 0. \end{aligned}$$

Next, use the approximation of the deconvolution operator, i.e., that $F^{-1}\tilde{w} \approx D\tilde{w}$, to get the rNS-$\alpha$ model, written in terms of w and p after renaming variables due to the approximation:

$$\begin{aligned} -\alpha^2\Delta w_t + w_t + (\nabla \times Dw) \times w + \nabla p - \nu\Delta Dw &= f, \\ \nabla \cdot w &= 0. \end{aligned}$$

2.1. Connections to Other Models

There are several models closely related to rNS-α that have been studied recently, including NS-α [9,11–13,18,19], NS–Voigt [20–22], and various other regularization models. It is important to make connections between the different models when possible, so existing results of various models can be fully utilized. It is clear from Equations (1)–(3) with $D = D_N$, that for any N, choosing $\alpha = 0$ recovers the Navier–Stokes equations. Thus, if a mesh sufficient for a DNS of Navier–Stokes is used with rNS-α and α is chosen on the order of the mesh width, then the solution is expected to match the Navier–Stokes solution closely.

Deconvolution theory from [2] shows that $N = \infty$ formally recovers NS-α, and the choice $N = 0$ recovers the NS–Voigt model [20–22]. Hence, one can interpret the rNS-α model as being 'in between' NS-α and NS–Voigt, in the sense of deconvolution order. These connections are summarized in the following Table 1.

Table 1. Connections summary.

α	N	rNS-α Model Reduces to
0	$[0, \infty]$	Navier–Stokes
>0	0	NS–Voigt
>0	∞	NS-α

A second important connection for rNS-α is to the generalized α-models studied in [23]. By replacing w with Fv and using the notation of [23], rNS-α fits the general form with $A = -\nu\Delta DF$, $M = F$, $N = DF$, and $\chi = 1$. Work done for these models reveals their strengths and weaknesses, and thus identifying connections between models is critical, as they can be extended to rNS-α. For example, the main arguments in [8] for well-posedness of rNS-α are similar to those made for NS–Voigt in [22,24].

When compared to other α-models, rNS-α has displayed better results on wall bounded turbulence, as demonstrated in [8,16,25]. A key difference between these models lies in the viscous term, which has the form $-\nu\Delta D_N w$ instead of the standard $-\nu\Delta w$. At the continuous level, the rNS-α viscosity provides no extra smoothing, but it does provide additional dispersion when compared to the standard viscous term. This is due to the fact that $\|D_N\| \geq 1$, and this can be seen at high wave numbers since here $\|D_N\| \approx (N+1)$ [2].

Further insight is gained by expanding the deconvolution operator of the viscous term. Following [26], one can write

$$\begin{aligned} -\nu\Delta D_N w &= -\nu\Delta w - \nu\Delta D_{N-1}(w - \overline{w}) &(7)\\ &= -\nu\Delta w - \nu\Delta D_{N-1}w' &(8)\\ &= -\nu\Delta w + \alpha^2\nu\Delta^2 D_{N-1}\overline{w}, &(9) \end{aligned}$$

where $w' := w - \overline{w}$ can be considered as fluctuations about the (filtered) mean velocity [1]. In Equation (8), the 'extra' term $-\nu\Delta D_{N-1}w'$ reveals a connection to variational multiscale models (VMMs). It takes the form of a viscosity for velocity fluctuations, which is in the same spirit as VMMs, but differs in that here fluctuations are defined with filtered quantities as means. Interestingly, similar dissipation/penalty/stabilizations for the advection equation [27] and in turbulence simulations [28] were recently considered. These were also constructed by using filtering to define means and fluctuations, and have proven to be very successful at increasing accuracy of simulations on coarse discretizations.

One can also observe the effect of the extra term on the energy balance. Assuming periodic or no-slip boundary conditions for both the velocity and filtered velocities and using Equation (9), test rNS-α with w to obtain

$$\frac{1}{2}\frac{d}{dt}\left(\alpha^2\|\nabla w\|^2+\|w\|^2\right)+\nu\|\nabla w\|^2+\nu\alpha^2(\Delta D_{N-1}^{1/2}\overline{w},\Delta D_{N-1}^{1/2}w)=(f,w), \tag{10}$$

and using the filter definition yields the energy equality (from regularity results in [8], all operations are justified)

$$\frac{1}{2}\frac{d}{dt}\left(\alpha^2\|\nabla w\|^2+\|w\|^2\right)+\nu\|\nabla w\|^2+\nu\alpha^2\|\Delta D_{N-1}^{1/2}\overline{w}\|^2+\nu\alpha^4\|\nabla\Delta D_{N-1}^{1/2}\overline{w}\|^2=(f,w). \tag{11}$$

This demonstrates the effect on the energy dissipation, which serves to dissipate higher order norms of $D_{N-1}^{1/2}\overline{w}$. Applying the filter definition once again, one can write

$$\frac{1}{2}\frac{d}{dt}\left(\alpha^2\|\nabla w\|^2+\|w\|^2\right)+\nu\|\nabla w\|^2+\nu\alpha^2\|D_{N-1}^{1/2}(w-\overline{w})\|^2+\nu\alpha^4\|\nabla D_{N-1}^{1/2}(w-\overline{w})\|^2=(f,w), \tag{12}$$

which more clearly shows that this extra viscous term acts to dissipate velocity fluctuations.

3. Analytical Results

The first analysis of rNS-α was performed in [8], where it was shown that the model is well-posed for a fixed end time, and that regularity of solutions depended on the regularity of the data. Global in time energy and regularity results were proven in [17], and are stated below, after some initial preliminaries. The treatment of energy by the model was studied in [16]. Results for the energy conservation of the model, energy spectra, and energy dissipation were all very good: an appropriate energy quantity is conserved, the model exhibits a $k^{-5/3}$ energy cascade on the large scales in the inertial range and at k^{-3} on smaller scales. Furthermore, energy is dissipated at the rate $\frac{U^3}{L}$, independent of the Reynolds number, which is consistent with true fluid flow.

Consider in this section a domain $\Omega \subset \mathbb{R}^d$, d = 2 or 3, which is a box. The notation $\|\cdot\|$ and $(\cdot,\cdot)$ denotes the $L^2(\Omega)$ norm and inner product, with all other norms being labeled clearly. Assume periodic boundary conditions on the box, which is the common setting for such energy studies, as it is typically the only case where well-posedness of solutions is known. For the energy balance and dissipation results, the results extend to the wall-bounded case if the well-posedness results also extend to the wall-bounded case, which is an open problem. The energy spectra results, however, rely heavily on the periodic setting, as they explicitly use Fourier decompositions. Most of the filtering and deconvolution results can be extended to the case of homogeneous Dirichlet boundary conditions, if filtered velocities satisfy no-slip boundary conditions. This is not widely accepted as a correct boundary condition for the filter (on the other hand, this is the boundary condition use in all successful turbulent flow simulations with rNS-α [8,16], and seems to work quite well).

Denote $(X, Q) := (H^1_{\#}(\Omega), L^2_{\#}(\Omega)) \subset (H^1(\Omega), L^2(\Omega))$ the periodic, zero-mean velocity and pressure spaces. The space $H^{-1}(\Omega)$ is the dual space of X.

3.1. Energy Bounds and Well-Posedness

Presented now are several long time energy bounds. The following was proven in [17], and shows the energy is trapped for all times if $f \in L^\infty((0,\infty); H^{-1}(\Omega))$ and $w_0 \in X$.

Lemma 1 (Energy trapping). *Suppose $f \in L^\infty((0,\infty); H^{-1}(\Omega))$ and $w_0 \in X$. Then, $\forall\, t > 0$,*

$$\alpha^2\|\nabla w(t)\|^2 + \|w(t)\|^2 \leq \max\left\{\frac{1}{\nu^2\beta}\|f\|^2_{L^\infty((0,\infty),H^{-1})}, \alpha^2\|\nabla w_0\|^2 + \|w_0\|^2\right\} := C_E,$$

where $\beta = \min\left\{\frac{1}{2\alpha^2}, \frac{C_\Omega^2}{2}\right\}$. Thus, the kinetic energy is contained in the ball $B(0, C_E)$ for all time.

Under the additional restriction that $f \in L^2((0,\infty); H^{-1}(\Omega))$, then energy must decay to 0. This is proven in [17], and stated in the following lemma.

Lemma 2 (Energy decay). *Suppose $f \in L^2((0,\infty); H^{-1}(\Omega))$ and $w_0 \in X$. Then, the energy decays to zero as $t \to \infty$:*

$$\lim_{t\to\infty}\left(\alpha^2\|\nabla w(t)\|^2 + \|w(t)\|^2\right) = 0.$$

Global well-posedness of the model was able to be proven, using the long time a priori energy bounds above [17].

Lemma 3 (Existence and uniqueness of weak solutions). *Suppose $f \in L^2((0,\infty); H^{-1}(\Omega))$ or $f \in L^\infty((0,\infty); H^{-1}(\Omega))$ and $w_0 \in X$. Then, there exists a global unique weak solution to Equations* (1)–(3).

Additional regularity of the data leads to global in time higher order regularity of solutions, and is proven in [17].

Lemma 4 (Higher order estimates). *Assume a solenoidal initial condition $w_0 \in H^2(\Omega) \cap X$. If $f \in L^\infty((0,\infty), L^2(\Omega))$, then $w \in L^\infty((0,\infty), H^2(\Omega))$. If $f \in L^2((0,\infty), L^2(\Omega))$, then $\|w\|_{H^2} \to 0$ as $t \to \infty$.*

Remark 1. *Assuming higher order regularity of the data, higher order regularity of solutions can be established using bootstrapping and the techniques used in [17].*

3.2. Energy and Helicity Balances

Integral invariants such as energy and helicity are fundamental for a priori estimates used in existence theorems. They can also provide physical insight into a model's behavior, as well as its physical relevance. It has long been established that these quantities are invariant in true fluid flow (i.e., in the Navier–Stokes equations), and as they are believed to be fundamental to the organization of flow structures, a good model should capture these quantities accurately.

The conserved model energy and helicity take the form:

$$\begin{aligned} E_{rNS\alpha}(t) &:= \frac{1}{2|\Omega|}\left(\alpha^2\|\nabla w(t)\|^2 + \|w(t)\|^2\right), \\ H_{rNS\alpha}(t) &:= \frac{1}{|\Omega|}\left(\alpha^2(\nabla\times D_N^{1/2}w(t), \nabla\times(\nabla\times D_N^{1/2}w(t))) + (D_N^{1/2}w(t), \nabla\times D_N^{1/2}w(t))\right). \end{aligned}$$

The energy of this model is the same as for NS–Voigt [22]. As w represents an approximation of $\overline{v}$, with v being the NS-α velocity, one can observe that the rNS-α energy is analogous to the NS-α energy $E_{NS\alpha} := \frac{1}{2|\Omega|}\left(\alpha^2\|\nabla\overline{v}(t)\|^2 + \|\overline{v}(t)\|^2\right)$ ([9,10]). By similar reasoning, the rNS-α helicity is connected to the NS–Voigt helicity and NS-α helicity.

It is stated below, and proven in [16], that the energy and helicity of the model are preserved. Furthermore, if the forcing is only spatially dependent, then the energy balance leads to a global in time bound on energy.

Theorem 1. *Let w be the solution to Equations (1)–(3) under periodic boundary conditions, with sufficiently smooth data. Then, the following model energy and helicity balances hold:*

$$E_{rNS\alpha}(T) + \frac{1}{|\Omega|}\nu \int_0^T \|\nabla D_N^{1/2} w(t)\|^2 \, dt = E_{rNS\alpha}(0) + \frac{1}{|\Omega|}\int_0^T (f(t), w(t)) \, dt,$$

$$H(T) + \frac{2}{|\Omega|}\nu \int_0^T (\nabla \times D_N w(t), \nabla \times (\nabla \times D_N w(t))) \, dt = H(0) + \frac{2}{|\Omega|}\int_0^T (f(t), \nabla \times D_N w(t)) \, dt.$$

In the case with vanishing viscosity and no external forcing, the model energy and helicity are exactly conserved. That is, for $v = 0$ and $f = 0$,

$$E_{rNS\alpha}(t) = E_{rNS\alpha}(0), \quad H_{rNS\alpha}(t) = H_{rNS\alpha}(0).$$

If the forcing is only spatially dependent, the energy balance from Theorem 1 can be further analyzed, revealing that $\epsilon_{rNS\alpha} := \frac{\nu}{|\Omega|}\|\nabla D_N^{1/2} w\|^2$ is the energy dissipation of the model. The following result is proven in [16].

Suppose that the forcing is solenoidal and only dependent on space, $f(x,t) = f(x) \in L^2(\Omega)$, and the initial condition $w_0 \in H^1(\Omega)$. Then,

$$\sup_{t\in(0,\infty)} E_{rNS\alpha}(t) \leq C(data) < \infty, \tag{13}$$

$$\frac{1}{T}\int_0^T \epsilon_{rNS\alpha}(t)\, dt = \frac{1}{T}\int_0^T \frac{\nu}{|\Omega|}\|\nabla D_N^{1/2} w\|^2 \, dt \leq \frac{1}{T}E_{rNS\alpha}(0) + \frac{1}{|\Omega|^{1/2}}\|f\| \left(\frac{1}{T}\int_0^T \frac{1}{|\Omega|}\|w\|^2\right)^{1/2}. \tag{14}$$

3.3. Energy Dissipation Rate

Consider now the rate of energy dissipation by the model, and assume a constant solenoidal forcing, i.e., $f = f(x) \in H^1(\Omega)$, and $\nabla \cdot f = 0$. The scale of the body force, large scale velocity, and length are defined as:

$$F := \left(\frac{1}{|\Omega|}\|f(x)\|^2\right)^{1/2},$$

$$U := \langle \frac{1}{|\Omega|}\|w(x,t)\|^2\rangle^{1/2},$$

$$L := \min\left(\frac{F}{\|\nabla f\|_\infty}, \frac{F}{\left(\frac{1}{|\Omega|}\|\nabla f\|^2\right)^{1/2}}\right),$$

where $\langle \phi \rangle := \limsup_{T\to\infty} \frac{1}{T}\int_0^T \phi(t)\, dt$ denotes the time average. It can be shown that each component of the definition for L has units of length.

The following theorem for the scaling of the time averaged energy dissipation is proven in [16]. Recall that N is assumed small, i.e., $N \leq 5$.

Theorem 2. *The time averaged energy dissipation of the rNS-α model is bounded as*

$$\langle \epsilon_{rNS\alpha}\rangle = \langle \frac{\nu}{|\Omega|}\|\nabla D_N^{1/2} w\|^2\rangle \leq \frac{C_N}{2}\left(3 + Re^{-1}\right)\frac{U^3}{L},$$

where C_N is a constant dependent only on N (e.g., $C_N \leq 6$ when $N = 2$).

Remark 2. *The above result is consistent with K41 phenomenology, which predicts time averaged energy dissipation to scale with $\frac{U^3}{L}$ [29], with the scalings derived from the Navier–Stokes equations directly [30–32], and for the usual NS-α model [33].*

3.4. Energy Spectra

It is a generally accepted theory in turbulence that energy is input at large scales, cascaded (but preserved) by the nonlinearity through the 'inertial range' of intermediate scales, and then dissipated exponentially fast by viscosity in the small scale range. For accurate computations, it is critical to resolve the flow to where viscosity takes over. For the Navier–Stokes equations, resolving all of these scales is a computationally intractable problem, and it is necessary to model. Studying the energy spectra of a model allows us to gain insight into its computability, as a successful model must have an inertial range that is shorter than that of the Navier–Stokes. The analysis and notation below follows studies done in [9,34–36].

Decomposing velocity into its Fourier modes yields the balance of energy

$$\frac{1}{2}\frac{d}{dt}\Big((w_k, w_k) + \alpha^2(-\Delta w_k, w_k)\Big) + \nu(-\Delta \widehat{D}_N(k) w_k, w_k) = T_k - T_{2k}, \tag{15}$$

where $\widehat{D}_N(k)$ is the Fourier transform of D_N, and

$$T_k = -((\nabla \times \widehat{D}_N(k) w_k^{<}) \times w_k, w_k) + ((\nabla \times \widehat{D}_N(k)(w_k + w_k^{>})) \times (w_k + w_k^{>}), w_k^{<}),$$

with

$$w_k^{<} = \sum_{|j|<k} w_j, \quad w_k^{>} = \sum_{|j|\geq 2k} w_j.$$

The quantity $T_k - T_{2k}$ represents net energy transferred into the wave numbers between $[k, 2k)$. Time averaging the balance Equation (15) gives the energy transfer equation

$$\langle \nu(-\Delta \widehat{D}_N(k) w_k, w_k)\rangle = \langle T_k\rangle - \langle T_{2k}\rangle. \tag{16}$$

Define the model energy of a size k^{-1} eddy, based on the rNS-α model energy $E_{rNS\alpha}$, as

$$E_{rNS\alpha}(k) = (1 + \alpha^2 k^2) \sum_{|j|=k} |\hat{w}_j|^2,$$

and combining with Equation (16) gives

$$\langle T_k\rangle - \langle T_{2k}\rangle \sim \frac{\nu k^3 \widehat{D}_N(k) E_{rNS\alpha}(k)}{1 + \alpha^2 k^2} \sim \frac{\nu k^3 (N+1) E_{rNS\alpha}(k)}{1 + \alpha^2 k^2},$$

with the last relation coming from [2]: $\widehat{D}_N(k) \sim (N+1)$ for sufficiently large k. This is a key estimate for determining the inertial range, since here one expects no leakage of energy through dissipation, and hence $\langle T_k\rangle \approx \langle T_{2k}\rangle$. This suggests that increasing N leads to a shorter inertial range, which is consistent with the energy spectrum computations in Section 4.

The inertial range's kinetic energy distribution is now considered. Begin by defining the average velocity of a size k^{-1} eddy, and then relate it to model energy:

$$U_k = \langle (w_k, w_k)\rangle^{1/2} \sim \left(\int_k^{2k} \frac{E_{rNS\alpha}(k)}{1 + \alpha^2 k^2}\right)^{1/2} \sim \left(\frac{k E_{rNS\alpha}(k)}{(1 + \alpha^2 k^2)}\right)^{1/2}.$$

From the Kraichnan energy cascade theory [37], one obtains that the eddy turnover time is

$$\tau_k := \frac{1}{kU_k} = \frac{(1+k^2\alpha^2)^{1/2}}{k^{3/2}E_{rNS\alpha}(k)^{1/2}},$$

and thus the model energy dissipation rate scales like

$$\epsilon_{rNS\alpha} \sim \frac{1}{\tau_k}\int_k^{2k} E_{rNS\alpha}(k) \sim \frac{k^{5/2}E_{rNS\alpha}(k)^{3/2}}{(1+k^2\alpha^2)^{1/2}}.$$

Solving for $E_{rNS\alpha}(k)$, one obtains the relation

$$E_{rNS\alpha}(k) \sim \frac{\epsilon_{rNS\alpha}^{2/3}(1+\alpha^2k^2)^{1/3}}{k^{5/3}},$$

and this yields a spectrum for kinetic energy ($E = \frac{1}{2}\|w\|^2$):

$$E(k) \sim \frac{E_{rNS\alpha}(k)}{(1+\alpha^2k^2)} \sim \frac{\epsilon_{rNS\alpha}^{2/3}}{k^{5/3}(1+\alpha^2k^2)^{2/3}}.$$

Split the spectrum into two parts: if $k\alpha > O(1)$, then

$$E(k) \sim \frac{\epsilon_{rNS\alpha}^{2/3}}{k^{5/3}(\alpha^2k^2)^{2/3}} \sim \frac{\epsilon_{rNS\alpha}^{2/3}}{k^3\alpha^{4/3}},$$

but if $k\alpha < O(1)$,

$$E(k) \sim \frac{\epsilon_{rNS\alpha}^{2/3}}{k^{5/3}}.$$

These scalings show that on larger inertial range scales, a $k^{-5/3}$ rolloff of energy is expected (which agrees with true fluid flow). However, for wave numbers bigger than $O(\alpha^{-1})$, a rolloff of k^{-3} is expected. These rolloffs are identical to usual NS-α on both the larger and smaller scales in the inertial range [9]. This implies that significantly less energy is held in higher wave numbers compared to a Navier–Stokes DNS, which makes the model more computable than the Navier-Stokes Equations (NSE). Numerical tests in [16] illustrate these scalings. Another important takeaway is that since the rolloffs mirror those of NS-α, one should expect computational cost (in terms of mesh width resolution) to be very similar NS-α, at least in this periodic setting. This means that total cost in the C^0 finite element setting should be much less for rNS-α compared to NS-α, since the former is more efficiently computable, as discussed above.

4. Sensitivity of the Model with Respect to the Filtering Radius α

The parameter α in rNS-α represents a filtering radius, and changing α in computations can lead to very different solutions. Error analysis of the model's discretization above suggests $\alpha = O(h)$ is a good choice, and choosing α much smaller than this would lead to no regularization, since such a discrete filter would effectively have 0 filtering radius. Throughout the literature, α is often chosen slightly larger than h, such as 2 h or 6 h (our experience makes us favor the choice of 2 h). Due to these various choices, it makes sense to study sensitivity of the model to changes in α, and in this section, a sensitivity equation for $\frac{\partial}{\partial\alpha}w$ is derived and analyzed. In later sections, efficient algorithms are proposed for computing it, and finally computations are performed. This section follows work done in [17].

The sensitivity equation for $\frac{\partial}{\partial\alpha}w$ is derived by applying $\frac{\partial}{\partial\alpha}$ to rNS-α, then setting $s = \partial w/\partial\alpha$ and $r = \partial p/\partial\alpha$. This gives

$$-\alpha^2\Delta s_t + s_t + \left(\nabla \times \frac{\partial}{\partial \alpha} D_N w\right) \times w + (\nabla \times \frac{\partial}{\partial \alpha} D_N w) \times s + \nabla r - \nu \Delta \frac{\partial}{\partial \alpha} D_N w = 2\alpha \Delta w_t, \quad (17)$$

$$\nabla \cdot s = 0, \quad (18)$$

$$s(0) = 0. \quad (19)$$

This system is reduced by eliminating variables for filtered velocities' sensitivities after evaluating $\frac{\partial}{\partial \alpha} D_N w$, and writing the system in terms of $s = \partial w / \partial \alpha$. The system depends on N, and for the first few values of N, one gets

N	$D_N w$	$\frac{\partial D w}{\partial \alpha}$
0	I	Ds
1	$2w - \overline{w}$	$Ds - 2\alpha\Delta\overline{\overline{w}}$
2	$3w - 3\overline{w} + \overline{\overline{w}}$	$Ds - 6\alpha\Delta\overline{\overline{w}} + 4\alpha\Delta\overline{\overline{\overline{w}}}$
3	$4w - 6\overline{w} + 4\overline{\overline{w}} - 3\overline{\overline{\overline{w}}}$	$Ds - 12\alpha\Delta\overline{\overline{w}} + 16\alpha\Delta\overline{\overline{\overline{w}}} - 18\alpha\Delta\overline{\overline{\overline{\overline{w}}}}$

Note that for general N, one can write $\frac{\partial D_N w}{\partial \alpha}$ can be written as $D_N s$ plus a linear combination of filtered w's,

$$\frac{\partial D_N w}{\partial \alpha} = D_N s + \alpha \Delta \sum_{i=1}^{N} \beta_i^{(N)} F^{i+1} w,$$

where the $\beta_i^{(N)}$'s are integers depending on N. Thus, the general system takes the form

$$-\alpha^2 \Delta s_t + s_t + (\nabla \times D_N s) \times w + (\nabla \times D_N w) \times s + \nabla r - \nu \Delta D_N s =$$

$$2\alpha \Delta w_t + \nu \alpha \Delta^2 \left(\sum_{i=1}^{N} \beta_i^{(N)} F^{i+1} w \right) - \alpha \left(\nabla \times \Delta \sum_{i=1}^{N} \beta_i^{(N)} F^{i+1} w \right) \times w, \quad (20)$$

$$\nabla \cdot s = 0, \quad (21)$$

$$s(0) = 0. \quad (22)$$

This system is proven in [17] to be well-posed in the periodic setting, and the result reads as follows:

Theorem 3 (Existence and uniqueness of weak sensitivity solutions). *Let $f \in L^\infty((0,\infty); L^2(\Omega))$ and $w_0 \in X$. Then, weak solutions to Equations (20)–(22) exist uniquely, and satisfy for $T < \infty$,*

$$\alpha^2 ||\nabla s(T)||^2 + ||s(T)||^2 + \nu \int_0^T ||\nabla s(t)||^2 dt \le C(data).$$

In the numerical tests section below, efficient numerical schemes for computing this sensitivity system are proposed and analyzed.

5. Numerical Schemes and Analysis

An efficient numerical scheme for approximating solutions to rNS-α using a C^0 finite element spatial discretization, together with an implicit-explicit BDF2 timestepping, is given. This time discretization linearizes the system at each time step and decouples the filtering/deconvolution equations from the momentum/mass system. After providing some preliminaries, the scheme and results for its stability, well-posedness, and convergence, are presented.

5.1. Problem Setting and Preliminaries

Assume a regular, conforming triangulation/tetrahedralization $\tau_h(\Omega)$, and associated inf-sup stable discrete velocity-pressure spaces $(X_h, Q_h) \subset (X, Q)$ that are piecewise polynomials on each element. Define the discretely divergence free subspace by

$$V_h := \{v_h \in X_h, \ (\nabla \cdot v_h, q_h) = 0 \, \forall q_h \in Q_h\}.$$

Discrete filtering is defined by the standard finite element discretization of the α filter: Given $\phi \in L^2(\Omega)$, find $F_h\phi = \overline{\phi}^h \in X_h$ satisfying

$$\alpha^2(\nabla\overline{\phi}^h, \nabla\chi_h) + (\overline{\phi}^h, \chi_h) = (\phi, \chi_h) \ \forall\chi_h \in X_h.$$

Discrete van Cittert deconvolution is analogous to the continuous case, but defined with the discrete filter:

$$D_N^h = \sum_{n=0}^{N} (I - F_h)^n.$$

An explicit filter operator is now given that filters known quantities exactly, but for unknown quantities, it instead filters a known second order approximation:

$$\widetilde{F}_h w_h^j = F_h w_h^j \text{ for } j=1,2,...,n, \textit{ and } \widetilde{F}_h w_h^{n+1} := F_h(2w_h^n - w_h^{n-1}).$$

This operator defines an explicit deconvolution operator

$$\widetilde{D_N^h} w_h^{n+1} = \sum_{n=0}^{N} (I - \widetilde{F}_h)^n.$$

Finally, the implicit-explicit (IMEX) BDF2, C^0 finite element scheme for the rNS-α model can be stated.

Algorithm 1: [BDF2] Given endtime $T > 0$, timestep $\Delta t > 0$, forcing $f \in L^\infty(0,T;H^{-1}(\Omega))$, and initial velocities w_h^{-1}, $w_h^0 \in V_h$, set $M = \frac{T}{\Delta t}$ and for $n = 1,2,...,M-1$, find $(w_h^{n+1}, q_h^{n+1}) \in (X_h, Q_h)$ satisfying for all $(v_h, r_h) \in (X_h, Q_h)$,

$$\frac{\alpha^2}{2\Delta t}\left(\nabla(3w_h^{n+1} - 4w_h^n + w_h^{n-1}), \nabla v_h\right) + \frac{1}{2\Delta t}\left(3w_h^{n+1} - 4w_h^n + w_h^{n-1}, v_h\right)$$
$$+ \left((\nabla \times D_N^h(2w_h^n - w_h^{n-1})) \times w_h^{n+1}, v_h\right) - (q_h^{n+1}, \nabla \cdot v_h)$$
$$+ \nu(\nabla\widetilde{D_N^h} w_h^{n+1}, \nabla v_h) = (f(t^{n+1}), v_h), \tag{23}$$
$$(\nabla \cdot w_h^{n+1}, r_h) = 0. \tag{24}$$

5.2. Analysis of the Scheme

Results of the stability and convergence of Algorithm 1 are now stated, and come from [8].

Lemma 5 (Stability). *Suppose that $\Delta t < \frac{\alpha^2}{4C(N)\nu}$. Then solutions of Algorithm 1 satisfy*

$$\alpha^2\|\nabla w_h^M\|^2 + \|w_h^M\|^2 + \sum_{n=0}^{M-1} \|w_h^{n+1} - 2w_h^n + w_h^{n-1}\|^2 + \nu\Delta t \sum_{n=1}^{M} \|\nabla w_h^n\|^2 \le C(data). \tag{25}$$

Moreover, solutions to Algorithm 1 exist uniquely since the scheme is linear at each timestep. Thus, the algorithm is well-posed.

Remark 3. *In most settings where large eddy simulation (LES) models such as rNS-α is used, it generally holds that $\nu \ll \alpha$. Moreover, typically $N \leq 3$, giving $C(N) \leq 6$. Thus, in practice, the timestep restriction will typically be mild.*

Convergence of the scheme is proven in [8] and is stated in the following theorem.

Theorem 4. *Suppose (X_h, Q_h) are chosen to be either (P_k, P_{k-1}) Taylor–Hood elements or (P_k, P_{k-1}^{disc}) Scott–Vogelius elements and suppose further that (w, p) is a solution of Equations (1) and (2) for a given $\alpha > 0$, $\nu > 0$, with forcing $f \in L^\infty(0,T;H^{-1}(\Omega))$ and initial velocity $w_0 \in V_h$, satisfying regularity criteria $w \in L^\infty(0,T;H^{k+1}(\Omega) \cap H^3(\Omega))$, $p \in L^\infty(0,T;H^k(\Omega))$, $w_t, w_{tt}, w_{ttt} \in L^\infty(0,T;H^{k+1}(\Omega) \cap H^3(\Omega))$. Suppose further that the stability criterion for the timestep size is satisfied. Then the error of Algorithm 1 in approximating solutions to the rNS-α model satisfies*

$$\left\| w(T) - w_h^M \right\| + \sum_{n=0}^{M-1} \left\| (w(t^{n+1}) - 2w(t^n) + w(t^{n-1})) - (w_h^{n+1} - 2w_h^n + w_h^{n-1}) \right\|$$
$$+\alpha^2 \left\| \nabla(w(T) - w_h^M) \right\| + \nu \left(\Delta t \sum_{n=1}^{M} \|\nabla(w(t^n) - w_h^n)\|^2 \right)^{1/2} \leq C\left(\Delta t^2 + h^k\right). \quad (26)$$

5.3. Numerical Scheme for rNS-α Sensitivity

An efficient numerical scheme is now detailed for the rNS-α sensitivity system to be used together with the IMEX BDF2 scheme above. The key feature for efficiency is that the discrete velocity/pressure and sensitivity systems will have *exactly* the same coefficient matrices. Thus, preconditioners for the velocity solve can be reused, and also one can take advantage of block solvers that work with multiple right-hand sides, since the proposed algorithms allows solving for s^n and w^{n+1} simultaneously. The discrete scheme is now presented. Step 1 is exactly the rNS-α scheme from above. Step 2 solves for velocity and pressure sensitivity, and is decoupled from Step 1.

Algorithm 2: [rNS-α with sensitivity] Given $w_h^0,\ w_h^1 \in V_h$, $s_h^0 = s_h^1 = 0 \in V_h$, $f \in L^\infty(0,T;L^2(\Omega))$, and timestep $\Delta t = \frac{T}{M}$, for $n = 2, 3, ..., M-1$,

Step 1: Find $w_h^{n+1} \in V_h$ satisfying:

$$\frac{1}{2\Delta t}\left(3w_h^{n+1} - 4w_h^n + w_h^{n-1}, v_h\right) + \frac{\alpha^2}{2\Delta t}\left(\nabla(3w_h^{n+1} - 4w_h^n + w_h^{n-1}), \nabla v_h\right)$$
$$+ \left((\nabla \times D_N^h(2w_h^n - w_h^{n-1})) \times w_h^{n+1}, v_h\right) + \nu\left(\nabla \widetilde{D_N^h w_h^{n+1}}, \nabla v_h\right) = \left(f(t^{n+1}), v_h\right) \quad (27)$$

Step 2: Find $s_h^{n+1} \in V_h$ satisfying:

$$\frac{1}{2\Delta t}\left(3s_h^{n+1} - 4s_h^n + s_h^{n-1}, v_h\right) + \frac{\alpha^2}{2\Delta t}\left(\nabla(3s_h^{n+1} - 4s_h^n + s_h^{n-1}), \nabla v_h\right)$$
$$+ \left((\nabla \times D_N^h(2w_h^n - w_h^{n-1})) \times s_h^{n+1}, v_h\right) + \nu\left(\nabla \widetilde{D_N^h s_h^{n+1}}, \nabla v_h\right)$$
$$= -\frac{\alpha}{\Delta t}\left(\nabla(3w_h^{n+1} - 4w_h^n + w_h^{n-1}), \nabla v_h\right) - \left((\nabla \times D_N^h(s_h^n - s_h^{n-1})) \times w_h^{n+1}, v_h\right)$$
$$- \alpha\left(\left(\nabla \times \Delta \sum_{i=1}^{N} \beta_i^{(N)} F_h^{i+1} w_h^{n+1}\right) \times w_h^{n+1}, v_h\right) - \nu\alpha\left(\nabla \Delta \sum_{i=1}^{N} \beta_i^{(N)} F_h^{i+1} w_h^{n+1}, \nabla v_h\right) \quad (28)$$

Stability and convergence of the Step 1 is stated above, and it is proven in [17] that Step 2 is stable and convergent. However, there is a slightly more restrictive condition on the time step size.

Theorem 5. *Let $w_0 \in H^1(\Omega)$, $f \in L^\infty([0,T], L^2(\Omega))$, and choose Δt such that $\Delta t < \min\{\frac{\alpha^2}{4C(N)\nu}, O(h^{2/3})\}$. Then,*

$$\left\|s_h^M\right\|^2 + \alpha^2 \left\|\nabla s_h^M\right\|^2 + \nu\Delta t \sum_{n=0}^{M-1} \left\|\nabla s_h^{n+1}\right\|^2 \leq C(data).$$

Furthermore, assuming sufficiently smooth data, the convergence estimate holds:

$$\left\|s(T) - s_h^M\right\|^2 + \nu\Delta t \sum_{n=0}^{M} \|\nabla(s(t^n) - s_h^n)\|^2 \leq C(\Delta t^2 + h^k).$$

6. Numerical Results

Results of computations for rNS-α are presented in this section. First, results of convergence rate tests are given, to illustrate the convergence results above. Next, two application problems are considered, 2D flow past a cylinder and 3D turbulent channel flow. For the two application problems, comparisons of the model solutions to the literature are made, and sensitivities are also computed and discussed.

6.1. Verification of Convergence Rates

In [8], convergence rate verification tests were performed for the IMEX scheme above for rNS-α. The test problem used was an exact rNS-α solution (i.e., it solves (1) and (2) with $f = 0$) given by

$$w = \begin{pmatrix} -\cos(n\pi x)\sin(n\pi y) \\ \sin(n\pi x)\cos(n\pi y) \end{pmatrix} e^{\frac{-2C_{D_N} n^2\pi^2\nu t}{1+2n^2\pi^2\alpha^2}}, \tag{29}$$

where C_D is defined for each N by $D_N w = C_{D_N} w$ (justification of the existence of such a constant is given in [8]).

Approximations to this exact solution were calculated in [8] on $\Omega = (0,1)^2$ and $0 \leq t \leq T = 0.1$, using Algorithm 1, with $f = 0$, $N = 1$, $n = 1$, $\nu = 1$, and $\alpha = \frac{1}{16}$ (with $C_{D_1} = 2 - \frac{1}{1+2\pi^2\alpha^2}$). Errors and rates for the scheme were computed, using successively refined uniform meshes and timesteps, with (P_2, P_1) Taylor–Hood elements. Based on Theorem 4, convergence is expected to satisfy

$$\|w - w_h\|_{2,1} := \|w - w_h\|_{L^2(0,T;H^1(\Omega))} = O(h^2 + \Delta t^2).$$

Thus, a convergence rate of 2 is predicted if the mesh width and the time step are reduced together, by a factor of 2 at each refinement, which is observed to be the case from Table 2.

Table 2. Velocity errors and rates for the convergence rate test are given in the table, and are consistent with the optimal theoretical rate of 2.

h	Δt	$\|w - w_h\|_{2,1}$	Rate
1/4	1/2	1.005×10^{-1}	-
1/8	1/4	6.553×10^{-2}	0.617
1/16	1/8	1.678×10^{-2}	1.965
1/32	1/16	4.089×10^{-3}	2.037
1/64	1/32	1.141×10^{-3}	1.841

6.2. 2D Channel Flow Past a Cylinder

Results of tests of the rNS-α model/scheme for 2D flow past a cylinder are now presented. This is a benchmark problem from [38–40], with the domain being a 2.2 $\times$ 0.41 rectangular channel with a radius = 0.05 cylinder, centered at $(0.2, 0.2)$, as shown in Figure 1. No slip boundary conditions are prescribed on the walls and cylinder, and the inflow and outflow profiles are given by

$$u_1(0,y,t) = u_1(2.2,y,t) = \frac{6}{0.41^2}\sin(\pi t/8)y(0.41-y),$$
$$u_2(0,y,t) = u_2(2.2,y,t) = 0.$$

The kinematic viscosity is set as $\nu = 10^{-3}$, the forcing $f = 0$, and the flow starts from rest. The correct behavior is for a vortex street to form by $t = 4$ and continue to $t = 8$. The flow is driven by an interaction of the fluid with the cylinder, which is an important scenario for industrial flows. This flow is not turbulent, but is still very challenging for most numerical models and methods, especially on coarser meshes.

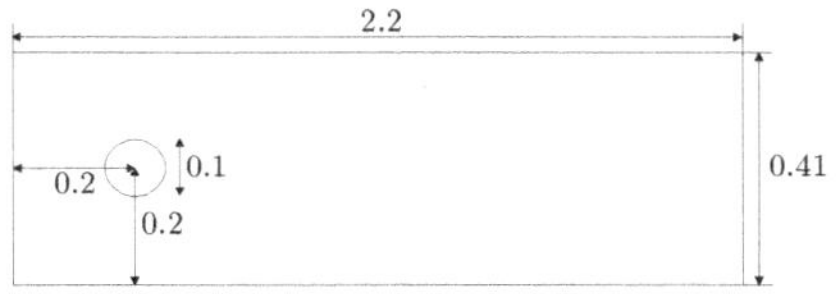

Figure 1. Shown above is the channel flow around a cylinder domain.

Important statistics for this flow are the predicted lift and drag forces. For a velocity field u and pressure p, lift and drag coefficients are defined by

$$c_d(t) = \frac{2}{\rho L U_{max}^2}\int_S \left(\rho\nu \frac{\partial u_{t_S}(t)}{\partial n} n_y - p(t) n_x\right) dS,$$
$$c_l(t) = \frac{2}{\rho L U_{max}^2}\int_S \left(\rho\nu \frac{\partial u_{t_S}(t)}{\partial n} n_x - p(t) n_y\right) dS,$$

where u_{t_S} is tangential velocity, $U_{max} = 1$ is the max velocity at the inlet, $L = 0.1$, $\rho = 1$, S is the cylinder, and $n = \langle n_x, n_y \rangle$ is the outward unit normal.

Solutions were computed using Algorithm 1 with $\Delta t = 0.002$ and $((P_2)^2, P_1^{disc})$ elements on a very coarse barycenter mesh that provided 5104 velocity degrees of freedom (dof). This choice of elements is made because the model is implemented using a Bernoulli-type pressure, which is significantly more complex than usual pressure and is known to affect the overall error if standard element choices are used [41], but not if divergence-free elements are used. The deconvolution parameter $N = 2$ was chosen, and tests were run with α varying from 0.010 to 0.016. To evaluate these solutions, values for the maximum drag $c_{d,max}$ and lift $c_{l,max}$ coefficients were calculated, and compared to those found in the resolved benchmark tests of [42]:

$$c_{d,max}^{ref} = 2.950918381, \quad c_{l,max}^{ref} = 0.47787543.$$

For comparison, the max lift and drag coefficient values were calculated on the same mesh, elements, and time step by the NSE discretized by the following extrapolated Crank–Nicolson finite element method:

$$\frac{1}{\Delta t}\left(u_h^{n+1}-u_h^n, v_h\right)+\left(\left(\nabla\times(\frac{3}{2}u_h^n-\frac{1}{2}u_h^{n-1})\right)\times u_h^{n+1/2}, v_h\right)$$
$$-\left(p_h^{n+1/2}, \nabla\cdot v_h\right)+\nu\left(\nabla u_h^{n+1/2}, \nabla v_h\right) = (f(t^{n+1/2}), v_h), \quad (30)$$
$$(\nabla\cdot u_h^{n+1}, q_h) = 0. \quad (31)$$

One observes from Table 3 that rNS-α provides much better max lift and drag coefficients than the coarse mesh NSE. The reference values come from upwards of 10^7 dof simulations, and it is remarkable the rNS-α gets this level of accuracy on such a coarse discretization.

Table 3. Shown in the table are max lift and drag coefficients for various models and parameters, for 2D flow past a cylinder.

Model	α	$c_{d,max}$	% Error	$c_{l,max}$	% Error
Navier–Stokes (NS)	-	3.393	15.0%	0.7499	56.9%
rNS-α	0.011	3.074	4.2%	0.511	6.9%
rNS-α	0.012	3.107	5.2%	0.506	5.9%
rNS-α	0.013	3.142	6.4%	0.501	4.8%
rNS-α	0.014	3.180	7.8%	0.497	4.0%
rNS-α	0.015	3.220	9.2%	0.492	3.0%
rNS-α	0.016	3.264	10.6%	0.487	1.9%

6.2.1. Sensitivity to α in Flow Past a Cylinder

For this test problem, sensitivity of the solution to α was considered, and Algorithm 2 was computed with $\alpha = 0.001$, $\Delta t = 0.001$, and (P_2, P_1^{disc}) Scott–Vogelius elements on a barycenter refined Delaunay triangulation, which provided 35,948 total dof. Plots of the sensitivity magnitude are shown in Figure 2 at various times. At early times, the sensitivity is created at the cylinder, and then as time progresses, it gets convected into the flow. By $t = 6$, values of $|s|$ are near 600. Creation of sensitivity at the solid boundary is not unexpected, since the sensitivity equation resembles a vorticity equation.

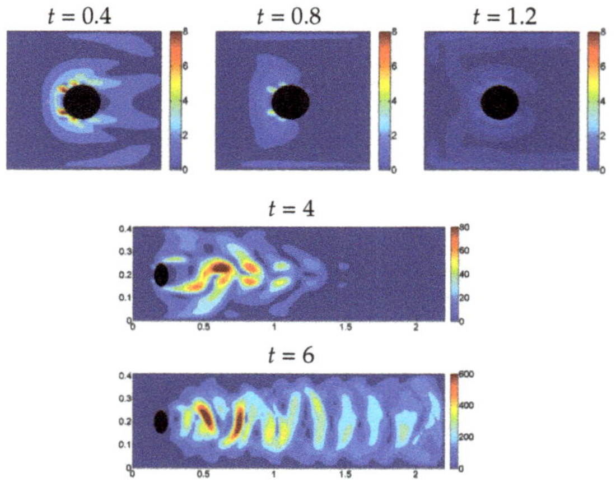

Figure 2. Sensitivity magnitudes at various times.

Lift and drag sensitivities are also calculated, defined by differentiating the lift and drag formulas above with respect to α. For the tests above, these quantities are also calculated:

$$s_d(t) = \frac{2}{\rho L U_{max}^2} \int_S \left(\rho\nu \frac{\partial s_{t_S}(t)}{\partial n} n_y - r(t) n_x \right) dS,$$
$$s_l(t) = \frac{2}{\rho L U_{max}^2} \int_S \left(\rho\nu \frac{\partial s_{t_S}(t)}{\partial n} n_x - r(t) n_y \right) dS.$$

Here, s_{t_S} denotes the tangential sensitivity. Shown in Figure 3 are plots of s_d and s_l versus time that came from this test. One observes very large sensitivity with the lift, and less so in the drag, but still significant. One also observes that the drag sensitivity is positive for most of the simulation, indicating that increasing α will increase drag.

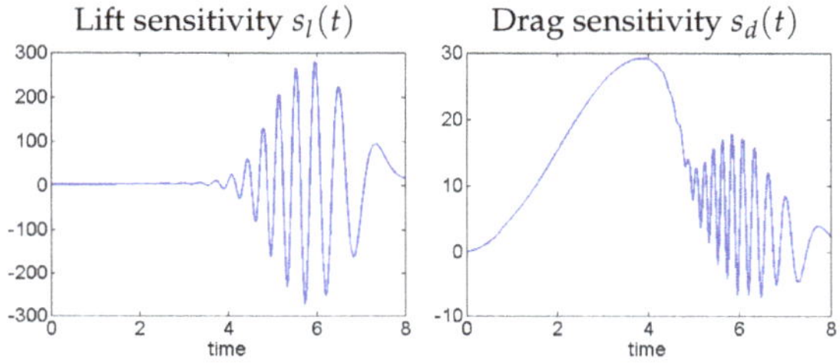

Figure 3. Sensitivities of lift and drag coefficients, plotted with time.

6.3. Turbulent Channel Flow at $Re_\tau = 590$

Results from rNS-α tests on a benchmark turbulent channel flow problem with friction Reynolds number $Re_\tau = 590$ from [16] are now discussed. The results below show that rNS-α does an outstanding job at matching the DNS mean velocity profile on a very coarse mesh. To our knowledge, no other model produces results this good on such a coarse mesh. For the widely cited DNS results [43], the number of degrees of freedom used was 113 million (our tests use just 82,821). For comparison, results on the same discretization for Leray-α, NS-α, and NS–Voigt models were computed on the same coarse mesh in [16] and shown to be much worse than rNS-α.

6.3.1. Problem Description and Results

The problem setup is given in detail in [8,16,44], and the interested reader is referred to those works for additional details. The domain is the box $\Omega = (-2\pi, 2\pi) \times (0,2) \times (-2\pi/3, 2\pi/3)$, with homogeneous Dirichlet boundary conditions enforced on the top and bottom walls $y = 0$ and $y = 2$, and periodic boundary conditions enforced on the remaining sides. The kinematic viscosity is set at $\nu = \frac{1}{Re_\tau}$. The initial condition is created by randomly perturbing the DNS average velocity profile pointwise. These perturbations create the turbulence as the simulation is run. The forcing is defined by $f = \langle 1,0,0 \rangle^T$, and is dynamically adjusted at each time step to maintain the bulk velocity. The simulations are run to $t = 60$, and the time and space average of the velocity is done from $t = 20$ to $t = 60$.

(P_3, P_2^{disc}) divergence-free Scott–Vogelius elements are used with a barycentric tetrahedral mesh that is weighted towards the walls, and provides 82,821 velocity dof. The time step size $\Delta t = 0.002$, deconvolution $N = 2$, α was varied. The value of $\alpha = 0.012$ gives an optimal (to within 0.0005) solution in the L^2 sense of best matching the DNS mean velocity profile; note that this value is close to the mesh resolution near the wall. Mean streamwise velocity profiles for rNS-α are shown in Figure 4, together with DNS data. Results this good are seemingly unique for any regularization model on such a coarse mesh, using a finite element discretization.

The $L^2(0,1)$ difference between the DNS mean streamwise velocity, and various model's mean streamwise velocity, is shown in Table 4 (the values of α are optimal to within 0.0005 in the interval [0,0.1] One immediately observes the rNS-α is much more accurate than Leray-α and NS–Voigt (which both barely improve on the 'no model', i.e., NSE on the coarse mesh). NS-α blows up for each choice of α (note the nonlinear problem was not solved at each time step; instead, a linearization on the filtered term in the nonlinearity is done, to make its efficiency the same as the other models (see [16])).

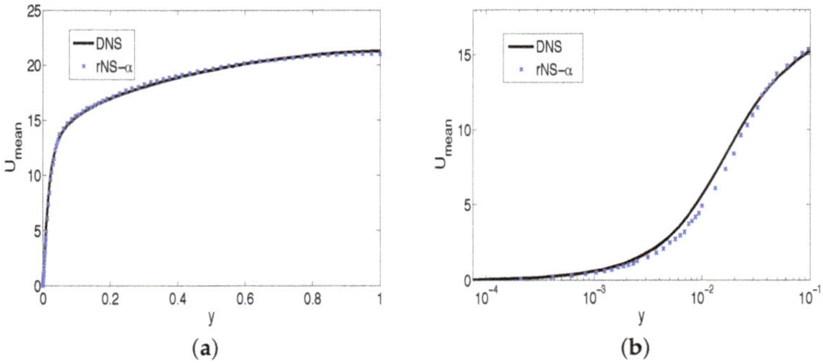

Figure 4. Shown above are the average velocity profiles of the coarse mesh reduced Navier–Stokes-α (rNS-α) simulation and direct numerical simulation (DNS) of the Navier-Stokes Equations (NSE), at $Re_\tau = 590$, shown in regular coordinates in (**a**), and near the wall in a log scale in (**b**).

Table 4. The $L^2(0,1)$ difference between the model's predicted velocity profile and the (cubic spline of) Moser, Kim, and Moin velocity profile for $Re_\tau = 590$, using results from the model with optimal α from 0.001 to 0.070, with increments of 0.001. No NS-α simulation with the implicit-explicit (IMEX) algorithm from the appendix remained stable for the entire simulation.

	$Re_\tau = 590$	
Model	**Optimal α**	$\Vert\langle u_{DNS}\rangle(y) - \langle u_{model}\rangle(y)\Vert_{L^2(0,1)}$
Navier-Stokes Equations (NSE) coarse mesh	-	1.409
rNS-α	0.012	0.184
NS-α	DNE	DNE
Leray-α	0.040	1.213
NS–Voigt	0.030	1.238

6.4. Model Sensitivity for Turbulent Channel Flow with $Re_\tau = 590$

rNS-α sensitivity was tested in [17] for this benchmark turbulent channel flow problem. Since the quantity of interest for this problem is the long time averaged velocity, a sensitivity study of rNS-α for this problem is now performed, to determine the long time averaged velocity sensitivity.

The same discretization as in [16] is used for this test: decoupled IMEX-BDF2-finite element discretization (step 1 of Algorithm 2), and (P_3, P_2^{disc}) Scott–Vogelius elements on a tetrahedral mesh that provided 82,281 velocity (and sensitivity) degrees of freedom. Both $N = 2$ and $N = 0$ (which is the NS–Voigt model) are tested, and $\alpha = 0.010$, 0.015 and 0.020. Starting from the T = 60, $N = 2$, $\alpha = 0.015$ (turbulent) solution as the initial condition (created using the setup above), the simulation is run to T = 70 s using a time step size of $\Delta t = 0.002$ (1000 total time steps). Algorithm 2 step 2 is used to calculate sensitivities at each time step, and time and space (x and z directions) averages are taken to get the mean streamwise velocity and sensitivity profiles.

From Figure 5, observe that rNS-α ($N = 2$) gives an excellent prediction of the mean streamwise velocity profile for each choice of α, significantly better than the NS–Voigt ($N = 0$) case. For the varying α, there is no visible change in the NS–Voigt solution, and a slight visible change in the rNS-α solution.

Figure 6 shows the mean streamwise sensitivity and relative sensitivity magnitudes, for both rNS-α and NS–Voigt. One observes that rNS-α is more sensitive than NS–Voigt to changes in α. More interesting is the observation that the sensitivity is largest near the boundary. Although this is perhaps expected due to (1) the boundary layer is critical for turbulent channel flow; and (2) the 'correct' boundary conditions for filtered velocities is an open problem for LES/α models (note the $N = 0$ model does not use filtered velocity boundary conditions).

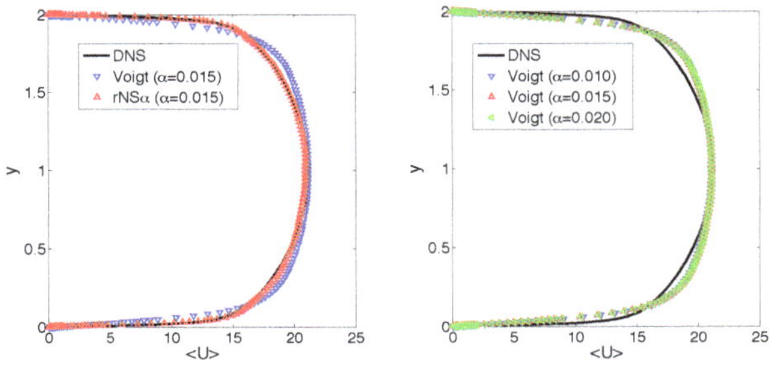

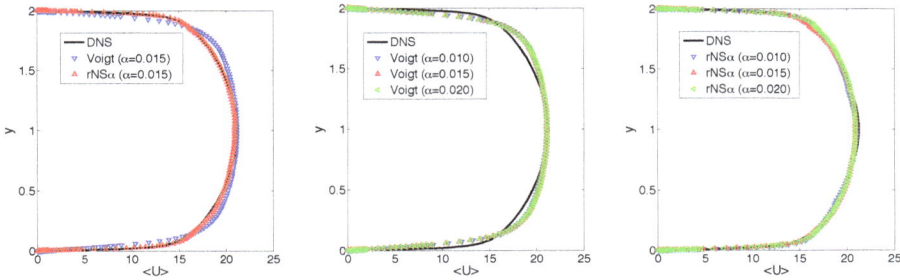

Figure 5. Average streamwise velocity profiles for DNS, rNS-α, and NS–Voigt models.

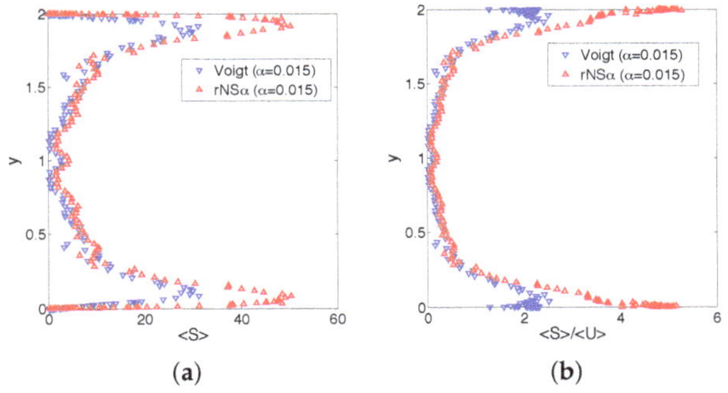

Figure 6. Average streamwise sensitivity (**a**) and relative streamwise sensitivity magnitude (**b**) for rNS-α and Voigt solutions with $\alpha = 0.015$.

7. Conclusions

The main results to date for rNS-α have been reviewed in this paper. This includes global well-posedness, energy spectrum analysis, and energy dissipation analysis. An efficient numerical discretization is reviewed, and corresponding stability and convergence results are given. Additionally, sensitivity equations for rNS-α are derived, and a numerical scheme for them is given. Finally, results of several numerical tests are given, which show excellent results on several benchmark problems.

There are several directions for future work. Further testing, further comparison to other LES models, and exploration of the model outside of the finite element framework should all be done. One could also immediately move this model into the multiphysics arena, for non-isothermal flow, or doubly diffusive flow, or even to magnetohydrodynamics. Additional numerical testing on difficult benchmark problems is also important for any newer models. An interesting study could likely be done for rNS-α's near-wall behavior; it remains unclear what makes rNS-α so much more accurate for turbulent channel flow, but one hypothesis is that it could be more accurate near the wall. Another interesting future direction of this work is to consider the rNS-α model, but remove all modeling except in the viscous term. That is, one could consider a simplified model to be

$$w_t + (\nabla \times w) \times w + \nabla p - \nu \Delta D_N w = f, \tag{32}$$

$$\nabla \cdot w = 0, \tag{33}$$

or possibly keep the Voigt term $-\alpha^2 \Delta w_t$. Proceeding as in Section 2, one can write the momentum equation of this reduced model as

$$w_t + (\nabla \times w) \times w + \nabla p - \nu \Delta w - \nu \Delta D_{N-1}(w - \overline{w}) = f.$$

Note that this system can be considered as Navier–Stokes with an additional term meant to dissipate velocity fluctuations. Our discussion in Section 2 shows that the altered viscous term is strongly related to the VMM approach to turbulence modeling, so it is possible that it is only this term that is to thank for the model's successes on the turbulent channel flow benchmark tests. One also notes that implementing the term would require very little extra work in a simulation, if the velocity term is treated implicitly and the filtered velocity is treated explicitly (by extrapolating from previous time steps).

Another possible direction for future work is choosing α locally and dynamically. Work in [45–47] has shown that carefully chosen local filtering radii can give large increases in accuracy in certain settings, and so extending these ideas to rNS-α seems like a logical direction.

Author Contributions: Abigail L. Bowers and Leo G. Rebholz did an equal amount of work on this manuscript, including the analysis, experiments, and writeup.

Conflicts of Interest: The authors declare no conflict of interest.

References

1. Berselli, L.; Iliescu, T.; Layton, W. *Mathematics of Large Eddy Simulation of Turbulent Flows*; Scientific Computation; Springer: Berlin, Germany, 2006.
2. Chacon-Rebello, T.; Lewandowski, R. *Mathematical and Numerical Foundations of Turbulence Models and Applications*; Springer: New York, NY, USA, 2014.
3. Layton, W.; Rebholz, L. *Approximate Deconvolution Models of Turbulence: Analysis, Phenomenology and Numerical Analysis*; Springer: Berlin, Germany, 2012.
4. Stolz, S.; Adams, N.; Kleiser, L. The approximate deconvolution model for large-eddy simulations of compressible flows and its application to shock-turbulent-boundary-layer interaction. *Phys. Fluids* **2001**, doi:10.1063/1.1397277.
5. Adams, N.; Stolz, S. A subgrid-scale deconvolution approach for shock capturing. *J. Comput. Phys.* **2002**, *178*, 391–426.
6. Adams, N.A.; Stolz, S. On the approximate deconvolution procedure for LES. *Phys. Fluids* **1999**, *2*, 1699–1701.
7. Stolz, S.; Adams, N.; Kleiser, L. An approximate deconvolution model for large-eddy simulations with application to incompressible wall-bounded flows. *Phys. Fluids* **2001**, *13*, 997–1015.
8. Cuff, V.; Dunca, A.; Manica, C.; Rebholz, L. The reduced order NS-α model for incompressible flow: Theory, numerical analysis and benchmark testing. *ESAIM Math. Model. Numer. Anal.* **2015**, *49*, 641–662.
9. Foias, C.; Holm, D.; Titi, E. The Navier–Stokes-alpha model of fluid turbulence. *Physica D* **2001**, *152*, 505–519.
10. Foias, C.; Holm, D.; Titi, E. The three dimensional viscous Camassa–Holm equations, and their relation to the Navier–Stokes equations and turbulence theory. *J. Dyn. Differ. Equ.* **2002**, *14*, 1–35.
11. Chen, S.; Foias, C.; Holm, D.; Olson, E.; Titi, E.; Wynne, S. The Camassa–Holm equations as a closure model for turbulent channel and pipe flow. *Phys. Rev. Lett.* **1998**, *81*, 5338–5341.
12. Chen, S.; Foias, C.; Olson, E.; Titi, E.; Wynne, W. The Camassa-Holm equations and turbulence. *Physica D* **1999**, *133*, 49–65.

13. Chen, S.; Foias, C.; Olson, E.; Titi, E.; Wynne, W. A connection between the Camassa-Holm equations and turbulent flows in channels and pipes. *Phys. Fluids* **1999**, *11*, 2343–2353.
14. Guermond, J.; Oden, J.; Prudhomme, S. An interpretation of the Navier–Stokes-alpha model as a frame-indifferent Leray regularization. *Physica D* **2003**, *177*, 23–30.
15. Cheskidov, A. Boundary layer for the Navier–Stokes-α model of fluid turbulence. *Arch. Ration. Mech. Anal.* **2004**, *172*, 333–362.
16. Rebholz, L.; Kim, T.Y.; Byon, Y.L. On an accurate α model for coarse mesh turbulent channel flow simulation. *Appl. Math. Model.* **2017**, *43*, 139–154.
17. Rebholz, L.; Zerfas, C.; Zhao, K. Global in time analysis and sensitivity analysis for the reduced NS-α model of incompressible flow. *J. Math. Fluid Mech.* **2016**, 1–23, doi:10.1007/s00021-016-0290-5.
18. Kim, T.Y.; Cassiani, M.; Albertson, J.; Dolbow, J.; Fried, E.; Gurtin, M. Impact of the inherent separation of scales in the Navier–Stokes-$\alpha\beta$ equations. *Phys. Rev. E* **2009**, *79045307*, 1–4.
19. Kim, T.Y.; Neda, M.; Rebholz, L.; Fried, E. A numerical study of the Navier–Stokes-$\alpha\beta$ model. *Comput. Methods Appl. Mech. Eng.* **2011**, *200*, 2891–2902.
20. Kalantarov, V.; Titi, E. Global attractors and determining modes for the 3D Navier–Stokes-Voight equations. *Chin. Ann. Math. Ser. B* **2009**, *30*, 697–714.
21. Levant, B.; Ramos, F.; Titi, E. On the statistical properties of the 3D incompressible Navier–Stokes-Voigt model. *Commun. Math. Sci.* **2010**, *8*, 277–293.
22. Larios, A.; Titi, E. On the higher-order global regularity of the inviscid Voigt-regularization of three-dimensional hydrodynamic models. *Discret. Contin. Dyn. Syst.* **2010**, *14*, 603–627.
23. Holst, M.; Lunasin, E.; Tsogtgerel, G. Analytical study of generalized α-models of turbulence. *J. Nonlinear Sci.* **2010**, *20*, 523–567.
24. Berselli, L.C.; Bisconti, L. On the structural stability of the Euler-Voight and Navier-Stokes-Voight models. *Nonlinear Anal.* **2012**, *75*, 117–130.
25. Abdi, N. TurBulence Modelling of the Navier–Stokes Equations Using The NS-α Approach. Master's Thesis, Freie Universitat, Berlin, Germany, 2015.
26. Dunca, A.; Epshteyn, Y. On the Stolz-Adams deconvolution model for the large-eddy simulation of turbulent flows. *SIAM J. Math. Anal.* **2005**, *37*, 1890–1902.
27. Dunca, A.; Neda, M. On the Vreman filter based stabilization for the advection equation. *Appl. Math. Comput.* **2015**, *269*, 379–388.
28. Vreman, A. The filtering analog of the variational multiscale method in large-eddy simulation. *Phys. Fluids* **2003**, *15*, 1–5.
29. Frisch, U. *Turbulence*; Cambridge University Press: Cambridge, UK, 1995.
30. Constantin, P.; Doering, C. Energy Dissipation in Shear Driven Turbulence. *Phys. Rev. Lett.* **1992**, *69*, 1648–1651.
31. Constantin, P.; Doering, C. Variational Bounds on Energy Dissipation in Incompressible Flows: Shear Flow. *Phys. Rev. E* **1994**, *49*, 4087–4099.
32. Doering, C.R.; Gibbon, J.D. *Applied Analysis of the Navier–Stokes Equations*; Cambridge University Press: Cambridge, UK, 1995.
33. Layton, W.; Rebholz, L.; Sussman, M. Energy and helicity dissipation rates of the NS-alpha and NS-alpha-deconvolution models. *IMA J. Appl. Math.* **2010**, *75*, 56–74.
34. Berselli, L.; Kim, T.Y.; Rebholz, L. Analysis of a reduced-order approximate deconvolution model and its interpretation as a NS-Voigt regularization. *DCDS-B* **2016**, *21*, 1027–1050.
35. Cao, Y.; Lunasin, E.; Titi, E. Global well-posedness of the three-dimensional viscous and inviscid simplified Bardina turbulence models. *Commun. Math. Sci.* **2006**, *4*, 823–848.
36. Cheskidov, A.; Holm, D.; Olson, E.; Titi, E. On a Leray-α model of turbulence. *Proc. R. Soc. A* **2005**, *461*, 629–649.
37. Kraichnan, R. Inertial-range transfer in two- and three-dimensional turbulence. *J. Fluid Mech.* **1971**, *47*, 525.
38. John, V. Reference values for drag and lift of a two dimensional time-dependent flow around a cylinder. *Int. J. Numerical Methods Fluids* **2004**, *44*, 777–788.
39. Schäfer, M.; Turek, S. The benchmark problem 'flow around a cylinder' flow simulation with high performance computers II. In *Notes on Numerical Fluid Mechanics*; Hirschel, E.H., Ed.; Vieweg: Braunschweig, Germany, 1996; Volume 52, pp. 547–566.
40. Hannasch, D.; Neda, M. On the accuracy of the viscous form in simulations of incompressible flow problems. *Numer. Methods Partial Differ. Equ.* **2012**, *28*, 523–541.

41. John, V.; Linke, A.; Merdon, C.; Neilan, M.; Rebholz, L.G. On the divergence constraint in mixed finite element methods for incompressible flows. *SIAM Rev.* **2017**, in press.
42. John, V.; Rang, J. Adaptive time step control for the incompressible Navier–Stokes equations. *Comput. Methods Appl. Mech. Eng.* **2010**, *199*, 514–524.
43. Moser, R.; Kim, J.; Mansour, N. Direct numerical simulation of turbulent channel flow up to $Re_\tau = 590$. *Phys. Fluids* **1999**, *11*, 943–945.
44. John, V.; Roland, M. Simulations of the turbulent channel flow at $Re_\tau = 180$ with projection-based finite element variational multiscale methods. *Int. J. Numer. Methods Fluids* **2007**, *55*, 407–429.
45. Galvin, K.; Rebholz, L.; Trenchea, C. Efficient, unconditionally stable, and optimally accurate FE algorithms for approximate deconvolution models. *SIAM J. Numer. Anal.* **2014**, *52*, 678–707.
46. Bowers, A.; Rebholz, L. Numerical study of a regularization model for incompressible flow with deconvolution-based adaptive nonlinear filtering. *Comput. Methods Appl. Mech. Eng.* **2013**, *258*, 1–12.
47. Layton, W.; Rebholz, L.; Trenchea, C. Modular nonlinear filter stabilization of methods for higher Reynolds numbers flow. *J. Math. Fluid. Mech.* **2012**, *14*, 325–354.

fluids

Article

Turbulence Intensity and the Friction Factor for Smooth- and Rough-Wall Pipe Flow

Nils T. Basse

Toftehøj 23, Høruphav, 6470 Sydals, Denmark; nils.basse@npb.dk

Academic Editor: William Layton
Received: 24 April 2017; Accepted: 8 June 2017; Published: 10 June 2017

Abstract: Turbulence intensity profiles are compared for smooth- and rough-wall pipe flow measurements made in the Princeton Superpipe. The profile development in the transition from hydraulically smooth to fully rough flow displays a propagating sequence from the pipe wall towards the pipe axis. The scaling of turbulence intensity with Reynolds number shows that the smooth- and rough-wall level deviates with increasing Reynolds number. We quantify the correspondence between turbulence intensity and the friction factor.

Keywords: turbulence intensity; Princeton Superpipe measurements; flow in smooth- and rough-wall pipes; friction factor

1. Introduction

Measurements of streamwise turbulence [1] in smooth and rough pipes have been carried out in the Princeton Superpipe [2–4] (Note that the author of this paper did not participate in making the Princeton Superpipe measurements.). We have treated the smooth pipe measurements as a part of [5]. In this paper, we add the rough pipe measurements to our previous analysis. The smooth (rough) pipe had a radius R of 64.68 (64.92) mm and a root mean square (RMS) roughness of 0.15 (5) μm, respectively. The corresponding sand-grain roughness is 0.45 (8) μm [6].

The smooth pipe is hydraulically smooth for all Reynolds numbers Re covered. The rough pipe evolves from hydraulically smooth through transitionally rough to fully rough with increasing Re. Throughout this paper, Re means the bulk Re defined using the pipe diameter D.

We define the turbulence intensity (TI) I as:

$$I(r) = \frac{v_{\mathrm{RMS}}(r)}{v(r)}, \tag{1}$$

where v is the mean flow velocity, v_{RMS} is the RMS of the turbulent velocity fluctuations and r is the radius ($r = 0$ is the pipe axis, $r = R$ is the pipe wall).

An overview of past research on turbulent flows over rough walls can be found in the pioneering work by Nikuradse [7] and a more recent review by Jiménez [8].

The development of predictive drag models has previously been carried out using both measurements [9] and direct numerical simulations (DNS) [10]. This work covered the transitionally and fully rough regimes and a variety of rough surface geometries.

The aim of this paper is to provide the fluid mechanics community with a scaling of the TI with Re, both for smooth- and rough-wall pipe flow. An application example is computational fluid dynamics (CFD) simulations, where the TI at an opening can be specified. A scaling expression of TI with Re is provided as Equation (6.62) in [11]. However, this formula does not appear to be documented, i.e., no reference is provided.

Our paper is structured as follows: in Section 2, we study how the TI profiles change over the transition from smooth to rough pipe flow. Thereafter, we present the resulting scaling of the TI with *Re* in Section 3. Quantification of the correspondence between the friction factor and the TI is contained in Section 4, and we discuss our findings in Section 5. Finally, we conclude in Section 6.

2. Turbulence Intensity Profiles

We have constructed the TI profiles for the measurements available (see Figure 1). Nine profiles are available for the smooth pipe and four for the rough pipe. In terms of *Re*, the rough pipe measurements are a subset of the smooth pipe measurements. Corresponding friction Reynolds numbers can be found in Table 1 in [3].

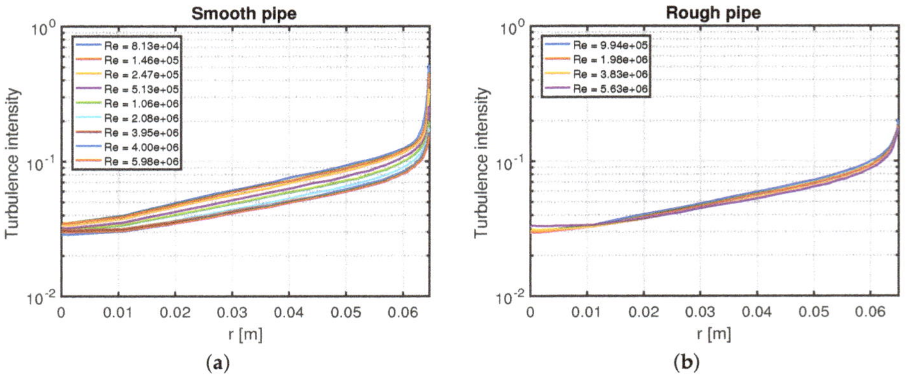

Figure 1. Turbulence intensity as a function of pipe radius, (**a**): smooth pipe; (**b**): rough pipe.

To make a direct comparison of the smooth and rough pipe measurements, we interpolate the smooth pipe measurements to the four *Re* values where the rough pipe measurements are done. Furthermore, we use a normalized pipe radius $r_n = r/R$ to account for the difference in smooth and rough pipe radii. The result is a comparison of the TI profiles at four *Re* (see Figure 2). As *Re* increases, we observe that the rough pipe TI becomes larger than the smooth pipe TI.

To make the comparison more quantitative, we define the turbulence intensity ratio (TIR):

$$r_{I,\text{Rough/Smooth}}(r_n) = \frac{I_{\text{Rough}}(r_n)}{I_{\text{Smooth}}(r_n)} = \frac{v_{\text{RMS,Rough}}(r_n)}{v_{\text{RMS,Smooth}}(r_n)} \times \frac{v_{\text{Smooth}}(r_n)}{v_{\text{Rough}}(r_n)}. \tag{2}$$

The TIR is shown in Figure 3. The left-hand plot shows all radii; prominent features are:

- The TIR on the axis is roughly one except for the highest *Re*, where it exceeds 1.1.
- In the intermediate region between the axis and the wall, an increase is already visible for the second-lowest *Re*, 1.98×10^6.

The events close to the wall are most clearly seen in the right-hand plot of Figure 3. A local peak of TIR is observed for all *Re*; the magnitude of the peak increases with *Re*. Note that we only analyse data to 99.8% of the pipe radius. Thus, the 0.13 mm closest to the wall is not considered.

The TIR information can also be represented by studying the TIR at fixed r_n vs. *Re* (see Figure 4). From this plot, we find that the magnitude of the peak close to the wall ($r_n = 0.99$) increases linearly with *Re*:

$$r_{I,\text{Rough/Smooth}}(r_n = 0.99) = 2.5137 \times 10^{-8} \times Re + 1.0161. \tag{3}$$

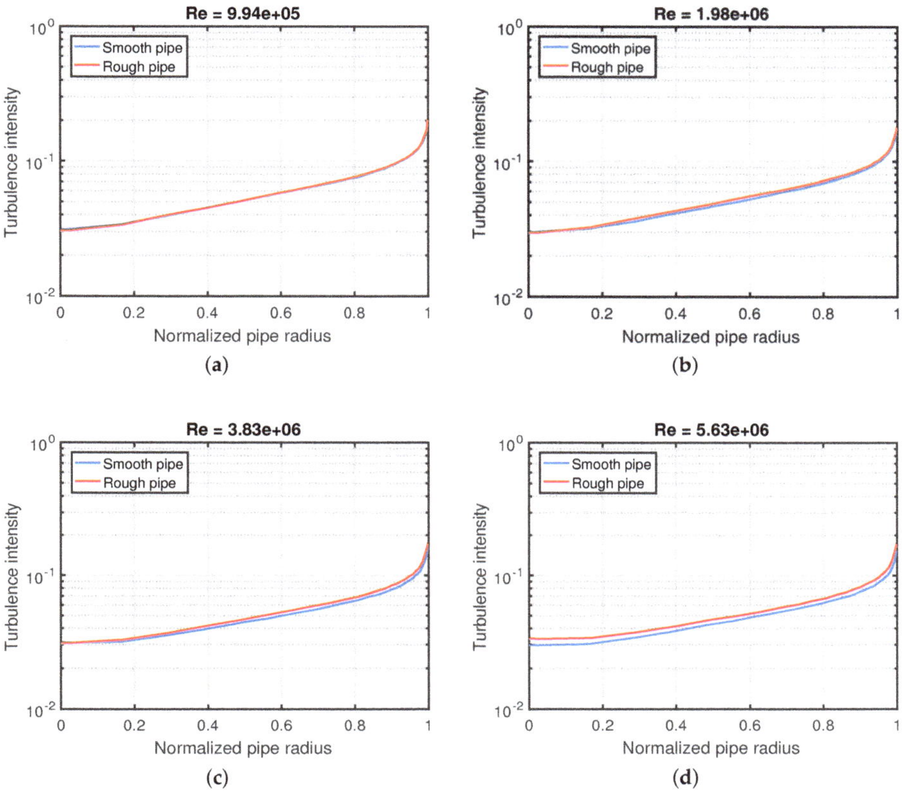

Figure 2. Comparison of smooth and rough pipe turbulence intensity (TI) profiles for the four *Re* values where the rough pipe measurements are done, (**a**): Re = 9.94e + 05; (**b**): Re = 1.98e + 06; (**c**): Re = 3.83e + 06; (**d**): Re = 5.63e + 06.

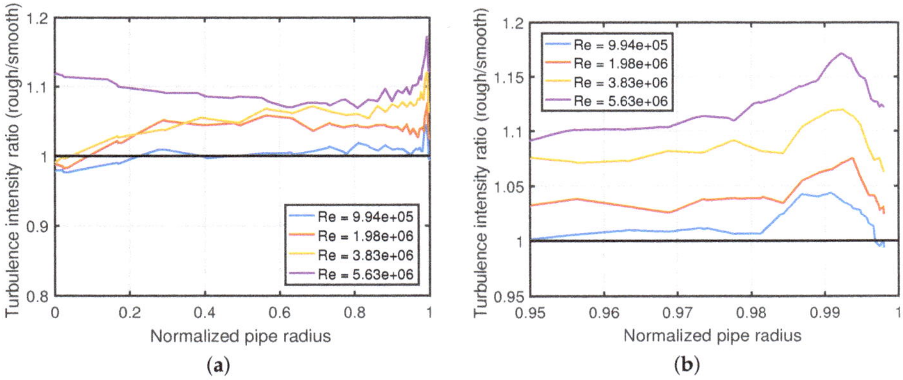

Figure 3. Turbulence intensity ratio (TIR); (**a**): all radii; (**b**): zoom to outer 5%.

Information on fits of the TI profiles to analytical expressions can be found in Appendix A.

Based on uncertainties in Table 2 in [3], the uncertainty of TI for the smooth (rough) pipe is 2.9% (3.5%), respectively. Note that we have used 4.4% instead of 4.7% for the uncertainty of $v_{\mathrm{RMS}}^2/v_\tau^2$ to derive the rough pipe uncertainty. The resulting TIR uncertainty is 4.5%.

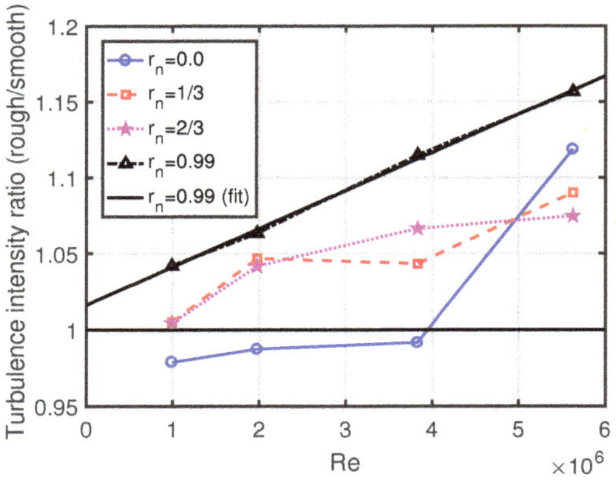

Figure 4. Turbulence intensity ratios for fixed r_n.

3. Turbulence Intensity Scaling

We define the TI averaged over the pipe area as:

$$I_{\text{Pipe area}} = \frac{2}{R^2} \int_0^R \frac{v_{\mathrm{RMS}}(r)}{v(r)} r \mathrm{d}r. \tag{4}$$

In [5], another definition was used for the TI averaged over the pipe area. Analysis presented in Sections 3 and 4 is repeated using that definition in Appendix B.

Scaling of the TI with *Re* for smooth- and rough-wall pipe flow is shown in Figure 5. For $Re = 10^6$, the smooth and rough pipe values are almost the same. However, when *Re* increases, the TI of the rough pipe increases compared to the smooth pipe; this increase is to a large extent caused by the TI increase in the intermediate region between the pipe axis and the pipe wall (see Figures 3 and 4). We have not made fits to the rough wall pipe measurements because of the limited number of datapoints.

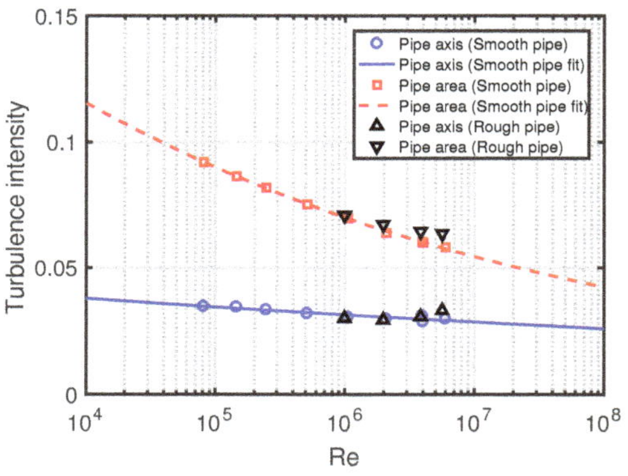

Figure 5. Turbulence intensity for smooth and rough pipe flow.

4. Friction Factor

The fits shown in Figure 5 are:

$$\begin{aligned} I_{\text{Smooth pipe axis}} &= 0.0550 \times Re^{-0.0407}, \\ I_{\text{Smooth pipe area}} &= 0.317 \times Re^{-0.110}. \end{aligned} \tag{5}$$

The Blasius smooth pipe (Darcy) friction factor [12] is also expressed as an *Re* power-law:

$$\lambda_{\text{Blasius}} = 0.3164 \times Re^{-0.25}. \tag{6}$$

The Blasius friction factor matches measurements best for $Re < 10^5$; the friction factor by e.g., Gersten (Equation (1.77) in [13]) is preferable for larger *Re*. The Blasius and Gersten friction factors are compared in Figure 6. The deviation between the smooth and rough pipe Gersten friction factors above $Re = 10^5$ is qualitatively similar to the deviation between the smooth and rough pipe area TI in Figure 5. For the Gersten friction factors, we have used the measured pipe roughnesses.

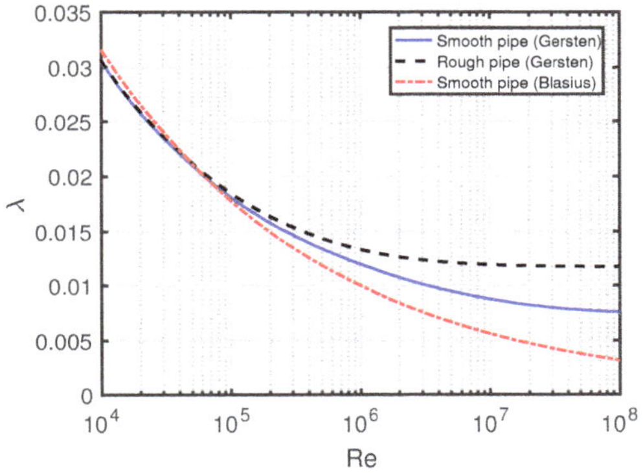

Figure 6. Friction factor.

For the smooth pipe, we can combine Equations (5) and (6) to relate the pipe area TI to the Blasius friction factor:

$$\begin{aligned} I_{\text{Smooth pipe area}} &= 0.526 \times \lambda_{\text{Blasius}}^{0.44}, \\ \lambda_{\text{Blasius}} &= 4.307 \times I_{\text{Smooth pipe area}}^{2.27}. \end{aligned} \tag{7}$$

The TI and Blasius friction factor scaling is shown in Figure 7.

For axisymmetric flow in the streamwise direction, the mean flow velocity averaged over the pipe area is:

$$v_m = \frac{2}{R^2} \times \int_0^R v(r) r dr. \tag{8}$$

Now, we are in a position to define an average velocity of the turbulent fluctuations:

$$\langle v_{\text{RMS}} \rangle = v_m I_{\text{Pipe area}} = \frac{4}{R^4} \int_0^R v(r) r dr \int_0^R \frac{v_{\text{RMS}}(r)}{v(r)} r dr. \tag{9}$$

The friction velocity is:

$$v_\tau = \sqrt{\tau_w / \rho}, \tag{10}$$

where τ_w is the wall shear stress and ρ is the fluid density.

The relationship between $\langle v_{\mathrm{RMS}} \rangle$ and v_τ is illustrated in Figure 8. From the fit, we have:

$$\langle v_{\mathrm{RMS}} \rangle = 1.8079 \times v_\tau, \tag{11}$$

which we approximate as:

$$\langle v_{\mathrm{RMS}} \rangle \sim \frac{9}{5} \times v_\tau. \tag{12}$$

Equations (11) and (12) above correspond to the usage of the friction velocity as a proxy for the velocity of the turbulent fluctuations [14]. We note that the rough wall velocities are higher than for the smooth wall.

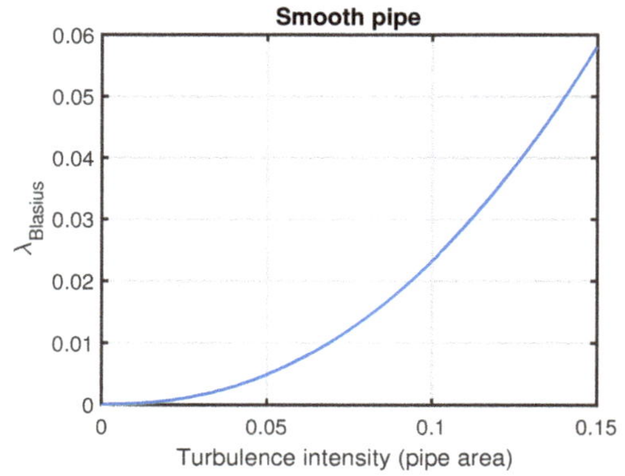

Figure 7. Relationship between pipe area turbulence intensity and the Blasius friction factor.

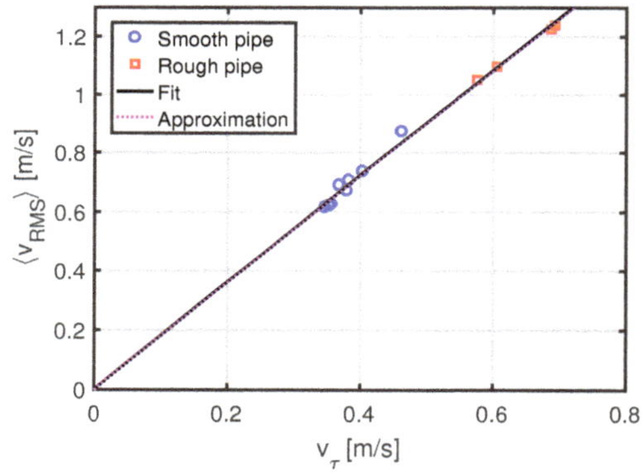

Figure 8. Relationship between friction velocity and the average velocity of the turbulent fluctuations.

Equations (9) and (12) can be combined with Equation (1.1) in [15]:

$$\lambda = \frac{4\tau_w}{\frac{1}{2}\rho v_m^2} = \frac{-(\Delta P/L)\,D}{\frac{1}{2}\rho v_m^2} = 8 \times \frac{v_\tau^2}{v_m^2} \sim \frac{200}{81} \times I_{\text{Pipe area}}^2, \tag{13}$$

where ΔP is the pressure loss and L is the pipe length. This can be reformulated as:

$$I_{\text{Pipe area}} \sim \frac{9}{10\sqrt{2}} \times \sqrt{\lambda}. \tag{14}$$

We show how well this approximation works in Figure 9. Overall, the agreement is within 15%.

We proceed to define the average kinetic energy of the turbulent velocity fluctuations $\langle E_{\mathrm{kin,RMS}} \rangle$ (per pipe volume V) as:

$$\begin{aligned} \langle E_{\mathrm{kin,RMS}} \rangle / V &= \tfrac{1}{2}\rho \langle v_{\mathrm{RMS}} \rangle^2 \sim -\tfrac{81}{50} \times (\Delta P/L)\, D/4 \\ &= \tfrac{81}{50} \times \tau_w = \tfrac{81}{50} \times v_\tau^2 \rho, \end{aligned} \tag{15}$$

with $V = L\pi R^2$, so we have:

$$\begin{aligned} \langle E_{\mathrm{kin,RMS}} \rangle &= \tfrac{1}{2} m \langle v_{\mathrm{RMS}} \rangle^2 \sim -\tfrac{81}{50} \times (\pi/2)\, R^3 \Delta P \\ &= \tfrac{81}{50} \times \tau_w V = \tfrac{81}{50} \times v_\tau^2 m, \end{aligned} \tag{16}$$

where m is the fluid mass. The pressure loss corresponds to an increase of the turbulent kinetic energy. The turbulent kinetic energy can also be expressed in terms of the mean flow velocity and the TI or the friction factor:

$$\langle v_{\mathrm{RMS}} \rangle^2 = v_m^2 I_{\text{Pipe area}}^2 \sim \frac{81}{200} \times v_m^2 \lambda. \tag{17}$$

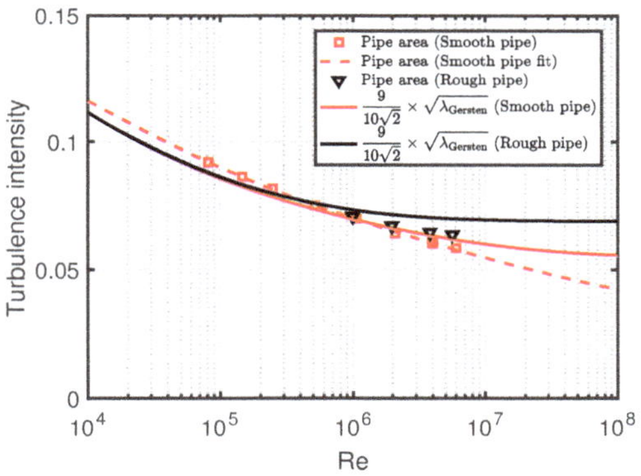

Figure 9. Turbulence intensity for smooth and rough pipe flow. The approximation in Equation (14) is included for comparison.

5. Discussion

5.1. The Attached Eddy Hypothesis

Our quantification of the ratio $\langle v_{\mathrm{RMS}} \rangle / v_\tau$ as a constant can be placed in the context of the attached eddy hypothesis by Townsend [16,17]. Our results are for quantities averaged over the pipe radius, whereas the attached eddy hypothesis provides a local scaling with distance from the wall. By proposing an overlap region (see Figure 1 in [18]) between the inner and outer scaling [19], it can be deduced that $\langle v_{\mathrm{RMS}} \rangle / v_\tau$ is a constant in this overlap region [20,21]. Such an overlap region has been shown to exist in [2,20]. The attached eddy hypothesis has provided the basis for theoretical work on e.g., the streamwise turbulent velocity fluctuations in flat-plate [22] and pipe flow [23] boundary layers. Work on the law of the wake in wall turbulence also makes use of the attached eddy hypothesis [24].

As a consistency check for our results, we can compare the constant 9/5 in Equation (12) to the prediction by Townsend:

$$\frac{v_{\mathrm{RMS,Townsend}}(r)^2}{v_\tau^2} = B_1 - A_1 \ln\left(\frac{R-r}{r}\right), \tag{18}$$

where fits have provided the constants $B_1 = 1.5$ and $A_1 = 1.25$. Here, A_1 is a universal constant, whereas B_1 is not expected to be a constant for different wall-bounded flows [25]. The constants are averages of fits presented in [3] to the smooth- and rough-wall Princeton Superpipe measurements. The Townsend-Perry constant A_1 was found to be 1.26 in [25]. Performing the area averaging yields:

$$\frac{\langle v_{\mathrm{RMS,Townsend}}\rangle^2}{v_\tau^2} = B_1 + \frac{3}{2} \times A_1 = 3.38. \quad (19)$$

Our finding is:

$$\frac{\langle v_{\mathrm{RMS}}\rangle^2}{v_\tau^2} \sim \left(\frac{9}{5}\right)^2 = 3.24, \quad (20)$$

which is within 5% of the result in Equation (19). The reason that our result is smaller is that Equation (18) is overpredicting the turbulence level close to the wall and close to the pipe axis. Equation (18) as an upper bound has also been discussed in [26].

5.2. The Friction Factor and Turbulent Velocity Fluctuations

The proportionality between the average kinetic energy of the turbulent velocity fluctuations and the friction velocity squared has been identified in [27] for $Re > 10^5$. This corresponds to our Equation (16).

A correspondence between the wall-normal Reynolds stress and the friction factor has been shown in [28]. Those results were found using DNS. The main difference between the cases is that we use the streamwise Reynolds stress. However, for an eddy rotating in the streamwise direction, both a wall-normal and a streamwise component should exist which connects the two observations.

5.3. The Turbulence Intensity and the Diagnostic Plot

Other related work can be found beginning with [29] where the diagnostic plot was introduced. In the following publications, a version of the diagnostic plot was brought forward where the local TI is plotted as a function of the local streamwise velocity normalised by the free stream velocity [30–32]. Equation (3) in [31] corresponds to our I_{Core} (see Equation (A1) in Appendix A).

5.4. Applicability of Turbulence Intensity Scaling with Friction Factor

The scaling of TI with the friction factor (Equation (14)) was found based on pipe flow measurements with two roughnesses. It is an open question whether our result holds in the fully rough regime. For the fully rough regime, the friction factor becomes a constant for high *Re*. As a consequence of our scaling expression, this should also be the case for the TI.

It is clear that the specific formula is not directly applicable for other wall-bounded flows, since B_1 takes different values. However, the basic behaviour, i.e., that the TI scales with the square root of the friction factor, may be universally valid.

6. Conclusions

We have compared TI profiles for smooth- and rough-wall pipe flow measurements made in the Princeton Superpipe.

The change of the TI profile with increasing *Re* from hydraulically smooth to fully rough flow exhibits propagation from the pipe wall to the pipe axis. The TIR at $r_n = 0.99$ scales linearly with *Re*.

The scaling of TI with *Re*—on the pipe axis and averaged over the pipe area—shows that the smooth- and rough-wall level deviates with increasing Reynolds number.

We find that $I_{\mathrm{Pipe\ area}} \sim \frac{9}{10\sqrt{2}} \times \sqrt{\lambda}$. This relationship can be useful to calculate the TI given a known λ, both for smooth and rough pipes. It follows that given a pressure loss in a pipe, the turbulent kinetic energy increase can be estimated.

Acknowledgments: We thank Alexander J. Smits for making the Superpipe data publicly available.

Conflicts of Interest: The authors declare no conflict of interest.

Appendix A. Fits to the Turbulence Intensity Profile

As we have done for the smooth pipe measurements in [5], we can also fit the rough pipe measurements to this function:

$$\begin{aligned} I(r_n) &= I_{\text{Core}}(r_n) + I_{\text{Wall}}(r_n) \\ &= \left[\alpha + \beta \times r_n^{\gamma}\right] + \left[\delta \times |\ln(1 - r_n)|^{\varepsilon}\right], \end{aligned} \tag{A1}$$

where α, β, γ, δ and ε are fit parameters. A comparison of fit parameters found for the smooth- and rough-pipe measurements is shown in Figure A1. Overall, we can state that the fit parameters for the smooth and rough pipes are in a similar range for $10^6 < Re < 6 \times 10^6$.

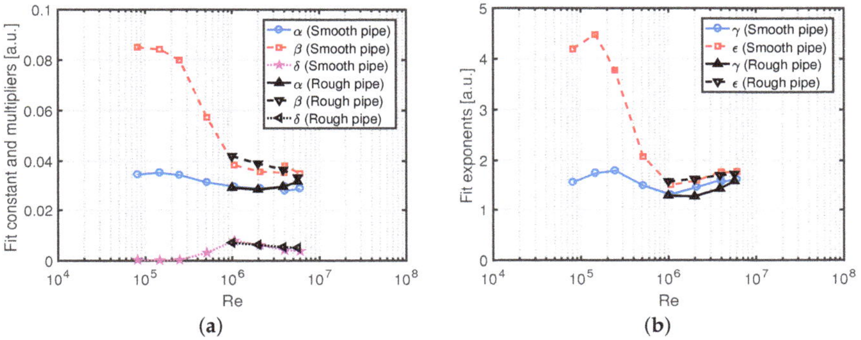

Figure A1. Comparison of smooth- and rough-pipe fit parameters, (**a**): fit parameters α, β and δ; (**b**): fit parameters γ and ε.

The min/max deviation of the rough pipe fit from the measurements is below 10%; see the comparison to the smooth wall fit min/max deviation in Figure A2.

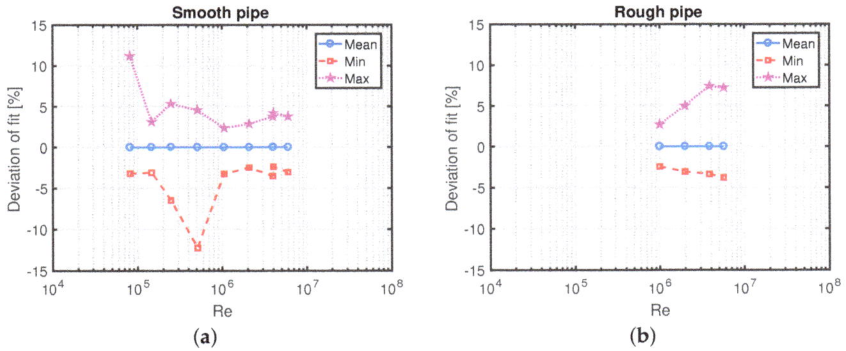

Figure A2. Deviation of fits to measurements; (**a**): smooth pipe, (**b**): rough pipe.

The core and wall fits for the smooth and rough pipe fits are compared in Figure A3.

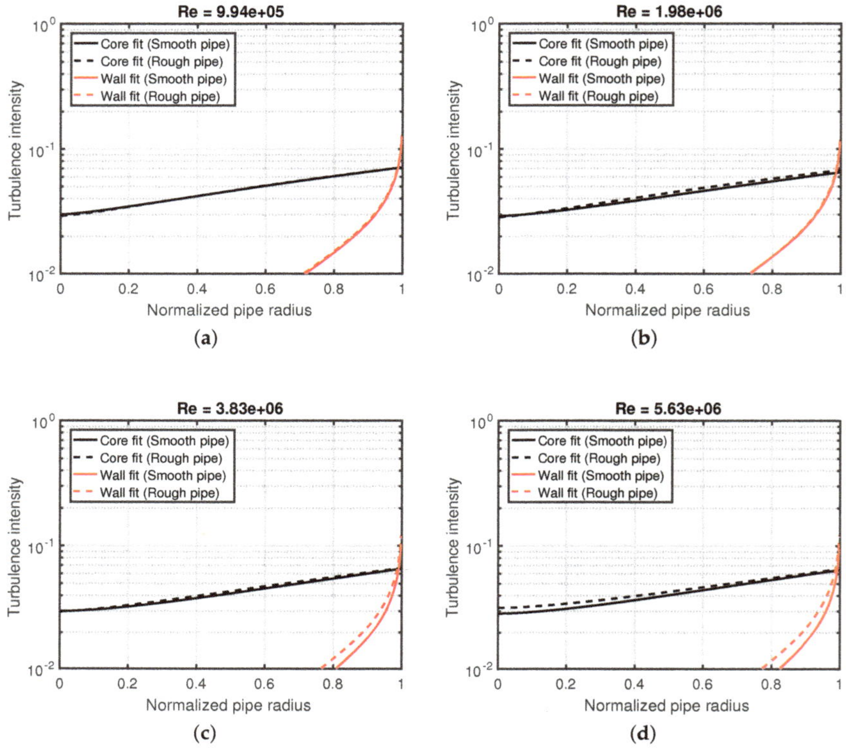

Figure A3. Comparison of smooth and rough pipe core and wall fits, (**a**): Re = 9.94e + 05; (**b**): Re = 1.98e + 06; (**c**): Re = 3.83e + 06; (**d**): Re = 5.63e + 06.

The position where the core and wall TI levels are equal is shown in Figure A4. This position does not change significantly for the rough pipe; however, the position does increase with *Re* for the smooth pipe: this indicates that the wall term becomes less important relative to the core term.

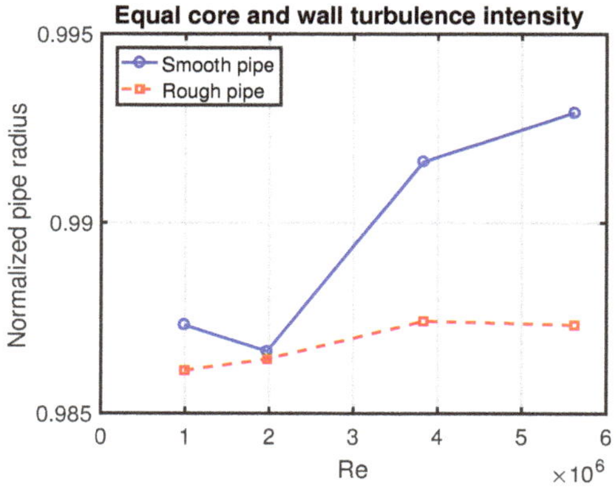

Figure A4. Normalised pipe radius where the core and wall TI levels are equal.

Appendix B. Arithmetic Mean Definition of Turbulence Intensity Averaged Over the Pipe Area

In the main paper, we have defined the TI over the pipe area in Equation (4). In [5], we used the arithmetic mean (AM) instead:

$$I_{\text{Pipe area, AM}} = \frac{1}{R}\int_0^R \frac{v_{\text{RMS}}(r)}{v(r)}\text{d}r. \tag{A2}$$

The AM leads to a somewhat different pipe area scaling for the smooth pipe measurements, which is illustrated in Figure A5. Compare to Figure 5.

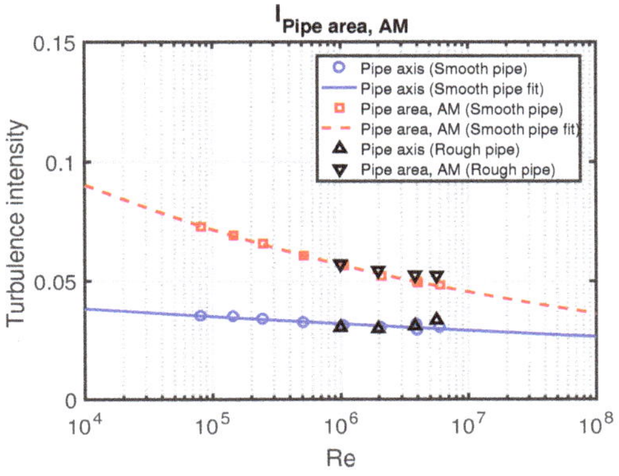

Figure A5. Turbulence intensity for smooth and rough pipe flow. The arithmetic mean (AM) is used for the pipe area TI.

The scaling found in [5] using this definition is:

$$I_{\text{Smooth pipe area, AM}} = 0.227 \times Re^{-0.100}. \tag{A3}$$

The AM scaling also has implications for the relationship with the Blasius friction factor scaling (Equation (7)):

$$\begin{aligned} I_{\text{Smooth pipe area, AM}} &= 0.360 \times \lambda_{\text{Blasius}}^{0.4}, \\ \lambda_{\text{Blasius}} &= 12.89 \times I_{\text{Smooth pipe area, AM}}^{2.5}. \end{aligned} \tag{A4}$$

We can now define the AM version of the average velocity of the turbulent fluctuations:

$$\langle v_{\text{RMS}} \rangle_{\text{AM}} = v_m I_{\text{Pipe area, AM}} = \frac{2}{R^3}\int_0^R v(r) r \text{d}r \int_0^R \frac{v_{\text{RMS}}(r)}{v(r)}\text{d}r. \tag{A5}$$

The AM definition can be considered as a first order moment equation for v_{RMS}, whereas the definition in Equation (9) is a second order moment equation.

Again, we find that the AM average turbulent velocity fluctuations are proportional to the friction velocity. However, the constant of proportionality is different than the one in Equation (11) (see Figure A6). The AM case can be fitted as:

$$\langle v_{\text{RMS}} \rangle_{\text{AM}} = 1.4708 \times v_\tau, \tag{A6}$$

which we approximate as:

$$\langle v_{RMS} \rangle_{AM} \sim \sqrt{\frac{2}{3}} \times \frac{9}{5} \times v_{\tau} \sim \sqrt{\frac{2}{3}} \times \langle v_{RMS} \rangle. \tag{A7}$$

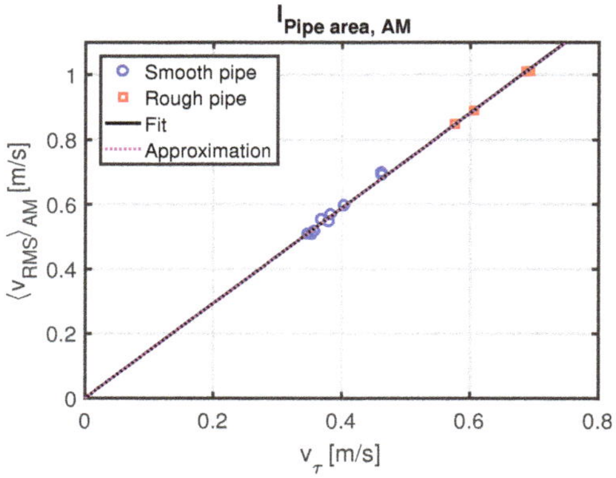

Figure A6. Relationship between friction velocity and the AM average velocity of the turbulent fluctuations.

As we did in Section 5, we can perform the AM averaging of Equation (18) (also done in [26]):

$$\frac{\langle v_{RMS,Townsend} \rangle^2_{AM}}{v_{\tau}^2} = B_1 + A_1 = 2.75, \tag{A8}$$

where we find:

$$\frac{\langle v_{RMS} \rangle^2_{AM}}{v_{\tau}^2} \sim \frac{2}{3} \times \left(\frac{9}{5}\right)^2 = 2.16. \tag{A9}$$

References

1. Marusic, I.; McKeon, B.J.; Monkewitz, P.A.; Nagib, H.M.; Smits, A.J.; Sreenivasan, K.R. Wall-bounded turbulent flows at high Reynolds numbers: Recent advances and key issues. *Phys. Fluids* **2010**, *22*, 065103.
2. Hultmark, M.; Vallikivi, M.; Bailey, S.C.C.; Smits, A.J. Turbulent pipe flow at extreme Reynolds numbers. *Phys. Rev. Lett.* **2012**, *108*, 094501.
3. Hultmark, M.; Vallikivi, M.; Bailey, S.C.C.; Smits, A.J. Logarithmic scaling of turbulence in smooth- and rough-wall pipe flow. *J. Fluid Mech.* **2013**, *728*, 376–395.
4. Princeton Superpipe. 2017. Available online: https://smits.princeton.edu/superpipe-turbulence-data/ (accessed on 7 June 2017).
5. Russo, F.; Basse, N.T. Scaling of turbulence intensity for low-speed flow in smooth pipes. *Flow Meas. Instrum.* **2016**, *52*, 101–114.
6. Langelandsvik, L.I.; Kunkel, G.J.; Smits, A.J. Flow in a commercial steel pipe. *J. Fluid Mech.* **2008**, *595*, 323–339.
7. Nikuradse, J. *Strömungsgesetze in Rauhen Rohren*; Springer-VDI-Verlag GmbH: Düsseldorf, Germany, 1933.
8. Jiménez, J. Turbulent flows over rough walls. *Annu. Rev. Fluid Mech.* **2004**, *36*, 173–196.
9. Flack, K.A.; Schultz, M.P. Review of hydraulic roughness scales in the fully rough regime. *J. Fluids Eng.* **2010**, *132*, 041203.
10. Chan, L.; MacDonald, M.; Chung, D.; Hutchins, N.; Ooi, A. A systematic investigation of roughness height and wavelength in turbulent pipe flow in the transitionally rough regime. *J. Fluid Mech.* **2015**, *771*, 743–777.
11. ANSYS Fluent User's Guide, Release 18.0; 2017. Available online: http://www.ansys.com/products/fluids/ (accessed on 7 June 2017).
12. Blasius, H. *Das Ähnlichkeitsgesetz bei Reibungsvorgängen in Flüssigkeiten*; Springer-VDI-Verlag GmbH: Düsseldorf, Germany, 1913; pp. 1–40.

13. Gersten, K. Fully developed turbulent pipe flow. In *Fluid Mechanics of Flow Metering*; Merzkirch, W., Ed.; Springer: Berlin, Germany, 2005.
14. Schlichting, H.; Gersten, K. *Boundary-Layer Theory*, 8th ed.; Springer: Berlin, Germany, 2000.
15. McKeon, B.J.; Zagarola, M.V.; Smits, A.J. A new friction factor relationship for fully developed pipe flow. *J. Fluid Mech.* **2005**, *538*, 429–443.
16. Townsend, A.A. *The Structure of Turbulent Shear Flow*, 2nd ed.; Cambridge University Press: Cambridge, UK, 1976.
17. Marusic, I.; Nickels, T.N.A.A. Townsend. In *A Voyage through Turbulence*; Davidson, P.A., Kaneda, Y., Moffatt, K., Sreenivasan, K.R., Eds.; Cambridge University Press: Cambridge, UK, 2011.
18. McKeon, B.J.; Morrison, J.F. Asymptotic scaling in turbulent pipe flow. *Phil. Trans. Royal Soc. A* **2007**, *365*, 771–787.
19. Millikan, C.B. A critical discussion of turbulent flows in channels and circular tubes. In Proceedings of the 5th International Congress for Applied Mechanics, New York, NY, USA, 12–16 September 1938.
20. Perry, A.E.; Abell, C.J. Scaling laws for pipe-flow turbulence. *J. Fluid Mech.* **1975**, *67*, 257–271.
21. Perry, A.E.; Abell, C.J. Asymptotic similarity of turbulence structures in smooth- and rough-walled pipes. *J. Fluid Mech.* **1977**, *79*, 785–799.
22. Marusic, I.; Kunkel, G.J. Streamwise turbulence intensity formulation for flat-plate boundary layers. *Phys. Fluids* **2003**, *15*, 2461–2464.
23. Hultmark, M. A theory for the streamwise turbulent fluctuations in high Reynolds number pipe flow. *J. Fluid Mech.* **2012**, *707*, 575–584.
24. Krug, D.; Philip, K.; Marusic, I. Revisiting the law of the wake in wall turbulence. *J. Fluid Mech.* **2017**, *811*, 421–435.
25. Marusic, I.; Monty, J.P.; Hultmark, M.; Smits, A.J. On the logarithmic region in wall turbulence. *JFM Rapids* **2013**, *716*, R3.
26. Pullin, D.I.; Inoue, M.; Saito, N. On the asymptotic state of high Reynolds number, smooth-wall turbulent flows. *Phys. Fluids* **2013**, *25*, 015116.
27. Yakhot, V.; Bailey, S.C.C.; Smits, A.J. Scaling of global properties of turbulence and skin friction in pipe and channel flows. *J. Fluid Mech.* **2010**, *652*, 65–73.
28. Orlandi, P. The importance of wall-normal Reynolds stress in turbulent rough channel flows. *Phys. Fluids* **2013**, *25*, 110813.
29. Alfredsson, P.H.; Örlü, R. The diagnostic plot—A litmus test for wall bounded turbulence data. *Eur. J. Mech. B Fluids* **2010**, *29*, 403–406.
30. Alfredsson, P.H.; Segalini, A.; Örlü, R. A new scaling for the streamwise turbulence intensity in wall-bounded turbulent flows and what it tells us about the "outer" peak. *Phys. Fluids* **2011**, *23*, 041702.
31. Alfredsson, P.H.; Örlü, R.; Segalini, A. A new formulation for the streamwise turbulence intensity distribution in wall-bounded turbulent flows. *Eur. J. Mech. B Fluids* **2012**, *36*, 167–175.
32. Castro, I.P.; Segalini, A.; Alfredsson, P.H. Outer-layer turbulence intensities in smooth- and rough-wall boundary layers. *J. Fluid Mech.* **2013**, *727*, 119–131.

fluids

MDPI

Article

High Wavenumber Coherent Structures in Low Re APG-Boundary-Layer Transition Flow—A Numerical Study

Weijia Chen and Edmond Y. Lo *

School of Civil and Environmental Engineering, Nanyang Technological University, Singapore 639798, Singapore; ch0003ia@e.ntu.edu.sg
* Correspondence: cymlo@ntu.edu.sg; Tel.: +65-6790-5268

Academic Editor: William Layton
Received: 27 February 2017; Accepted: 18 April 2017; Published: 28 April 2017

Abstract: This paper presents a numerical study of high wavenumber coherent structure evolution in boundary layer transition flow using recently-developed high order Combined compact difference schemes with non-uniform grids in the wall-normal direction for efficient simulation of such flows. The study focuses on a simulation of an Adverse-Pressure-Gradient (APG) boundary layer transition induced by broadband disturbance corresponding to the experiment of Borodulin et al. (Journal of Turbulence, 2006, 7, pp. 1–30). The results support the experimental observation that although the coherent structures seen during transition to turbulence have asymmetric shapes and occur in a random pattern, their local evolutional behaviors are quite similar. Further calculated local wavelet spectra of these coherent structures are also very similar. The wavelet spectrum of the streamwise disturbance velocity demonstrates high wavenumber clusters at the tip and the rear parts of the Λ-vortex. Both parts are imbedded at the primary Λ-vortex stage and spatially coincide with the spike region and high shear layer. The tip part is associated with the later first ring-like vortex, while the rear part with the remainder of the Λ-vortex. These observations help to shed light on the generation of turbulence, which is dominated by high wavenumber coherent structures.

Keywords: combined compact difference; numerical simulation; boundary layer transition; coherent structures; vortex dynamics; wavelet analysis

1. Introduction

The transition of the boundary layer from laminar to turbulent flow has attracted research interest for more than a century as it plays an important role in both fundamental studies of fluid mechanics and in engineering applications [1]. In this transition process, relatively small environmental disturbances introduced into the flow may amplify, interact with other nonlinear modes and lead to laminar flow breakdown. During the initial receptivity stage, the small amplitude disturbances are transformed into internal unstable Tollmien–Schlichting (TS) waves within the boundary layer [2]. If the TS waves amplify to sufficient amplitude [3], strong non-linear effects become significant and generate a series of two-dimensional (2D) harmonic waves resonantly [4]. As an enhancement of non-linear interaction, the harmonic waves are unstable to infinitesimal three-dimensional (3D) broadband disturbances that are always present both naturally and in experiments [5]. Based on their interacting mechanism, the normal transition can be subdivided into two kinds: fundamental transition first observed in [6] or subharmonic transition first discovered in [7].

In fundamental resonance, the TS wave interacts with the 3D disturbance wave of a given spanwise periodicity, resulting in an overall 3D structure with a "peak and valley distribution" in the spanwise direction for the streamwise disturbance velocity [6]. Λ-vortices in an aligned pattern appearing with tips at the peak positions when the 3D disturbance amplitudes have attained a magnitude of comparable order as the 2D TS wave.

Λ-vortices also occur in the subharmonic resonance, but are formed in a different way. The 3D structures are initiated by a 2D fundamental TS wave with frequency β_1 and its spanwise subharmonic wave with frequency $\beta_{1/2} \approx \beta_1/2$ [8]. The subharmonic wave rapidly amplifies to reach a magnitude of the order of the TS wave and leads to distortion in the spanwise direction [9]. Then, groups of weak Λ-vortices are formed in a staggered pattern.

In [9], the authors showed experimentally that, in the late stage of transition flow, the local behavior of the disturbances in the vicinity of Λ-structures is very similar with regard to both fundamental and subharmonic resonances. The main common features of the transition process include: (a) the formation of Λ-structures; (b) the appearance of spikes (large negative streamwise disturbance velocity in the time series) near the tips of Λ-structures; and (c) the formation of ring-like vortices departing away from the Λ-vortex and moving upward in the external part of the boundary layer. Although the physical observations of transition have been widely recorded in the literature [6,9,10], little consensus has been reached on a common mechanism behind the transition. In the experiments of [11], an Adverse-Pressure-Gradient (APG) boundary layer transition is induced by a TS wave plus initially weak broadband disturbances. The observed Λ-vortices, intensive Λ-shape high shear (HS) layers, Ω-vortices, ring-like vortices and associated spikes seen in time-traces and other phenomena are distributed randomly in time and space and have somewhat distorted shapes. However, it is reported that their general properties are qualitatively similar across sub-harmonic, harmonic and broadband disturbances.

Since the local behaviors of the vortex are quite similar in the late stage of transition, there is a strong interest in a theory to explain the common features of vortex evolution. In [10], the authors found that the first sign of randomness in the transition process was observed at a position corresponding to the tip of the Λ-vortex. In an experimental study of [12], a 3D Soliton-like Coherent Structures (SCS), essentially a wave packet accompanied with a high shear layer, is proposed to be the building block of the vortical coherent structures, which leads to turbulent burst [11]. Similarly, the Direct Numerical Simulation (DNS) results of [13] indicated that turbulence is not generated by vortex breakdown, but rather by positive and negative spikes and consequent high shear layers. A follow-on study by [14] further proposed that the shear layer instability is the "mother of turbulence". Both studies proposed theories suggesting a universal mechanism for turbulence generation and sustainment in [12,14]. Though different theories exist, the transition processes are essentially based on the existence of coherent flow structures and the formation of high wavenumber components, which is the precursor of turbulence. Therefore, the present study directly applies a local wavenumber detector, wavelet analysis to investigate the relationship between the typical vortical structure and the high wavenumber components, which is captured by a high order combined compact difference scheme in numerical simulations [15,16]. It can be shown that the local high wavenumber components in the streamwise direction mainly cluster inside the high shear layer region. The authors observe such behavior across fundamental, subharmonic resonance and transition induced by the broadband disturbance. The last case is the most general scenario, and hence, it is the focus of the present study.

Numerical simulation provides a helpful tool to investigate how spatial disturbance signals evolve temporally. Such studies have provided significant results to the study of boundary layer transition [10,17–20], but only a few of them present a spatial spectrum of the disturbance as in this study. The present study uses recently-developed high order combined compact difference schemes with non-uniform grids for numerical simulations of the Navier–Stokes (NS) equations for flat plate boundary layer transitions [15,16,21]. While these schemes are "DNS-like", they however are highly optimized for boundary layer transitions, i.e., having (a) low-numerical dispersion/dissipation to preserve wave speed and amplitude, (b) non-uniform gridding in the wall-normal direction to capture the high velocity gradients near the wall and (c) reported good comparisons with the experiments of [22] for Zero-Pressure-Gradient (ZPG) and [23] for APG for Re_{δ_1} (Reynolds number based on displacement thickness δ_1) up to 1200. In the present study, simulation results of the experiments from [11] on the APG boundary layer transition with Re_{δ_1} up to 1130 are analyzed in detail. Furthermore the Continuous Wavelet Transform (CWT) is used as the signal identifier to demonstrate the local high wavenumber spectrum in the boundary layer transition. This technique has been used to investigate turbulence by many researchers, e.g., in [24], the authors used wavelet analysis to demonstrate the Richardson cascade in turbulent flow.

The organization of this paper is as follows: Section 2 gives a brief description of the Combined Compact Difference (CCD)-based numerical model used to simulate the experiment of [11] on APG boundary layer transition. Section 3 makes a qualitative comparison between the simulation results and the experiment. It demonstrates the common features of the vortex in the boundary layer across fundamental and subharmonic transition also existing in broadband disturbance. Section 4 discusses the application of CWT to analyze transition signals demonstrating that CWT is a good identifier of high wavenumber components in transition flow. Section 5 tracks the evolution of a typical Λ-vortex showing that the high wavenumber components are imbedded in the associated spike region, and high shear layers developed in transition boundary layer flow. Section 6 provides the conclusion.

2. Numerical Model

In the numerical model of the boundary layer flow used here, all variables in the Navier–Stokes equation are expressed in non-dimensional form and related to their dimensional counterparts, denoted by bars or uppercase, as follows:

$$\begin{aligned} x &= \frac{X}{\overline{L}},\ y = \frac{Y\sqrt{Re}}{\overline{L}},\ z = \frac{Z}{\overline{L}},\ Re = \frac{\overline{U}_\infty \overline{L}}{\overline{\nu}} \\ t &= \frac{\overline{U}_\infty T}{\overline{L}},\ u = \frac{\overline{u}}{\overline{U}_\infty},\ v = \frac{\overline{v}\sqrt{Re}}{\overline{U}_\infty},\ w = \frac{\overline{w}}{\overline{U}_\infty} \end{aligned} \tag{1}$$

where x, y and z denote the coordinates of streamwise, wall-normal and spanwise direction, respectively, and u, v, w are the corresponding velocity components in each direction, $\overline{L}$ is the characteristic length, $\overline{U}_\infty$ is the freestream velocity, $\overline{\nu}$ is the kinematic viscosity and Re is a reference Reynolds number. X, Y, Z and T are used to denote the corresponding dimensional spatial and temporal coordinates. The NS equations are solved using a vorticity-velocity formulation with corresponding vorticity components in non-dimensional form given as:

$$\omega_x = \frac{1}{Re}\frac{\partial v}{\partial z} - \frac{\partial w}{\partial y},\ \omega_y = \frac{\partial w}{\partial x} - \frac{\partial u}{\partial z},\ \omega_z = \frac{\partial u}{\partial y} - \frac{1}{Re}\frac{\partial v}{\partial x} \tag{2}$$

In addition, the total flow field $(\boldsymbol{V}, \boldsymbol{\Omega})$ is decomposed into a steady 2D base flow $(\boldsymbol{V_B}, \boldsymbol{\Omega_B})$ and an unsteady 3D disturbance flow $(\boldsymbol{V\prime}, \boldsymbol{\Omega\prime})$, which is expressed as:

$$\boldsymbol{V}(t,x,y,z) = \boldsymbol{V_B}(x,y,z) + \boldsymbol{V}'(t,x,y,z) \tag{3}$$

$$\boldsymbol{\Omega}(t,x,y,z) = \boldsymbol{\Omega_B}(x,y,z) + \boldsymbol{\Omega}'(t,x,y,z) \tag{4}$$

with:

$$\boldsymbol{V_B} = \{u_B, v_B, 0\},\ \boldsymbol{\Omega_B} = \{0, 0, \omega_{z_B}\} \tag{5}$$

$$\boldsymbol{V}' = \{u', v', w'\},\ \boldsymbol{\Omega}' = \left\{\omega_x', \omega_y', \omega_z'\right\} \tag{6}$$

2.1. Computational Domain

The computational domain was setup to closely match the experimental environment of [11]. The physical system is that of an APG boundary layer flow (with Hartree parameter $\beta_H = -0.115$) over a flat plate, which is disturbed by a blowing and suction strip at a certain upstream location. A comparison sketch of the distance from leading edge, disturbance source location and main region of interest along the streamwise direction are sketched in Figure 1. The rectangular computational domain is shown schematically in Figure 2. The computational domain is 500 mm × 25.3 mm × 48 mm in the streamwise (x), wall-normal (y) and spanwise (z) directions, respectively, and started at 250 mm from the leading edge. The blowing and suction strip was located over 287 mm to 308 mm to simulate the disturbance sources used by [11]. The amplitude of the disturbances were calibrated to that at location X = 450 mm using available experimental results. The main region of interest was over X of 450 mm to 530 mm where experimental data are reported, even though the experimental section continued to 1490 mm. The present computational domain, though smaller, covers this region of interest and extends to X = 580 mm, after which a buffer domain was used until X = 753 mm.

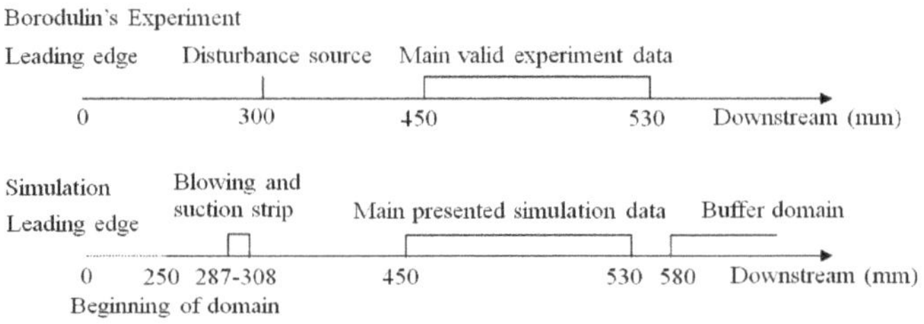

Figure 1. Comparison sketch of the investigation domain in streamwise direction between the experiments of [11] and the present simulation.

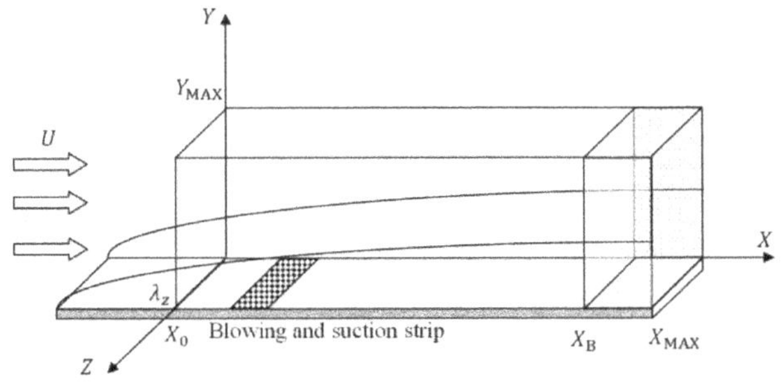

Figure 2. Computational domain.

The length of the domain in the wall-normal direction in the simulation covers about 12-times boundary layer displacement thickness δ_1, with typical turbulent structures extending upwards to $5\delta_1$. Thus, there is no suppression of turbulence development. The measured disturbance signals in the experiment repeats every 48 mm in the spanwise direction; hence, the spanwise domain size used here is one disturbance wavelength, $\overline{\lambda}_z$ of 48 mm.

2.2. Simulation of Base Flow

The steady base flow solution $(\boldsymbol{V_B}, \boldsymbol{\Omega_B})$ was first obtained from numerical integration of the NS equations to a steady state:

$$\frac{\partial \omega_{zB}}{\partial t} + u_B \frac{\partial \omega_{zB}}{\partial x} + v_B \frac{\partial \omega_{zB}}{\partial y} = \frac{1}{Re}\frac{\partial^2 \omega_{zB}}{\partial x^2} + \frac{\partial^2 \omega_{zB}}{\partial y^2} \tag{7}$$

$$\frac{1}{Re}\frac{\partial^2 v_B}{\partial x^2} + \frac{\partial^2 v_B}{\partial y^2} = -\frac{\partial \omega_{zB}}{\partial x} \tag{8}$$

$$\frac{\partial^2 u_B}{\partial x^2} = -\frac{\partial^2 v_B}{\partial x \partial y} \tag{9}$$

At the inflow boundary, the solution of the Falkner–Skan equation was used to specify the base flow variables. No-slip and no-penetration conditions were used at the wall. The wall vorticity was calculated from the velocity field considering the continuity condition. The vorticity vanished at the freestream boundary where $u_B(x, y_{MAX})$ was prescribed as:

$$u_B(x, y_{MAX}) = U_e(x) \tag{10}$$

where the form of $U_e(x)$ is given later in Section 3.

At the outflow boundary, all equations were solved dropping the second x-derivative terms, and u_B is calculated from:

$$\frac{\partial^2 u_B}{\partial y^2} = \frac{\partial \omega_{z_B}}{\partial y} \tag{11}$$

A more detailed description of these boundary conditions in a Blasius boundary layer flow is found in [15].

2.3. Simulation of Base Flow

The NS equations consist of three vortex transport equations:

$$\frac{\partial \omega'_x}{\partial t} + \frac{\partial a}{\partial y} - \frac{\partial c}{\partial z} = \frac{1}{Re}\frac{\partial^2 \omega'_x}{\partial x^2} + \frac{\partial^2 \omega'_x}{\partial y^2} + \frac{1}{Re}\frac{\partial^2 \omega'_x}{\partial z^2} \tag{12}$$

$$\frac{\partial \omega'_y}{\partial t} - \frac{\partial a}{\partial x} + \frac{\partial b}{\partial z} = \frac{1}{Re}\frac{\partial^2 \omega'_y}{\partial x^2} + \frac{\partial^2 \omega'_y}{\partial y^2} + \frac{1}{Re}\frac{\partial^2 \omega'_y}{\partial z^2} \tag{13}$$

$$\frac{\partial \omega'_z}{\partial t} + \frac{\partial c}{\partial x} - \frac{\partial b}{\partial y} = \frac{1}{Re}\frac{\partial^2 \omega'_z}{\partial x^2} + \frac{\partial^2 \omega'_z}{\partial y^2} + \frac{1}{Re}\frac{\partial^2 \omega'_z}{\partial z^2}, \tag{14}$$

where:

$$a = v'\omega'_x - u'\omega'_y + v_B'\omega'_x - u_B\omega'_y \tag{15}$$

$$b = w'\omega'_y - v'\omega'_z - v_B\omega'_z - v'\omega_{z_B} \tag{16}$$

$$c = u\omega'_z - w\omega'_x + u_B\omega'_z + u'\omega_{z_B} \tag{17}$$

and three velocity Poisson Equations [19,25]:

$$\frac{\partial^2 u'}{\partial x^2} + \frac{\partial^2 u'}{\partial z^2} = -\frac{\partial \omega'_y}{\partial z} - \frac{\partial^2 v'}{\partial x \partial y} \tag{18}$$

$$\frac{1}{Re}\frac{\partial^2 v'}{\partial x^2} + \frac{\partial^2 v'}{\partial y^2} + \frac{1}{Re}\frac{\partial^2 v'}{\partial z^2} = \frac{\partial \omega'_x}{\partial z} - \frac{\partial \omega'_z}{\partial x} \tag{19}$$

$$\frac{\partial^2 w'}{\partial x^2} + \frac{\partial^2 w'}{\partial z^2} = \frac{\partial \omega'_y}{\partial x} - \frac{\partial^2 v'}{\partial y \partial z} \tag{20}$$

All disturbance flow variables vanished at the inflow boundary. Potential flow held at the freestream boundary so that the vorticity components vanished. Equations (18) and (20) were solved for u' and w' at the freestream boundary. For v', the local wall-normal gradient was prescribed to impose an exponential decay [26]:

$$\frac{\partial v'}{\partial y} = -\frac{\alpha^*}{\sqrt{Re}}v' \tag{21}$$

where $\alpha^* = \sqrt{\alpha^2 + \gamma^2}$ and α and γ are the streamwise and spanwise wavenumbers, respectively.

At the wall boundary, no-slip conditions were used. The vorticity components were calculated from their relationship with velocity [19,26,27]. A buffer domain method was used at the outflow boundary to allow disturbances to travel out of the computational domain without upstream reflection. Periodicity conditions were employed at the spanwise boundaries at $z = 0$ and $z = \lambda_z$. Centerline symmetry is not assumed here, unlike, e.g., other reported work [19,26,27], so that the asymmetric disturbance waves could develop freely.

2.4. Numerical Method

Since the unsteady disturbance flow was assumed to be spanwise periodic, the disturbance flow variables $(\boldsymbol{V\prime}, \boldsymbol{\Omega\prime})$ and the terms (a, b, c) were expanded using Fourier modes:

$$(\boldsymbol{V}', \boldsymbol{\Omega}', a, b, c) = \sum_{k=-K}^{K} (\boldsymbol{V}'_k, \boldsymbol{\Omega}'_k, a_k, b_k, c_k) exp(Ik\gamma_z) \tag{22}$$

where $I = \sqrt{-1}$ and γ_z is the lowest spanwise Fourier mode, which is related to the spanwise wavelength λ_z by:

$$\gamma_z = \frac{2\pi}{\lambda_z} \tag{23}$$

Substituting Equation (22) into Equations (12) to (20) gave a set of K governing equations for each (x, y)-plane integration domain. The nonlinear terms a, b and c of the vorticity transport equations were evaluated pseudo-spectrally using fast Fourier transform for conversion from Fourier space to physical space and back with the $3/2$ rule used for de-aliasing [28].

Combined Compact Difference (CCD) schemes up to 12th-order accuracy were used in the streamwise and wall-normal directions as reported in [15,16]. The spectral resolution of these schemes can be demonstrated by the modified wavenumber analysis following [29], wherein the scaled wavenumber $w = kh$ where k is the physical wavenumber and h is the grid spacing. The modified scaled wavenumber w', i.e., the one represented by the finite difference scheme, is typically complex. The closer $w\prime$ is to w, the more accurate is the scheme. The modified wavenumber analysis for each scheme used in the simulation is presented in Figure 3. The upwind CCD scheme co-optimized with fourth order 5 to 6 alternating stage Runge–Kutta (RK) scheme (denoted as uniform upwind CCD12RK56 in Figure 3a) was used for the $\frac{\partial}{\partial x}$ term in Equations (12) to (14) and (18) to (20) in order to preserve low-dispersion (related to the real part of w_1) and low-dissipation errors for wave propagation (related to the imaginary part of w_1). This scheme also suppressed numerical oscillations introduced from the buffer domain, because of the very negative $(w_1)_I$ near $w = 3$. The $\frac{\partial^2}{\partial x^2}$ terms in Equations (12) to (14) and (18) to (20) are discretized with a centered difference scheme (denoted as uniform CCD12 in Figure 3b) [15]. In the wall-normal y direction, the CCD schemes were directly constructed on a non-uniform grid without the use of coordinate transform. The spectral resolution for the non-uniform CCD12 scheme used for $\frac{\partial}{\partial y}$ is given in Figure 3a, and that for $\frac{\partial^2}{\partial y^2}$ is given in Figure 3b. A second-order 5 to 6 alternating stage Strong Stability Preserving (SSP) RK scheme was used for the term $\frac{\partial^2}{\partial y^2}$ to allow a larger time step. All of the schemes presented in Figure 3 have a high spectral resolution up to $w = 2.5$, and detailed analyses of these schemes are given in [15,16]. A multigrid method is used to solve the velocity Poisson Equation (19). A stable stretching grid [16] was used in the y direction to obtain fine resolution in the near-wall region.

In the present simulations, 1000 uniform grid points with $\Delta X = 0.5$ mm were used in the streamwise direction, which represented a resolution of about 61 grid points per TS wavelength. Since 12th-order CCD schemes was used in this direction, a scaled wavenumber as high as 2.5 could be accurately simulated, implying that harmonic waves with the wavenumber roughly 25-times the TS wave could be accurately captured [15]. Considering that the TS scaled wavenumber in the present simulation is 0.103 (TS wavenumber $\times$ $\Delta X = 205.9$ rad/m $\times$ 0.5×10^{-3} m), the highest wavenumber that can be accurately resolved is thus about 5000 rad/m (205.9 rad/m $\times$ 2.5/0.103). In the wall-normal direction, 120 grid points were used. A stable stretching grid was used in the near-wall region with grid sizes ranging from 0.01 to 0.62 mm. This had been demonstrated to successfully resolve the large velocity gradient of the high shear layer in the boundary layer transition flow [16]. Sixteen Fourier modes were used in the spanwise direction. The total number of grid points used in this study was 3.84×10^6. ΔT was selected to be as small as 6.56×10^{-6} s so that the maximum CFL number in the streamwise direction ($\frac{U_e \Delta T}{\Delta X}$) ≤ 0.12, which satisfied the scheme optimization condition given in [15]. This ΔT also satisfied the stability criterion in the wall-normal direction as detailed in [16]. As all variables were non-dimensionalized in Equation (1), the needed kinematic viscosity $\overline{\nu}$ used was 1.526×10^{-5} m^2/s, as in the experiments by [11]. The reference velocity was selected to be $\overline{U}_\infty$ = 30.52 m/s. The reference length $\overline{L}$ was set as 0.05 m, so that the reference Reynolds number Re is 105. The local Reynolds number based on displacement thickness δ_1 ranged from 820 to 1130 over the computational domain.

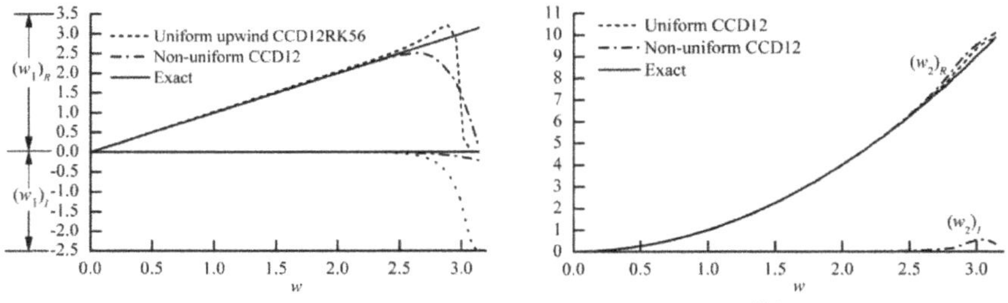

Figure 3. Modified scaled wavenumber $w1$ and $w2$ plotted against scaled wavenumber w for the finite difference schemes on a non-uniform grid. (**a**) $w1$ for the first derivative. (**b**) $w2$ for the second derivative. CCD, Combined Compact Difference.

It is noted that the number of grid points in the present simulation (1000 streamwise $\times$ 120 wall normal $\times$ 32 spanwise) would usually be insufficient to resolve the full-scale vortex motion even in this low-Reynolds number turbulent boundary layer when using lower resolution schemes. As a comparison, a similar study on fundamental resonance using Direct Numerical Simulation (DNS) in [19] on the ZPG transition flow up to the stage where the Ω-vortex was formed used fourth order finite difference schemes on a grid of 3000 (streamwise) $\times$ 65 (wall-normal) $\times$ 32 (spanwise). The highest wavenumber disturbances dominated in the streamwise direction necessitating finer grid point in this direction. In [30] where a second order finite difference scheme in each direction was used, typically 4096 $\times$ 400 $\times$ 128 grids were needed to simulate a fully-developed turbulent boundary layer. Here, the use of 12th order CCD schemes allowed the grid points in the streamwise direction to be significantly reduced. In particular for the experiment of [11], which the present results are being compared to, the highest disturbance frequency recorded is around 1100 Hz (Figure 19 in [11]). Since the TS wave frequency is 109.1 Hz, this means about 10 harmonics of TS wave exist. As the TS wave number is 205.9 rad/m, hence the wavenumber of the 10th harmonic is 2059 rad/m. Considering that the amplitude of broadband disturbance is comparable with the harmonics of TS wave, the disturbance should thus be resolved by the present numerical scheme with resolution up to 5000 rad/m. The flow parameters, grid resolution and size of the computational domain are summarized in Table 1.

Table 1. Parameters of flow and computational domain. TS, Tollmien–Schlichting.

Streamwise Grid Points	1200	ΔX Interval Per TS Wave Length	61
ΔX (mm)	5	Wall-normal grid points	120
ΔY_{max} (mm)	0.01	Spanwise grid points	32
Spanwise modes	16	ΔZ (mm)	1.4
ΔT (s)	6.56×10^{-6}	ΔT interval per TS wave period	1400
TS wave frequency (Hz)	109.1	TS wave number (rad/m)	205.9

The preceding analysis shows that the present resolution can capture the major vortical motion during the transition. Further as the present study is on an APG boundary layer and also focuses on mid-stage including ring-like vortices formed after Ω-vortices, the velocity gradients in the wall-normal direction are larger than in [19], necessitating the stretching grid with higher resolution in the near-wall region. The comparison with experiments from [11] as given below demonstrates that grid used was sufficient to resolve the dynamics of the most energetic vortices, enabling the numerical simulation to realistically model the vortex evolution in the transition flow experiment.

2.5. Continuous Wavelet Transform

The CWT used to investigate the evolution of coherent structures is described next. The CWT is defined as:

$$C(a,b) = \int_{-\infty}^{\infty} \frac{1}{\sqrt{a}} \psi\left(\frac{x-b}{a}\right) u(x)dx \tag{24}$$

where ψ is the mother wavelet and a is the scale of the transform and can be determined by:

$$a = \frac{2\pi F_{\psi}}{\alpha \Delta x} \tag{25}$$

F_{ψ} is the centered frequency of the wavelet; α is the wavenumber spectrum of the signal; and Δx is the grid space or sampling period.

The most commonly-used wavelet is the complex-valued Morlet wavelet because it gives information on the phase and has good resolution for the high wavenumber components [31]. The complex-valued Morlet wavelet used here is given by:

$$\psi(x) = \frac{1}{\pi^4} exp\left(I\omega_0 x - \frac{x^2}{2}\right) \tag{26}$$

where ω_0 is the non-dimensional frequency, taken to be six, as in [32].

2.6. Excitation of Disturbance Waves

The disturbance waves in experiment of [11] were produced by blowing and suction of air through a slit with point-like sources on the flat-plate surface perpendicular to the flow direction and located at distance X = 300 mm from the leading edge. The disturbance signals consisted of two main components: the signal corresponding to the fundamental 2D TS wave and broadband 3D pseudorandom small-amplitude signals ("noise" of the TS wave). The broadband signals have approximately the same amplitudes in the frequency spectrum, but different phases for the spectral modes. The broadband signals repeat in the spanwise direction with a period of 48 mm.

In the present simulation, the disturbances are introduced from the wall-normal velocity component $\hat{v}\prime(x,y,k)$ in the spanwise spectral space. The TS wave is generated by:

$$\hat{v}'(x,0,0) = A_{2D} f(x) \sin(\beta t) + s_0(t) \tag{27}$$

where β is the disturbance frequency and is 109.1 Hz, $s_0(t)$ is the random signal, discussed later, and A_{2D} is used to adjust the amplitude to make the r.m.s. streamwise velocity u (maximum in y) equal to around 4% of the freestream velocity at X = 450 mm. It is noted that this is a relatively large amplitude so that the broadband disturbance wave can develop nonlinearly even at the initial stage and trigger vortex formation in a shorter distance downstream. This is so that a smaller computational domain can be used. However this amplitude should not be too large as to distort the base flow. This selection of the initial amplitude follows that used in the experiment (see Section 2.8 for a validation) where similar arguments are used to justify the initial amplitude.

The spatial dependence was introduced via the disturbance distribution function $f(x)$ developed in the present study. It is selected to satisfy the following three conditions:

1. Disturbances start and end with zero amplitude: $f(x_1) = f(x_2) = 0$.
2. No discontinuity for the first derivative at the beginning and the end of the strip, i.e.,

$$\frac{df}{dx}(x_1) = \frac{df}{dx}(x_2) = 0$$

3. The form is conservative, so that $\int_{x_1}^{x_2} f(x)dx = 0$.

Conditions (1) and (2) make the transition from disturbance source to the wall smooth, while Condition (3) enforces the law of mass conservation to the imposed disturbances. One possible choice that is used here is:

$$f(x) = \begin{cases} 24.96\xi^6 - 56.16\xi^5 + 31.2\xi^4 & \xi = \frac{x-x_1}{x_{st}-x_1} \quad x_1 < x < x_{st} \\ -24.96\xi^6 + 56.16\xi^5 - 31.2\xi^4 & \xi = \frac{x_1-x}{x_2-x_{st}} \quad x_{st} < x < x_2 \end{cases} \tag{28}$$

The shape is shown in Figure 4.

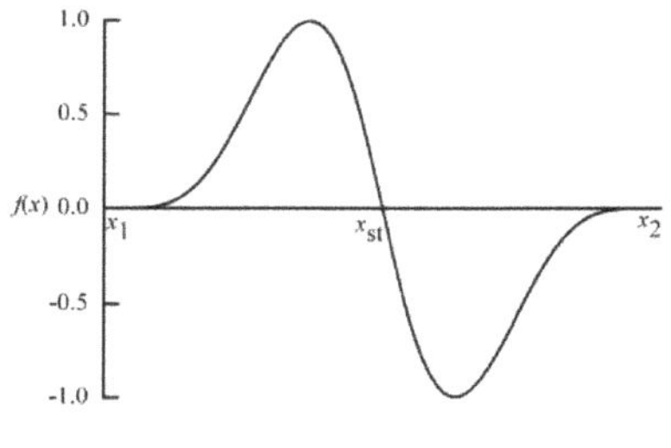

Figure 4. The distribution of the two-dimensional disturbance source.

The randomized disturbance $s_0(t)$ in Equation (27) is first generated in the Fourier space with the frequency band up to six-times the TS wave. The inverse Fourier transform is used to recover the signal in the time space as shown in Figure 5. This broad band signal has randomized phases, but the same normalized amplitude, as shown in Figure 6 for its spectrum. This signal amplitude is about 2% of A_{2D} of the fundamental TS wave (refer to Figure 6 in [11]). Besides $s_0(t)$ in Equation (27), an additional 14 different signals were imposed on the real and imaginary parts of the Fourier modes $\hat{v}'(x, 0, k)$, where $k = \pm1, \pm2, \pm3, ..., \pm7$. Thus, by construction, all of these spanwise Fourier modes have the same spectrum shown in Figure 6.

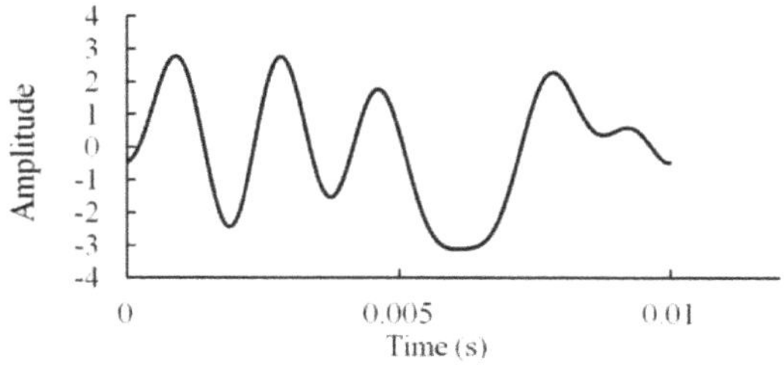

Figure 5. Random signal in time space.

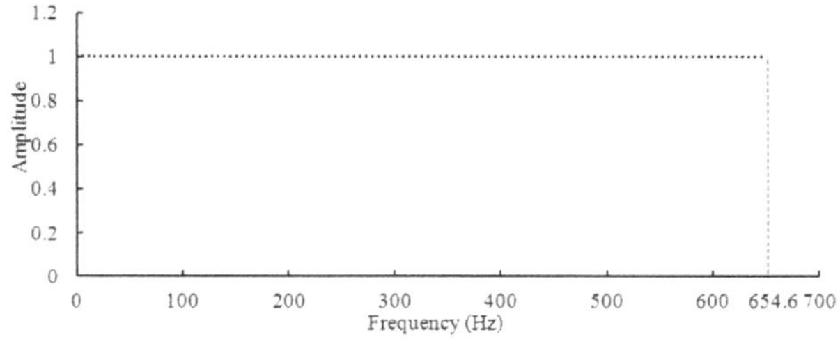

Figure 6. Top hat frequency spectrum of the random signal $s_0(t)$. The vertical axis is the Fourier amplitude while the horizontal axis show frequency range up to 6 times the TS frequency of 109.1 Hz.

2.7. Code Validation

Since randomized disturbances are used in the present simulation of the experiment of [11], a direct quantitative comparison between the results is not possible. However, the subharmonic resonance experiment of [23] is very similar to the present case with the only difference being the source of disturbance. The comparison was reported earlier in [15] where the TS wave interacted non-linearly with the subharmonic wave, and the amplitude of the latter amplified dramatically along the downstream direction, as shown in Figure 7, where the streamwise disturbance amplitudes of different components are plotted against Reynolds number Re_x (the length scale x refers to the distance from the leading edge). Although the simulated amplification of the subharmonic wave is slightly weaker than in the experiment, the agreement was generally good.

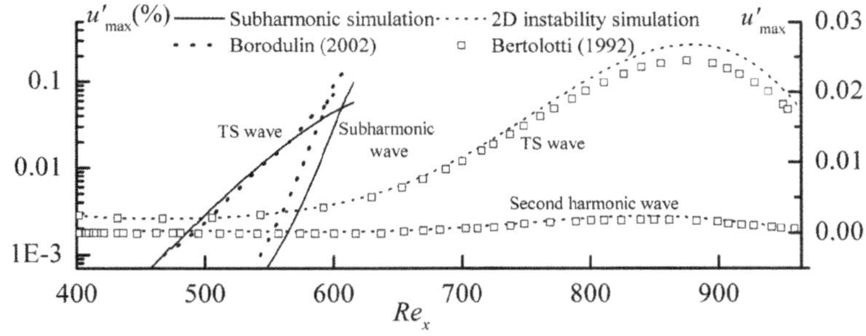

Figure 7. Amplification of streamwise disturbance amplitude in resonant simulation. Left vertical axis: wave amplitude of subharmonic resonance with comparison of [23]. Right vertical axis: wave amplitude in 2D instability simulation with comparison of [33].

In addition, the CCD12RK56 scheme on the uniform grid is validated by simulating 2D amplification of the TS wave in [33]. As shown in Figure 7, the initial amplitude of TS wave is small, but as it amplifies, its second harmonic wave is triggered. The results compare well with [33]. More details are found in [15].

2.8. Comparison of Settings between Simulation and Experiment

Before further investigation of the coherent structures via the simulation, the simulation results are first compared with the experiment to calibrate for the initial conditions and base flow. The non-dimensional freestream velocity in Equation (10) is prescribed as:

$$U_e(x) = Cx^{\frac{\beta_H}{2-\beta_H}} \tag{29}$$

where β_H is the Hartree parameter fixed as -0.115 and C is selected as 8.295 to match the freestream velocity gradient in the experiments of [11]. Detailed comparisons of the base flow are not presented here, but are found in [21], including comparisons of the streamwise distribution of freestream velocity, base flow velocity and the boundary layer displacement thickness.

In [11], the disturbance flow is first measured at the initial section X = 450 mm. The initial amplitudes of the TS wave and broadband disturbances are adjusted to match the experimental condition at this location. The wall-normal profile of the TS wave amplitude at X = 450 mm is shown in Figure 8 along with the simulation showing that a good match was achieved.

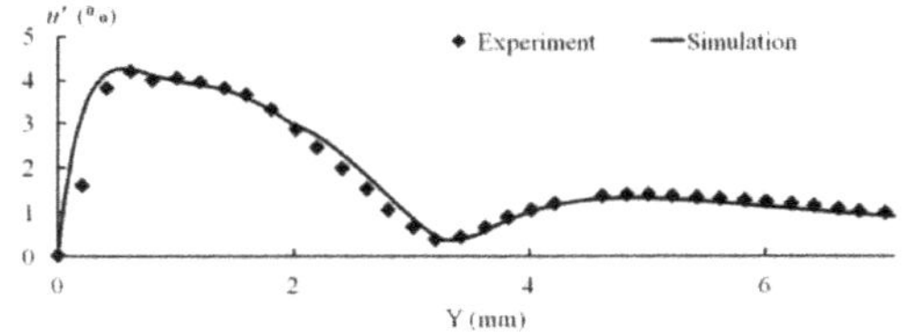

Figure 8. Wall-normal profile of boundary layer disturbance amplitude at X = 450 mm and Z = 14 mm.

The wave amplitude spectrum plotted against frequency and spanwise Fourier modes is shown in Figure 9. The TS wave amplitude is dominant at a frequency of 109 Hz and spanwise Fourier Mode 0. The disturbance wave amplitudes decay with both the frequency and Fourier mode.

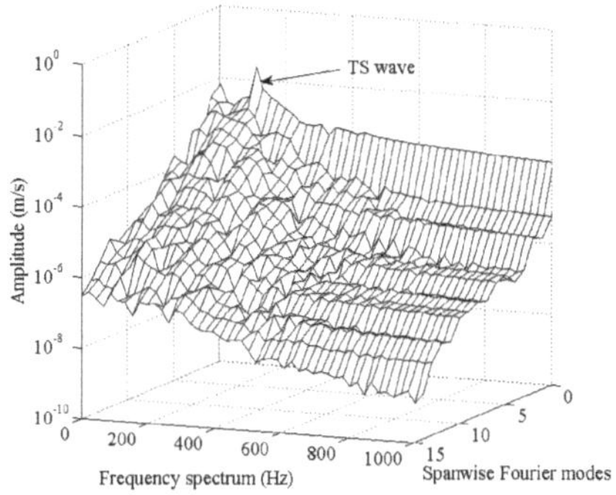

Figure 9. Wave amplitude against frequency and spanwise Fourier modes at X = 450 mm and Y = 1 mm.

3. Overall Properties of Coherent Structures

Common shapes of the typical coherent structures in broadband disturbances are presented in this section to demonstrate the common features of coherent structures by streamwise velocity disturbance $u\prime$ and the Q-criterion. The structures in broadband disturbances are further compared with those seen in the experimental results of [11].

3.1. Coherent Structures in Broadband Disturbance Flow

The process of coherent structures' formation in broadband disturbance flow is shown in Figure 10a,b, which are contour plots of the streamwise disturbance velocity $u\prime$ at $Y = 1.2$ mm with Figure 10a being at an early time stage and (b) at a later stage after the vortices have broken down. There is a weak spanwise flow distortion before $X = 0.4$ m, and localized structures are formed around $X = 0.45$ m. Both structures at $X = 0.52$ m (Figure 10a) and at $X = 0.5$ m (Figure 10b) show a Λ shape that resemble the well-known Λ-vortex observed in [10,19]. However, these structures arise in a random order, though the distances in the streamwise direction between these structures are roughly one TS wavelength. In Figure 10a, the structures are roughly arranged in an aligned pattern, which is the characteristic of fundamental resonance (see [8]). In Figure 10b, these structures are in a staggered pattern like the ones in subharmonic resonance (see [8]).

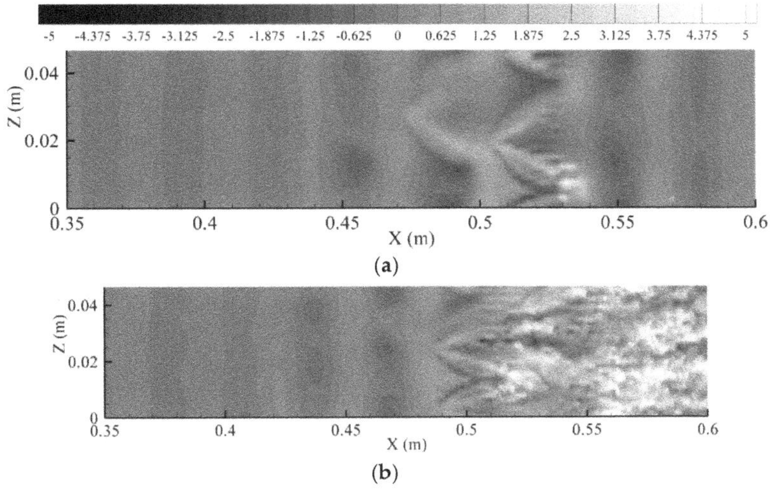

Figure 10. Instantaneous contour of streamwise disturbance $u\prime$ (m/s) at $Y = 1.2$ mm. (**a**) $T = 0.059$ s; (**b**) $T = 0.072$ s.

Visualization of vortices is performed using the iso-surface of the Q-criterion following [34], which is defined as a second invariant of the velocity gradient. While it focuses only on the rotational part of the vorticity at the expense of the strain, it is widely used for visualizing vortex structures in early-to mid-stage transition [30,34–36]. Chakraborty et al. [35] further compared several prevailing vortex identification criteria including the Q-criterion and concluded that all resulted in remarkably similar looking structures. The λ_2-isosurface as an improved criterion was used in a recent transition study by Liu et al. [14], though Chakraborty et al. [35] indicated that the λ_2-isosurface only captures the strain on a specific plane [37]. Since late stage vortex breakdown, after the formation of ring-like vortices, is not the focus here, using the Q-criterion for comparison of vortex development with experiments is sufficient, as the observed vortices are driven mainly by the rotational part. The strain is further examined via the wall region shear $\frac{\partial u\prime}{\partial y}$.

Figure 11 visualizes the vortices at two different times T showing that the vortices develop with different shapes and patterns. In Figure 11a, the vortex centered at $Z = 0.01$ m has developed into an Ω-shape, while the one centered at $Z = 0.04$ m is still a primary Λ-vortex. Similarly, in Figure 11b, a mature Λ-vortex centered at $Z = 0$ m has formed and started to evolve to an Ω-vortex, while the one centered at $Z = 0.025$ m has formed a ring-like vortex at its tip.

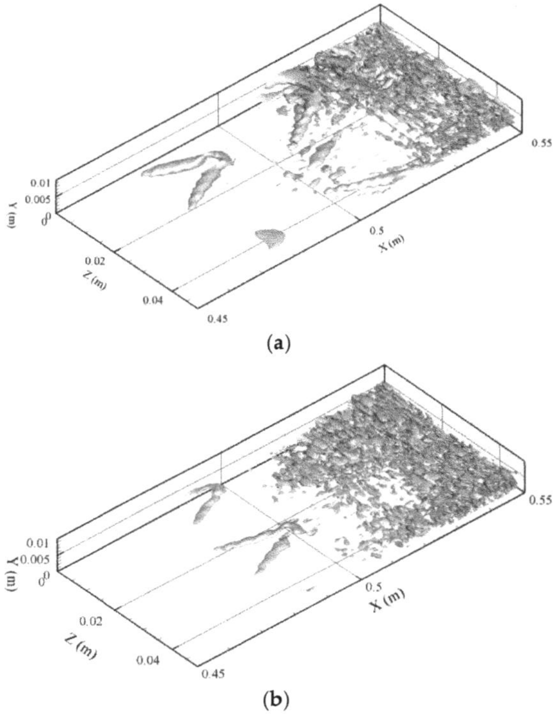

Figure 11. Vortex visualization by Q-criterion at (**a**) $T = 0.066$ s; (**b**) $T = 0.105$ s.

The vortex features shown in Figures 10 and 11, and similar plots (not shown), indicate that broadband disturbances do not qualitatively change the typical Λ-shape of the coherent structures. The effects of broadband disturbances on coherent structure formation are summarized as: (1) the coherent structures are spatially arranged in random order; and (2) at the same spatial location, the coherent structures may be in different evolution stages.

3.2. Spatial Shapes of Coherent Structures

The typical spatial shapes of several individual coherent structures observed at around $X = 0.49$ m in the simulation are shown in Figure 12 by plotting the iso-surface of $u\prime = -8\%$ of freestream velocity. They have the common features of the coherent structures reported for fundamental and subharmonic resonance [9,11]: they are close to a Λ-shape having two legs in the rear and one tip in the front. However, most of these structures are asymmetric due to the effects of broadband disturbances. In Figure 12a,d, one leg is longer than the other. There, the two structures are partially overlapped and merged in Figure 12b, while the coherent structure is close to symmetric in Figure 12c. The structure in Figure 12d is only at the beginning of its formation.

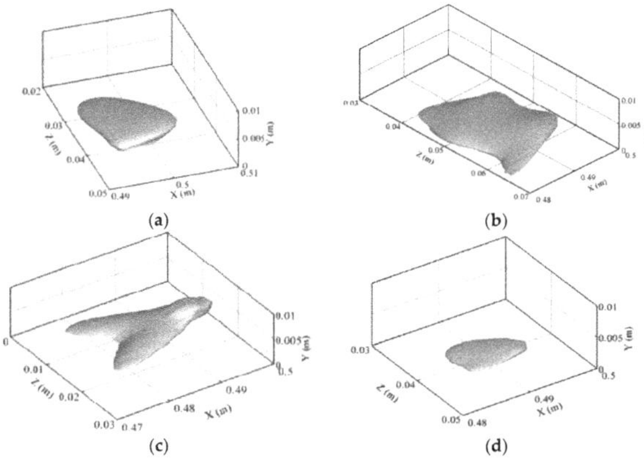

Figure 12. Iso-surfaces of the streamwise disturbance velocities by plotting $u\prime = -8\%$, at (**a**) $T = 0.053$ s, (**b**) $T = 0.059$ s, (**c**) $T = 0.067$ s, (**d**) $T = 0.069$ s.

The further downstream development of the coherent structures at around X = 0.53 m is shown in Figure 13. It can be seen that the structures at this location have evolved to shapes qualitatively similar to the typical Λ-vortex. Compared with Figure 12, these structures have stretched, becoming sharper in the streamwise direction and spread in the spanwise direction. The separation of the two legs has become clear with the tip starting to swell. Figure 13a shows a structure very similar to the typical symmetric Λ-structure. The structures in Figure 13b,c evolve from those in Figure 12a,b, respectively. The asymmetry of both structures has become more obvious. The left leg of the structures has developed further than the right leg in Figure 13b (note that the part at Z = 0.045 m is the leg of another structure). The two legs of the two close structures have merged to a tail in Figure 13c,d showing a structure that only developed its left leg.

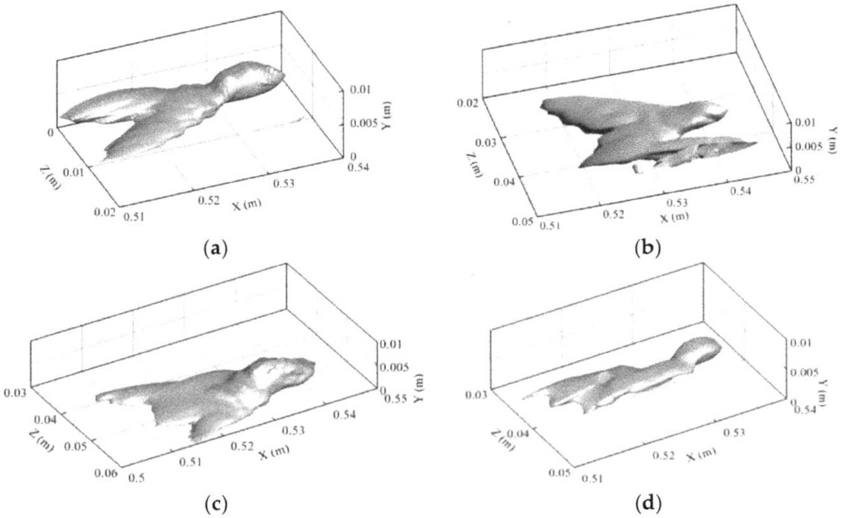

Figure 13. Iso-surfaces of the streamwise disturbance velocities by plotting $u\prime = -12\%$, at (**a**) $T = 0.059$ s, (**b**) $T = 0.062$ s, (**c**) $T = 0.067$ s, (**d**) $T = 0.078$ s.

Figures 12 and 13 fully demonstrate the polymorphism of coherent structures generated by broadband disturbances. However, all of the structures display qualitatively the same shape as the typical Λ-vortex observed in boundary layer transition flow. This observation indicates that there may be common mechanisms that govern the coherent structure evolution.

Although the simulation results are not quantitatively comparable with the experiments of [11], the structure properties, including disposition and spatial shape, match the experimental results well (see Figures 17 and 21 in [11]). The spatial shapes of vortical structure seen in the experiment are qualitatively similar to the ones found in the present simulation shown in Figures 12 and 13. Therefore, it is concluded that the present simulation is able to sufficiently capture the effects of broadband disturbances on coherent structure evolution allowing further investigations using CWT on the computed results.

4. Characteristic Signal Pattern of Coherent Structures

The signal pattern from a transition flow experiment is usually analyzed using the time series of streamwise disturbance velocity [8,12], as it is experimentally easier to record the velocity at a certain location with a point probe than to record the entire velocity field. However, this is not a constraint in simulations. To further investigate the disturbance signal evolution, it is more insightful to analyze the spatial-temporal development of the disturbance signal, which can demonstrate how the high wavenumber components develop as the vortex evolves.

Flow structures with different shapes are shown in Figure 14 via the u' contour at $Y = 1.18$ mm and at $T = 0.056$ s. This particular time is chosen for discussion as the coherent structures exhibit a whole range of evolution behavior with random spatial locations and asymmetric shapes. Structures 1 and 4 are in their primary stage with rhombus-like shapes, where Structures 2, 3, 5 and 6 are in the developed stage with quasi-Λ-shapes.

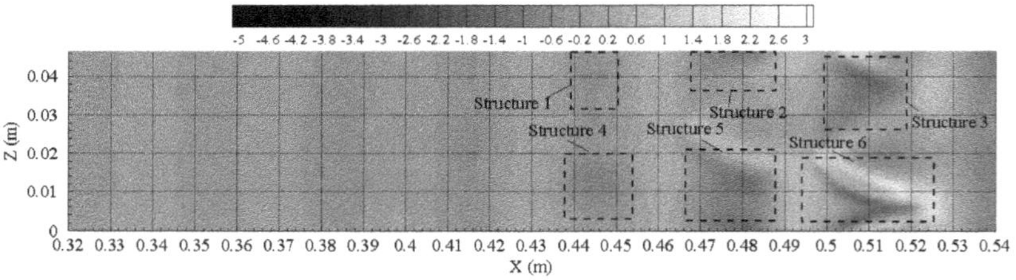

Figure 14. Instantaneous contour of streamwise disturbance velocity u' (m/s) at $Y = 1.18$ mm and $T = 0.056$ s.

Figure 15 presents the evolution of vortex visualized by the Q-criterion at $T = 0.056$ s. Structure 1 in Figure 14 has not developed a vortex, while Structure 4 forms a very weak primary Λ-vortex, which is a precursor of the well-known Λ-vortex. Meanwhile, Structures 2 and 5 have developed to the stage of a quasi-Λ-vortex. Structure 3 has formed a Λ-vortex with two asymmetric legs. Structure 6 has developed to a later stage when the first ring-like vortex has started to separate.

CWT is next performed on u' in the downstream direction. The highest wavenumber that CWT can resolve depends on the spectral resolution of the numerical scheme used. The present CCD scheme resolves the effective scaled wavenumber is up to 2.5, i.e., up to 5000 rad/m, as discussed earlier.

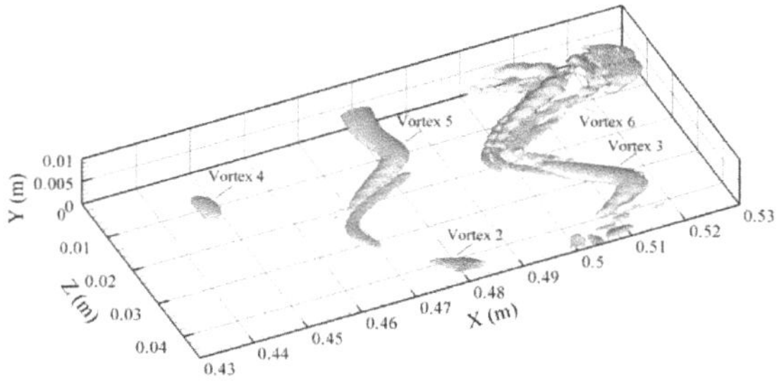

Figure 15. Vortex visualization by Q-criterion at T = 0.056 s.

Figure 16a plots the streamwise disturbance velocity u' at T = 0.056 s, Y = 1.18 mm, Z = 13.5 mm, which goes through the spanwise centers of Structures 4, 5 and 6 in Figure 14. Figure 16b plots the corresponding wavelet contour in absolute amplitude (m/s) on the $\log_{10}$ scale so that the small-scale wave amplitudes are also seen. The energy at wavenumbers smaller than 500 rad/m mainly corresponds to the fundamental TS wave, subharmonic wave and the secondary harmonic wave with the wave energy continuously increasing along the downstream direction. The figure also shows energy peak structures at wavenumbers larger than 1000 rad/m, indicating a discontinuous energy distribution in this region. The peak structures are also amplifying with downstream distance. After X = 0.5 m, the energy at the high wavenumber is almost the same as in the low wavenumber.

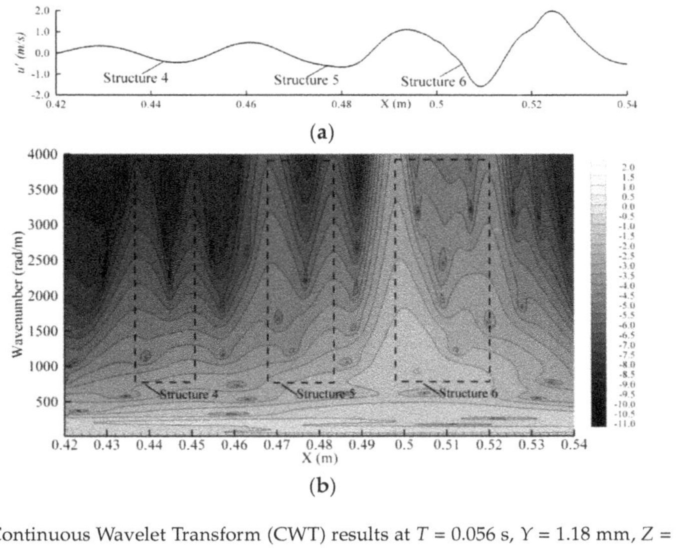

Figure 16. Continuous Wavelet Transform (CWT) results at T = 0.056 s, Y = 1.18 mm, Z = 13.5 mm, (**a**) streamwise disturbance signal and (**b**) wavelet transform of the streamwise disturbance velocity u' ($\log_{10}$ (m/s)).

The twin-peak region (demarcated by dashed lines in Figure 16) for Structure 4 extends from X = 0.436 m to 0.454 m, for Structure 5 from X = 0.468 m to 0.486 m and for Structure 6 from X = 0.498 m to 0.522 m. A comparison with Figures 14 and 15 shows that these twin-peak regions coincided with the spatial region of the coherent structures. The two peaks of Structures 4 and 5 in Figure 16 are physically located very close so that the two peaks occurred within the one complete structure seen in Figure 15. For Structure 6, the two peaks are somewhat separated in Figure 16, and Vortex 6 in

Figure 15 is just at the stage when the first ring-like vortex separates from Λ-vortex. The two peaks with larger amplitude are spatially associated with these two vortices. Thus, the high wavenumber components in the wavelet spectrum are as expected spatially coincident with the coherent structures with a clustering of energetic high wavenumber components at the streamwise edge locations of the coherent structures.

The above discussion shows that as the coherent structures develop further downstream, the energy at high wavenumber becomes more prominent. Moreover, it indicates that prior to the formation of Λ-vortex, the high wavenumber structure has already arisen upstream and is composed of two wave packets corresponding to the twin-peak CWT signals of the coherent structures in Figure 16. As the primary coherent structure evolves to the Λ-vortex, the two wave packets are still spatially connected though amplified. At a further stage when the front wave packet becomes strong enough, it separates from the rear wave packet. In physical space, this corresponds to the separation of the first ring-like vortex from the Λ-vortex. Here, we also note that the CWT results reflect the presence of subharmonic waves, specifically via a slightly high amplitude (lighter) band at a wavenumber around 100 rad/m over X= 0.46 to 0.54 m, while the harmonics of the TS wave are always present (the TS wave here has a wavenumber of 205.9 rad/m). Thus, both subharmonic and harmonic waves are generated.

Figure 17 plots the contour of the shear $\partial u'/\partial y$ at Z = 0.0135 m, i.e., the spanwise location is just around the centers of Structures 4, 5 and 6 shown in Figure 14. It can be seen that the high shear structures, especially for Structures 4 and 5 (the dark regions at round Y = 1 mm), coincide with the wavelet structure patterns in the streamwise direction shown in Figure 16. Therefore, the high wavenumber components are imbedded in the high shear layer structures near the wall. Details about the high shear structures are discussed in the next section. It is also noted that the high shear $\frac{\partial u'}{\partial y}$ is the dominant component of spanwise disturbance vorticity ω'_z during the transition stage discussed in this paper; the subsequent discussion focuses on u' and $\frac{\partial u'}{\partial y}$.

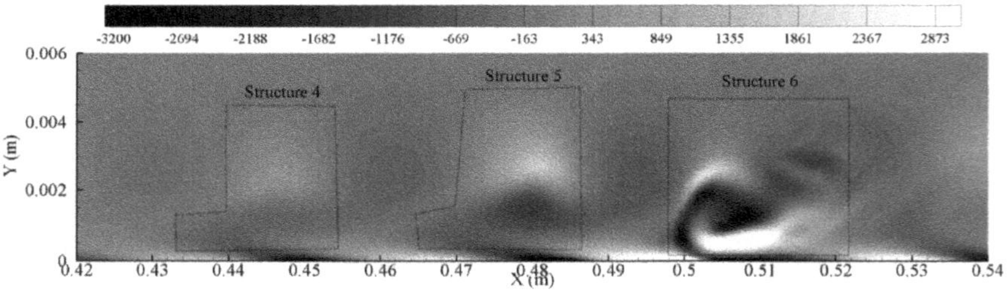

Figure 17. Computed evolution of $\frac{\partial u'}{\partial y}$ (1/s) at Z = 0.0135 m, T = 0.059 s.

5. Coherent Structure Evolution

We now follow the full evolution process of a quasi-Λ-vortex, spike region, high shear layer and the high wavenumber components in the transition flow. Spike regions and high shear layers have been considered as important structures associated with turbulence generation [12,14]; hence, the wavelet spectrum is analyzed with these two structures. It is seen below that high wavenumber components have already developed at the primary stage of the Λ-vortex, spike region and high shear layer and evolve together with them. Further in the transition stage up to the formation of the first ring-like vortex, the high shear layer dominates the spanwise vorticity. To demonstrate this, a slightly asymmetric vortical structure is selected for convenience of comparison with the typical evolution process in subharmonic and fundamental resonance, as observed in [10,19,21,38], but without loss of generality.

5.1. Spike Region and Primary Λ-Vortex

As shown in Figure 16, the high wavenumber components mainly concentrate on the local minimum of the streamwise disturbance velocity, which is commonly referred to as a spike. Here, the spike region of Figure 11 is visualized in 3D by plotting the iso-surface of the streamwise disturbance velocity $u\prime$ with about 60% of the local minimum around the coherent structure during a certain period. The 60% threshold value is chosen to have a clear 3D view of the local velocity distribution.

Figure 18a shows a visualization of the evolution of the spike region at its initial stage by plotting the iso-surface of $u\prime = -0.4$ m/s (57% of local minimum) at time $T = 0.053$ s along with the primary vortex visualized by the Q-criterion. In this early stage, the spike region comprising of a flat, rhombus-like structure located very close to the wall at around $Y = 1.5$ mm is seen. On top of the structure is a primary vortex, which evolves to the mature Λ-vortex further downstream location (see Figure 19).

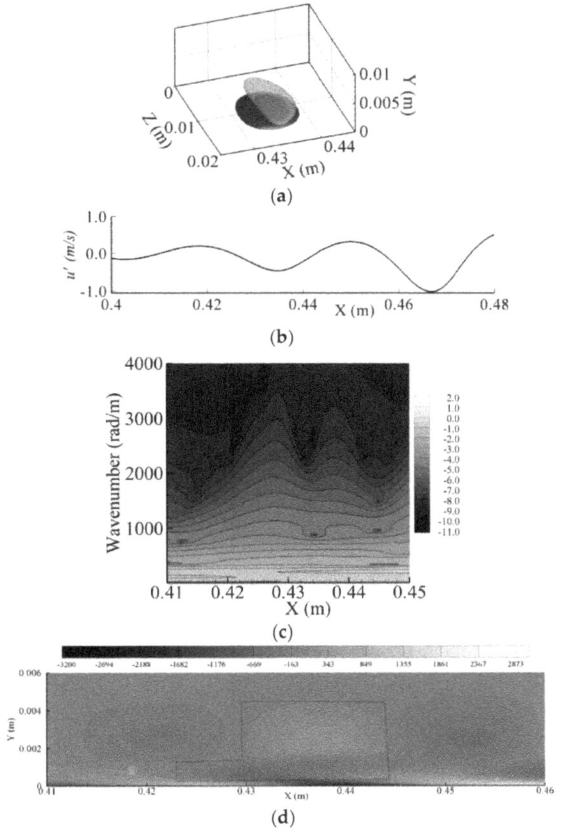

Figure 18. (**a**) Visualized evolution of rhombus-like structure (dark gray) by $u\prime = -0.4$ m/s (57% of local minimum) and primary vortex (transparent light gray) by the Q-criterion. (**b**) Streamwise disturbance $u\prime$ at $Z = 0.0135$ m, $Y = 1.4$ mm. (**c**) Wavelet transform of the streamwise disturbance velocity $u\prime$ ($\log_{10}$ (m/s)) at $Z = 0.0135$ m, $Y = 0.0014$ m. (**d**) $\frac{\partial u'}{\partial y}$ contours at $Z = 0.0135$ m. (a) to (d) are plotted at $T = 0.053$ s.

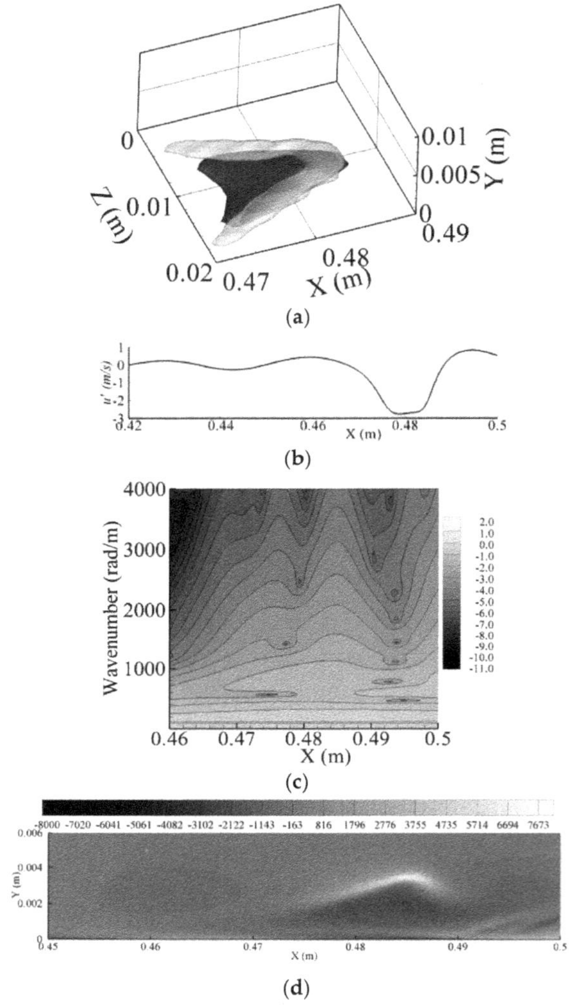

Figure 19. (**a**) Evolution of the spike region (dark gray) with $u\prime = -1.5$ m/s (60% of local minimum) and a Λ-vortex (transparent light gray) according to the Q-criterion. (**b**) Corresponding streamwise disturbance velocity $u\prime$ at $Z = 0.0135$ m, $Y = 2$ mm. (**c**) Wavelet transform of the streamwise disturbance velocity $u\prime$ ($\log_{10}$ (m/s)) at $Z = 0.0135$ m, $Y = 2$ mm. (**d**) $\frac{\partial u'}{\partial y}$ at $Z = 0.0135$ m. (a) to (d) are plotted at $T = 0.0649$ s.

It is illustrative to examine the spectral components of this structure to shed light on the development of later high wavenumber components. Figure 18b plots the disturbance $u\prime$ though the center of the primary Λ-vortex, which spread from $X = 0.425$ to 0.442 m. The corresponding wavelet transform in Figure 18c over this X-range shows the expected two separated high wavenumber bursts at around $X = 0.435$ m, which is the spike location in Figure 18b. The energy of these two bursts is mainly concentrated at wavenumbers below 1000 rad/m.

The CWT results clearly show that the spike region defined here actually contains two wave packets even during this early stage. It is consistent with observation of Kachanov and Borodulin's experiments [8,39] where two groups of high-frequency secondary fluctuations in the streamwise disturbance velocity time series were seen. The upper packet further from the wall corresponds to the first spike. Its early spectral view is shown in Figure 18c, and its physical view can be seen at a later time (in Figure 20c). The other packet (the rear peak in Figure 18c) close to the wall has velocity practically equal to the velocity of the high shear layer. This observation demonstrates the existence of the twin-wave packet structure at the stage before the formation of a strong spike. The present CWT result, consistent with experimental observations, shows the origin of these two wave packets as developing from the local valley of the approximately sine wave behavior of $u\prime$. At the present stage, even though the first spike that as typically associated with the first ring-like vortex has not yet fully developed, its characteristic spectrum has already formed via the front peak in the wavelet transform.

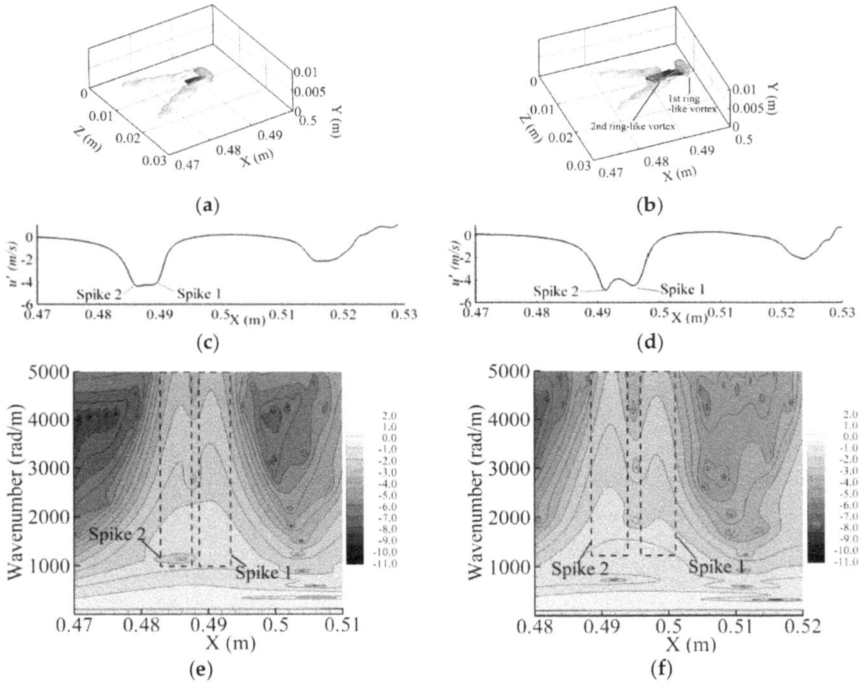

Figure 20. Evolution of the spike region (dark gray) with $u\prime = -4.0$ m/s (66% of local minimum) and a Λ-vortex (transparent light gray) according to the Q-criterion at (**a**) $T = 0.066$ s, (**b**) $T = 0.0675$ s. Corresponding streamwise disturbance velocity $u\prime$ at $Z = 0.0135$ m, $Y = 3.12$ mm and (**c**) $T = 0.066$ s; (**d**) $T = 0.0675$ s. Wavelet transform of the streamwise disturbance velocity $u\prime$ ($\log_{10}$ (m/s)) at $Z = 0.0135$ m, $Y = 3.12$ mm and (**e**) $T = 0.066$ s; (**f**) $T = 0.0675$ s.

The last subplot in Figure 18 for $\frac{\partial u'}{\partial y}$ shows the distribution of the high shear layer of this structure. $\frac{\partial u'}{\partial y}$ turns from negative to positive as its distance increases from the wall. Near the wall, it stretches from $X = 0.425$ to 0.445 m, which approximately matches the distribution of the twin-peak distribution of the high wavenumber components in Figure 18c. Therefore, the high wavenumber components are already developed at the location of the wall high shear layer at this primary stage where the Λ-vortex is not yet mature.

5.2. Λ-Vortex Evolution and First Spike Formation

After the primary Λ-vortex is formed, it continues to stretch, and its two legs become more slender as they propagate downstream. Figure 19 shows this development and the corresponding wavelet patterns. As shown in Figure 19a, the mature Λ-vortex has started to transform into an Ω-vortex.

In Figure 19b, the spike starts to become blunt with one peak starting to separate into two. The two packets that are close enough to generate a sharp spike shown in Figure 18b start to separate. This spike formation is seen to arise from the interaction effects of the two wave packets when they are sufficiently close and have large enough energies. Figure 19c shows the twin-peak structure still present in the spectrum; but the two peaks spread to higher wavenumber, and their amplitude becomes stronger.

Compared with Figure 18d, the high shear layer in Figure 19d at this stage becomes much stronger, moves further away from the wall and forms a hump near its front tip at $X = 0.485$ m. The streamwise length of the near wall high shear layer still matches that of the high wavenumber spectrum well.

5.3. Formation of Ring-Like Vortex

Figure 20 shows the detachment process of the ring-like vortex from the remaining part of the Λ-vortex. Figure 20a shows that an Ω-vortex has formed at $T = 0.066$ s. The two legs of the Ω-vortex are more slender than those of the Λ-vortex. At this stage, a ring-like vortex is about to form, and it tends to separate from the remaining of Ω-vortex. The disturbance $u\prime$ signal in Figure 20c shows the situation with two faint peaks seen. The wavelet transform results in Figure 20e are consistent with these observations with the two wavenumber peaks starting to separate. Here, it is noted that the energy at the highest wavenumbers though small is approaching the resolution of the numerical scheme, and accuracy beyond this stage may be affected.

Figure 20b at $T = 0.0675$ s shows that the first ring-like vortex has separated from its mother vortex, and the two legs of the Ω-vortex have reconnected with two thin links. At this stage, a second spike in the $u\prime$ signal is formed (Figure 20d) resulting in two clear spikes, i.e., the second spike stage. The wavelet transform in Figure 20f shows a clear separation of the front and rear packets. It is seen that the front packet is essentially associated with the first ring-like vortex with the upper and lower wave packets first described in Borodulin and Kachanov's experimental results [8,39] and which is seen clearly in physical space. Here, they are linked to the first and the second spike respectively. However, before their physical separation, these two distinct structures already exist in the primary Λ-vortex stage, which is revealed by the wavelet transform in Figure 18c.

The pattern of the high shear layer $\frac{\partial u'}{\partial y}$ is shown in Figure 21a. The first ring-like vortex forms at the head of Λ-vortex along with a hump being generated at the front of the high shear layer $\frac{\partial u'}{\partial y}$. It is seen that the tip of the $\frac{\partial u'}{\partial y}$ pattern keeps growing strongly, which corresponds to the first ring-like vortex in Figure 20b. The tail part of the $\frac{\partial u'}{\partial y}$ pattern corresponds to the second ring-like vortex. These patterns are similar to the ones reported for subharmonic resonance and fundamental resonance [10,16,19]. Contours of $\frac{\partial v'}{\partial x}$, the other part of spanwise disturbance vorticity ω_z', are shown in Figure 21b. Although the plot shows a similar "hook" shape in the streamwise direction as $\frac{\partial u'}{\partial y}$, its magnitude is only around 30% of $\frac{\partial u'}{\partial y}$, indicating the high shear $\frac{\partial u'}{\partial y}$ is the main contributor to the spanwise disturbance vorticity ω_z' up to the present stage.

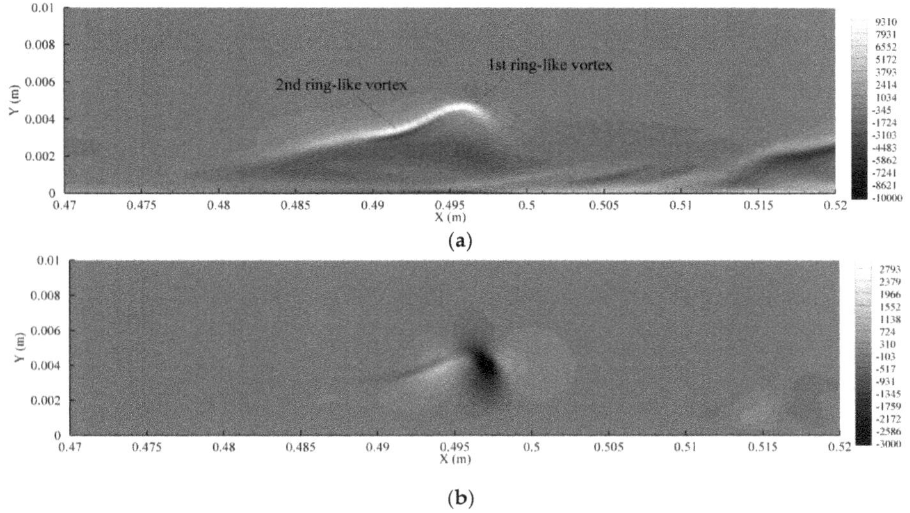

Figure 21. (**a**) High shear layer $\frac{\partial u'}{\partial y}$ (1/s), (**b**) $\frac{\partial v'}{\partial x}$ (1/s) at Z = 0.0135 m, T = 0.0675 s.

Figure 21 also shows the stretching of the high shear layer. Though the two spike signals in Figure 20d have already separated, the high shear layer is still connected. The front high wavenumber component peak matches its tip at X = 0.498 m and the rear one at X = 0.492 m. As in the spike signal and the wavelet spectrum, the high shear layer tends to split into two parts (one goes with the first ring-like vortex and the other with the remaining Λ-vortex), therefore, combining the observations in Figures 20 and 21, the high wavenumber components of the first ring-like vortex come from the "front peak" in the wavelet spectrum stemming from the one in Figure 18c, where the Λ-vortex is still in its primary stage. The "rear peak" in the wavelet spectrum in Figure 18c contributes to the continuing evolution of the remaining Λ-vortex. Further, both peaks are imbedded in the near wall high shear layer and in the spike region in the primary Λ-vortex stage.

6. Conclusions

A high order accuracy numerical model is used to simulate the experiments of [11] on an APG boundary layer transition induced by broadband disturbances. The numerical simulation results demonstrate that there may be common features in the boundary layer transition induced by small amplitude disturbances and particularly in the development of coherent structures. A wavelet analysis further shows that the local spectral components of the Λ-vortex and its associated high shear layer contain two wave packets during most of its life.

Due to the use of initial broadband disturbances in the simulations here, the coherent structures that arise in the transition process are distributed randomly both spatially and temporally. The coherent structures have different shapes, and most of them are asymmetric. However, they are qualitatively similar to those observed in the typical fundamental and subharmonic resonances. They all have a tip at their head and one or two legs at the rear.

Although the coherent structures exhibit different shapes and spatial distributions in the simulation of broadband disturbances, their spectrum are quite similar. They always contain a twin-peak structure, which forms prior to the maturity of the Λ-vortex. As the Λ-vortex evolves to the Ω-vortex and at a later stage when the first ring-like vortex forms, the two peaks of the high wavenumber components separate. The front one is related to the first spike or the first ring-like vortex and the rear one to the second spike or the other ring-like vortex. Thus, the results reveal the source location of high wavenumber disturbance in boundary layer transition flow: they are originally imbedded in the spike region and wall-region high shear layer and evolve with them.

Acknowledgments: The contributions of Jim C. Chen in this study are gratefully acknowledged.

Author Contributions: Weijia Chen designed and performed the numerical experiments; Weijia Chen and Edmond Y. Lo co-analyzed the data and co-wrote the paper.

Conflicts of Interest: The authors declare no conflict of interest.

References

1. Cebeci, T.; Cousteix, J. *Modeling and Computation of Boundary-Layer Flows: Laminar, Turbulent and Transitional Boundary Layers in Incompressible and Compressible Flows*; Horizon Publishing: Long Beach, CA, USA, 2005.
2. Schubauer, G.B.; Skramstad, H.K. *Laminar-boundary-Layer Oscillations and Transition on a Flat Plate*; National Bureau of Standards: Washington, DC, USA, 1948; pp. 251–292.
3. Morkovin, M.V. On the many faces of transition. In *Viscous Drag Reduction*; Wells, C.S., Ed.; Plenum Press: New York, NY, USA, 1969; pp. 1–31.
4. Kachanov, Y.S. On the resonant nature of the breakdown of a laminar boundary-layer. *J. Fluid Mech.* **1987**, *184*, 43–74. [CrossRef]
5. Orszag, S.A.; Patera, A.T. Secondary instability of wall-bounded shear flows. *J. Fluid Mech.* **1983**, *128*, 347–385. [CrossRef]
6. Klebanoff, P.S.; Tidstrom, K.D.; Sargent, L.M. The three-dimensional nature of boundary layer instability. *J. Fluid Mech.* **1962**, *12*, 1–34. [CrossRef]
7. Kachanov, Y.S.; Kozlov, V.V.; Levchenko, V.Y. Nonlinear development of a wave in a boundary layer. *Fluid Dyn.* **1977**, *12*, 383–390. [CrossRef]
8. Kachanov, Y.S. Physical mechanisms of laminar-boundary-layer transition. *Annu. Rev. Fluid Mech.* **1994**, *26*, 411–482. [CrossRef]
9. Bake, S.; Fernholz, H.H.; Kachanov, Y.S. Resemblance of K- and N-regimes of boundary-layer transition at late stages. *Eur. J. Mech. B-Fluids* **2000**, *19*, 1–22. [CrossRef]
10. Bake, S.; Meyer, D.G.W.; Rist, U. Turbulence mechanism in klebanoff transition: A quantitative comparison of experiment and direct numerical simulation. *J. Fluid Mech.* **2002**, *459*, 217–243. [CrossRef]
11. Borodulin, V.I.; Kachanov, Y.S.; Roschektayev, A.P. Turbulence production in an APG-boundary-layer transition induced by randomized perturbations. *J. Turbul.* **2006**, *7*, 1–30. [CrossRef]
12. Lee, C.B.; Li, R.Q. Dominant structure for turbulent production in a transitional boundary layer. *J. Turbul.* **2007**, *8*, 1–34. [CrossRef]
13. Liu, C.; Lu, P. DNS study on physics of late boundary layer transition. In Proceedings of the 50th AIAA Aerospace Sciences Meeting, Nashville, TN, USA, 9–12 January 2012.
14. Liu, C.; Yan, Y.; Lu, P. Physics of turbulence generation and sustenance in a boundary layer. *Comput. Fluids* **2014**, *102*, 353–384. [CrossRef]
15. Chen, W.; Chen, J.C. Combined compact difference method for solving the incompressible navier-stokes equations. *Int. J. Numer. Methods Fluids* **2012**, *68*, 1234–1256. [CrossRef]
16. Chen, W.; Chen, J.C.; Lo, E.Y. An interpolation based finite difference method on non-uniform grid for solving navier-stokes equations. *Comput. Fluids* **2014**, *101*, 273–290. [CrossRef]
17. Fasel, H. Investigation of the stability of boundary layers by a finite-differencee model of the navier-stokes equations. *J. Fluid Mech.* **1976**, *78*, 355–383. [CrossRef]
18. Fasel, H. Numerical investigation of the three-dimensional development in boundary-layer transition. *AIAA J.* **1990**, *28*, 29–37. [CrossRef]
19. Rist, U.; Fasel, H. Direct numerical simulation of controlled transition in a flat-plate boundary layer. *J. Fluid Mech.* **1995**, *298*, 211–248. [CrossRef]
20. Yeo, K.S.; Zhao, X.; Wang, Z.Y.; Ng, K.C. DNS of wavepacket evolution in a blasius boundary layer. *J. Fluid Mech.* **2010**, *652*, 333–372. [CrossRef]
21. Chen, W. Numerical Simulation of Boundary Layer Transition by Combined Compact Difference Method. Ph.D. Thesis, Nanyang Technological University, Nanyang Ave, Singapore, 2013.

22. Kachanov, Y.S.; Levchenko, V.Y. The resonant interaction of disturbances at laminar turbulent transition in a boundary-layer. *J. Fluid Mech.* **1984**, *138*, 209–247. [CrossRef]
23. Borodulin, V.I.; Kachanov, Y.S.; Koptsev, D.B. Experimental study of resonant interactions of instability waves in a self-similar boundary layer with an adverse pressure gradient: I. Tuned resonances. *J. Turbul.* **2002**, *3*, 62. [CrossRef]
24. Argoul, F.; Ameodo, A.; Grasseau, G.; Gagne, Y.; Hopfinger, E.; Frisch, U. Wavelet analysis of turbulence reveals the multifractal nature of the richardson cascade. *Nature* **1989**, *338*, 51–53. [CrossRef]
25. Meitz, H.L. *Numerical Investigation of Suction in a Transitional Flat-Plate Boundary Layer*; The University of Arizona: Tucson, AZ, USA, 1996.
26. Meitz, H.L.; Fasel, H.F. A compact-difference scheme for the navier–stokes equations in vorticity–velocity formulation. *J. Comput. Phys.* **2000**, *157*, 371–403. [CrossRef]
27. Liu, C.H.; Maslowe, S.A. A numerical investigation of resonant interactions in adverse-pressure-gradient boundary layers. *J. Fluid Mech.* **1999**, *378*, 269–289. [CrossRef]
28. Peyret, R. *Spectral Methods for Incompressible Viscous Flow*; Springer: Berlin, Germany, 2002.
29. Hirsch, C. *Numerical Computation of Internal and External Flows: Fundamentals of Computational Fluid Dynamics*, 2nd ed.; Butterworth-Heinemann: Oxford, UK, 2007.
30. Wu, X.H.; Moin, P. Direct numerical simulation of turbulence in a nominally zero-pressure-gradient flat-plate boundary layer. *J. Fluid Mech.* **2009**, *630*, 5–41. [CrossRef]
31. Shaikh, F.N. Investigation of transition to turbulence using white-noise excitation and local analysis techniques. *J. Fluid Mech.* **1997**, *348*, 29–83. [CrossRef]
32. Torrence, C.; Compo, G.P. A practical guide to wavelet analysis. *Bull. Am. Meteorol. Soc.* **1998**, *79*, 61–78. [CrossRef]
33. Bertolotti, F.P.; Herbert, T.; Spalart, P.R. Linear and nonlinear stability of the blasius boundary-layer. *J. Fluid Mech.* **1992**, *242*, 441–474. [CrossRef]
34. Hunt, J.C.R.; Wray, A.A.; Moin, P. Eddies, Stream, and Convergence Zones in Turbulent Flows. Available online: https://ntrs.nasa.gov/archive/nasa/casi.ntrs.nasa.gov/19890015184.pdf (accessed on 17 April 2017).
35. Chakraborty, P.; Balachandar, S.; Adrian, R.J. On the relationships between local vortex identification schemes. *J. Fluid Mech.* **2005**, *535*, 189–214. [CrossRef]
36. Ovchinnikov, V.; Choudhari, M.M.; Piomelli, U. Numerical simulations of boundary-layer bypass transition due to high-amplitude free-stream turbulence. *J. Fluid Mech.* **2008**, *613*, 135–169. [CrossRef]
37. Jeong, J.; Hussain, F. On the identification of a vortex. *J. Fluid Mech.* **1995**, *285*, 69–94. [CrossRef]
38. Borodulin, V.I.; Gaponenko, V.R.; Kachanov, Y.S.; Meyer, D.G.W.; Rist, U.; Lian, Q.X.; Lee, C.B. Late-stage transitional boundary-layer structures: Direct numerical simulation and experiment. *Theor. Comput. Fluid Dyn.* **2002**, *15*, 317–337. [CrossRef]
39. Borodulin, V.I.; Kachanov, Y.S. Role of the mechanism of local secondary instability in K-breakdown of boundary layer. *Soviet J. Appl. Phys.* **1988**, *3*, 70–81.

fluids

MDPI

Article

Resolution and Energy Dissipation Characteristics of Implicit LES and Explicit Filtering Models for Compressible Turbulence

Romit Maulik and Omer San *

School of Mechanical and Aerospace Engineering, Oklahoma State University, Stillwater, OK 74078, USA; romit.maulik@okstate.edu

* Correspondence: osan@okstate.edu; Tel.: +1-405-744-2457; Fax: +1-405-744-7873

Academic Editor: William Layton

Received: 7 March 2017; Accepted: 1 April 2017; Published: 6 April 2017

Abstract: Solving two-dimensional compressible turbulence problems up to a resolution of $16,384^2$, this paper investigates the characteristics of two promising computational approaches: (i) an implicit or numerical large eddy simulation (ILES) framework using an upwind-biased fifth-order weighted essentially non-oscillatory (WENO) reconstruction algorithm equipped with several Riemann solvers, and (ii) a central sixth-order reconstruction framework combined with various linear and nonlinear explicit low-pass spatial filtering processes. Our primary aim is to quantify the dissipative behavior, resolution characteristics, shock capturing ability and computational expenditure for each approach utilizing a systematic analysis with respect to its modeling parameters or parameterizations. The relative advantages and disadvantages of both approaches are addressed for solving a stratified Kelvin-Helmholtz instability shear layer problem as well as a canonical Riemann problem with the interaction of four shocks. The comparisons are both qualitative and quantitative, using visualizations of the spatial structure of the flow and energy spectra, respectively. We observe that the central scheme, with relaxation filtering, offers a competitive approach to ILES and is much more computationally efficient than WENO-based schemes.

Keywords: compressible turbulence; large eddy simulations; implicit LES; WENO schemes; relaxation filtering; Euler equations

1. Introduction

The use of nonlinearly weighted and upwind-biased schemes (such as weighted essentially non-oscillatory (WENO) reconstructions) is considered a state-of-the-art numerical tool for the simulation of turbulent compressible flows in the presence of shock waves [1–3]. These numerical schemes have an inherent numerical dissipation mechanism due to their asymmetric stencil reconstructions. They are also extensively used in the implicit large eddy simulation framework for turbulent flow computation and represent a promising approach to meeting the degree of accuracy required in high precision engineering requirements [4–8]. On the other hand, non-dissipative central (symmetric) linear reconstruction schemes can be used in a relaxation filtering framework where an explicit low-pass spatial filtering procedure is utilized to remove high-frequency contents of the simulation [9–11]. The main idea of this paper is to examine the relative advantages and disadvantages of these approaches for solving canonical turbulent compressible flow systems including a stratified shear layer instability problem to demonstrate the evolution of linear perturbations into a transition to nonlinear two-dimensional hydrodynamic turbulence.

The stratified Kelvin-Helmholtz instability (KHI) is a famous problem which manifests itself when there is a velocity difference at the interface between two fluids of different densities [12].

It can commonly be observed through experimental observation, numerical simulation and it is also visible in many natural phenomena such as for example in situations with wind flow over bodies of water causing wave formation and in the planet Jupiter's atmosphere between atmospheric bands moving at different speeds [13]. The study of this instability in a benchmark formulation reveals key information about the transition to turbulence for two fluids moving at different speeds. For these practical applications, it is common to choose a dual shear layer (DSL) setup to simulate the formation of KHI in a two-dimensional framework of the Euler equations for the validation of numerical methods [14]. The evolution of the instability can be compared across simulations with different underlying numerical schemes qualitatively (through visual examination of variable contours) and through statistical tools such as the measurement of kinetic energy spectra, the total energy in the flow and the rate of dissipation of total energy. Our goal is to categorize the differences in the averaged kinetic energy spectral scaling capture of the different numerical methods used in this investigation and pronounce conclusions about their advantages and disadvantages. We also aim to examine field plots in the form of density contours to visually assess the extent of spatial filtering. Next we study another test case of a Riemann shock interaction problem given by [15] (hereby denoted RSI) which will be used primarily to assess the shock capturing ability of the investigated numerical methods. It must be noted that KHI type vortical structures are also observed in this test case [16]. Through this investigation, we hope to connect the choice of solver to the choice of physics that needs to be captured for the numerical methods (and broader fluid mechanics) community.

The numerical methods utilized in this investigation are from two chief fields of thought (with regards to modeling approach). We use the implicit large eddy simulation methodology [5,17–22] where weighted non-oscillatory (WENO) upwinding schemes [23–25] are coupled with certain Riemann solvers in an implicit large eddy simulation (or ILES) framework. The term implicit refers to the addition of numerical dissipation by the upwinding scheme rather than any explicit dissipation term [26]. It is no surprise, therefore, that the characterization of the dissipative behavior of various WENO type schemes is an important area of interest in the computational fluid dynamics community [27–29]. Although the primary motivation for the development of non-oscillatory upwinding schemes was for the simulation of hyperbolic conservation laws with sharp discontinuities in the solution field [30–34], ILES has emerged as a popular alternative to the more formal techniques for subgrid scale modeling in turbulent flows [35–39] since it is convenient to let the numerical dissipation inherent to these schemes substitute the subgrid scale dissipation that would otherwise require artificial modeling [40–43]. We remark that controlling this numerical dissipation is not a trivial task, but the computational efficiency of this approach (as it does not require any explicit turbulence model) makes it attractive.

We shall also utilize the concept of relaxation filtering [44–46] (which employs low-pass spatial filters for removing high frequency content) and utilize a central scheme for the reconstruction of conserved quantities at cell interfaces. In this framework, the flow field variables are filtered after a certain number of timesteps to dissipate the amount of energy related to residual stresses and thus the dissipative effects of the unresolved scales on the resolved scales are modeled. We aim to utilize the dissipative behavior of this framework to prevent Gibbs oscillations which arise due to the use of central schemes in governing laws with dominating hyperbolic behavior. Relaxation filtering has been used with success for the large eddy simulation of homogeneous isotropic turbulence [47] and represents a computationally inexpensive approach to subgrid scale modeling [48]. Various filters have also been utilized for the relaxation filtering methodology [49,50] and the choice of free filtering parameters is crucial for this method. In addition, the relaxation filtering approach has been observed to have the desirable quantity of being grid independent [51].

We note that the primary purpose of both these approaches is the addition of dissipation in coarse grained simulations so as to capture kinetic energy spectra scaling as well as the formation of shocks. We remind the reader that the addition of artificial dissipation is crucial for the accurate simulations of hyperbolic conservation laws. The aforementioned numerical methods are implemented using the message passing interface (MPI) approach for distributing a physical domain among multiple processes for concurrent computation [52]. Through this, exceptional high fidelity simulations (HFS) are obtained for the DSL and RSI problems as benchmark 'true' numerical experiments for the purpose of determining the viability of a numerical method for coarse grained usage. On a thorough comparison of density contours, kinetic energy spectra capture and total energy measurement, conclusions are drawn about the suitability of numerical methods for the purpose of numerical simulation using the two-dimensional Euler equations. We reiterate that the main idea of this work is to provide the reader an exhaustive comparison of the well studied WENO-Riemann methods and the symmetric (central) reconstruction based explicit filtering approach through the aforementioned conclusions combined with a detailed discussion devoted to the mathematical development of these numerical approaches. The domain decomposition used in our MPI methodology is also reported briefly for the purpose of reproducibility. We aspire to provoke questions in our readers' mind about the performance of the numerical schemes discussed in representing accurate physics for two-dimensional flows given by the DSL and RSI test cases which are a simpler representation for many real world applications. For the convenience of the reader, we can draw up a list of points to summarize the major questions this investigation attempts to address:

1. How does a central scheme augmented with a relaxation filter compare with the performance of the conventional upwind-biased approach to numerical dissipation in ILES?
2. What reduction in computational expense can we expect through the utilization of this new explicit filtering framework?
3. Can we quantify the effect of the free modeling parameters that accompany low-pass spatial filters on the statistical and instantaneous features of our solution field?
4. What parameters dominate the dissipative nature of a WENO-ILES scheme for our test cases?

The rest of our work will be devoted to the development of the mathematics and a documentation of results to address these important questions.

2. Governing Equations

As mentioned in the previous section, we shall utilize the two-dimensional Euler equations in their conservation form as our underlying partial differential equations. These can be expressed as [53,54]

$$\frac{\partial q}{\partial t} + \frac{\partial F}{\partial x} + \frac{\partial G}{\partial y} = 0 \tag{1}$$

where

$$q = \begin{bmatrix} \rho \\ \rho u \\ \rho v \\ \rho e \end{bmatrix}, \quad F = \begin{bmatrix} \rho u \\ \rho u^2 + P \\ \rho u v \\ \rho u H \end{bmatrix}, \quad G = \begin{bmatrix} \rho v \\ \rho u v \\ \rho v^2 + P \\ \rho v H \end{bmatrix}, \tag{2}$$

where

$$H = e + P/\rho, \ \ P = \rho(\gamma - 1)\left(e - \frac{1}{2}(u^2 + v^2)\right). \tag{3}$$

Here ρ, P, u and v are the density, pressure, horizontal and vertical components of velocity respectively. Also, e and H denote the total energy and total enthalpy per unit mass with γ being the ratio of specific heats. Before proceeding further, we develop the eigensystem of the system of equations which will then be used in devising hyperbolic conservation laws. The convective flux Jacobian matrices for our system of equations are

$$A = \frac{\partial F}{\partial q} = \begin{bmatrix} 0 & 1 & 0 & 0 \\ \phi^2 - u^2 & (3-\gamma)u & -(\gamma-1)v & \gamma - 1 \\ -uv & v & u & 0 \\ (\phi^2 - H)u & H - (\gamma-1)u^2 & -(\gamma-1)uv & \gamma u \end{bmatrix} \tag{4}$$

and

$$B = \frac{\partial G}{\partial q} = \begin{bmatrix} 0 & 0 & 1 & 0 \\ -uv & v & u & 0 \\ \phi^2 - v^2 & -(\gamma-1)u & (3-\gamma)v & \gamma - 1 \\ (\phi^2 - H)v & -(\gamma-1)uv & H - (\gamma-1)v^2 & \gamma v \end{bmatrix} \tag{5}$$

where $\phi^2 = \frac{1}{2}(\gamma-1)(u^2+v^2)$. A similarity transform of the above flux Jacobian matrices can be shown as

$$\begin{aligned} LAR = \Lambda \quad &\Rightarrow R\Lambda L = A \\ SBT = \Psi \quad &\Rightarrow T\Psi S = B \end{aligned} \tag{6}$$

where Λ and Ψ are the diagonal matrices comprising of the real eigenvalues of A and B respectively; L and S represent 4×4 matrices whose columns are the eigenvectors of this eigendecomposition. They can be expressed as [55]

$$R = \begin{bmatrix} 1 & 0 & \beta & \beta \\ u & 0 & \beta(u+a) & \beta(u-a) \\ v & -1 & \beta v & \beta v \\ \frac{\phi^2}{(\gamma-1)} & -v & \beta(H+ua) & \beta(H-ua) \end{bmatrix}$$

$$L = \begin{bmatrix} 1 - \frac{\phi^2}{a^2} & (\gamma-1)\frac{u}{a^2} & (\gamma-1)\frac{v}{a^2} & -\frac{\gamma-1}{a^2} \\ v & 0 & -1 & 0 \\ \phi^2 - ua & a - (\gamma-1)u & -(\gamma-1)v & \gamma-1 \\ \phi^2 + ua & -a - (\gamma-1)u & -(\gamma-1)v & \gamma - 1 \end{bmatrix}$$

$$T = \begin{bmatrix} 1 & 0 & \beta & \beta \\ u & 1 & \beta u & \beta u \\ v & 0 & \beta(v+a) & \beta(v-a) \\ \frac{\phi^2}{(\gamma-1)} & u & \beta(H+va) & \beta(H-va) \end{bmatrix}$$

$$S = \begin{bmatrix} 1 - \frac{\phi^2}{a^2} & (\gamma-1)\frac{u}{a^2} & (\gamma-1)\frac{v}{a^2} & -\frac{\gamma-1}{a^2} \\ -u & 1 & 0 & 0 \\ \phi^2 - va & -(\gamma-1)u & a - (\gamma-1)v & \gamma-1 \\ \phi^2 + va & -(\gamma-1)u & -a - (\gamma-1)v & \gamma - 1 \end{bmatrix}$$

$$\Lambda = \begin{bmatrix} u & 0 & 0 & 0 \\ 0 & u & 0 & 0 \\ 0 & 0 & u+a & 0 \\ 0 & 0 & 0 & u-a \end{bmatrix}, \quad \Psi = \begin{bmatrix} v & 0 & 0 & 0 \\ 0 & v & 0 & 0 \\ 0 & 0 & v+a & 0 \\ 0 & 0 & 0 & v-a \end{bmatrix}$$

where a is the speed of sound and can be calculated through the relation $a^2 = \gamma P/\rho$. The parameter β is given by $1/(2a^2)$. We must reiterate that this is one of several possible eigendecompositions possible for this system and there may be slight variations in results for certain algorithms according to one's choice [56]. The solvers we evaluate in this paper are independent of the choice of eigenvector matrices.

3. Numerical Methods

3.1. Finite Volume Framework

The semi-discrete form of the governing equations can be written as

$$\frac{dq_{i,j}}{dt} + \frac{1}{\Delta x}(F_{i+1/2,j} - F_{i-1/2,j}) + \frac{1}{\Delta y}(G_{i,j+1/2} - G_{i,j-1/2}) = 0 \tag{7}$$

with $q_{i,j}$ being the cell-averaged vector of dependant variables, $F_{i\pm1/2,j}$ representing the fluxes at the right and left cell boundaries and $G_{i,j\pm1/2}$ representing the fluxes at the top and bottom boundaries of the cell. We use the method of lines to represent our system of PDE's as an ODE through time (so as to implement a Runge-Kutta scheme for time integration)

$$\frac{dq_{i,j}}{dt} = \pounds(q_{i,j}) \tag{8}$$

with the right hand side of the above equation representing the combined effect of all the spatial derivatives in the governing equations. The above ODE representation can be advanced to obtain the solution field at a future time step $n+1$ given the current time step n by a total variation diminishing third-order Runge Kutta scheme [57] as follows

$$\begin{aligned} q_{i,j}^{(1)} &= q_{i,j}^{(n)} + \Delta t \pounds(q_{i,j}^{(n)}) \\ q_{i,j}^{(2)} &= \frac{3}{4}q_{i,j}^{(n)} + \frac{1}{4}q_{i,j}^{(1)} + \frac{1}{4}\Delta t \pounds(q_{i,j}^{(1)}) \\ q_{i,j}^{(n+1)} &= \frac{1}{3}q_{i,j}^{(n)} + \frac{2}{3}q_{i,j}^{(2)} + \frac{2}{3}\Delta t \pounds(q_{i,j}^{(2)}) \end{aligned} \tag{9}$$

where a time step Δt is prescribed through a CFL criterion as

$$\Delta t = \min\left(\eta\frac{\Delta x}{\max(|\Lambda|)}, \eta\frac{\Delta y}{\max(|\Psi|)}\right). \tag{10}$$

In our work, we have defined the parameter $\eta = 0.5$ for all simulations.

3.2. Upwind-Biased ILES Schemes

3.2.1. WENO Reconstructions

The left and right states of the cell boundaries are reconstructed using the weighted essentially non-oscillatory (WENO) approach which was first introduced in [58]. The order of accuracy of these reconstructions depends on the length of the stencil chosen and affect the solution (through a different dissipative behavior). In this work, we shall focus on the 5th order accurate WENO scheme (WENO-5) which utilizes a five point stencil. The reader is directed to [14] for a thorough examination of the effect of WENO stencil sizes on the Kelvin-Helmholtz instability test case. In what follows (and for the rest of this document), we shall introduce stencil expressions only for the x direction of our domain. We note, for the purpose of clarity, that it is the conserved variables which are being reconstructed in this formulation. We utilize a modified implementation of the WENO-5 reconstruction [59] which can be given as

$$q^L_{i+1/2} = w_0(\frac{1}{3}q_{i-2} - \frac{7}{6}q_{i-1} + \frac{11}{6}q_i) + w_1(-\frac{1}{6}q_{i-1} + \frac{5}{6}q_i + \frac{1}{3}q_{i+1}) + w_2(\frac{1}{3}q_i + \frac{5}{6}q_{i+1} - \frac{1}{6}q_{i+2}) \quad (11)$$

$$q^R_{i-1/2} = w_0(\frac{1}{3}q_i + \frac{5}{6}q_{i-1} - \frac{1}{6}q_{i-2}) + w_1(-\frac{1}{6}q_{i+1} + \frac{5}{6}q_i + \frac{1}{3}q_{i-1}) + w_2(\frac{1}{3}q_{i+2} - \frac{7}{6}q_{i+1} + \frac{11}{6}q_i) \quad (12)$$

where the nonlinear weights are defined by

$$w_k = \frac{\alpha_k}{\alpha_0 + \alpha_1 + \alpha_2}, \quad \alpha_k = \frac{d_k}{(\beta_k + \epsilon)^p} \quad (13)$$

with the smoothness indicators defined as

$$\beta_0 = \frac{13}{12}(q_{i-2} - 2q_{i-1} + q_i)^2 + \frac{1}{4}(q_{i-2} - 4q_{i-1} + 3q_i)^2 \quad (14)$$

$$\beta_1 = \frac{13}{12}(q_{i-1} - 2q_i + q_{i+1})^2 + \frac{1}{4}(q_{i-1} - q_{i+1})^2 \quad (15)$$

$$\beta_2 = \frac{13}{12}(q_i - 2q_{i+1} + q_{i+2})^2 + \frac{1}{4}(3q_i - 4q_{i+1} + q_{i+2})^2. \quad (16)$$

In this study, we compute the smoothness indicators for each conserved variables. The optimal linear weighting coefficients are $d_0 = 1/10$, $d_1 = 3/5$ and $d_2 = 3/10$ in Equation (11) and $d_0 = 3/10$, $d_1 = 3/5$ and $d_2 = 1/10$ in Equation (12). The presence of any discontinuity thus leads to an adaptation in the weights for order reduction and increased dissipation in the stencil corresponding to the discontinuity. This expression for the calculation of the nonlinear weights of the smoothness indicator for the WENO-5 interfacial flux reconstruction will be referred to as WENO-JS. We also investigate the effect of choosing a different approach for the calculation of the nonlinear weights (hereby denoted as WENO-Z) which are given by [60]

$$\alpha_k = d_k \left(1 + \frac{|\beta_0 - \beta_2|}{(\beta_k + \epsilon)^p}\right). \quad (17)$$

Before proceeding, we must observe that there are many variants of the WENO reconstructions described above (for example in [61–64]) and this investigation is by no means exhaustive. The different variants of the WENO reconstructions have all been developed as a response to demands of either different orders of accuracy [65,66] or reduced computational expense. The interested reader is directed to [33] for an excellent discussion on the effect of the order of reconstruction on a classical two-dimensional configuration utilizing the Euler equations.

We remark, for clarity, that the left (L) and right (R) states correspond to reconstructions considering the possibility of advection from both directions. The value of ϵ is set to 10^{-6} for the WENO-JS approach and 10^{-20} for the WENO-Z approach in order to prevent any errors by division of zero. We note that the power of the denominator in α_k is usually set to $p = 2$ in most utilizations of the WENO-5 approach. In this investigation, we aim to perform a sensitivity analysis for p to quantify its effect on our solutions. A mathematical analysis of the role of the smoothness indicator and the parameter ϵ can be found in [67].

3.2.2. Riemann Solvers

Once the left and right states have been constructed using WENO-5, we may now use Riemann solvers to calculate the fluxes at these boundaries. We utilize four celebrated Riemann solvers for our numerical simulations which are as follows. We emphasize that the selection of these solvers was motivated by the desire to span a wide range of dissipative behaviors as reported in literature [16].

3.2.3. Rusanov Scheme

The Rusanov solver [68] utilizes information from the maximum local wave propagation speed to give us the follow expression for the flux

$$F_{i+1/2} = \frac{1}{2}(F^R + F^L) - \frac{c_{i+1/2}}{2}\left(q^R_{i+1/2} - q^L_{i+1/2}\right) \tag{18}$$

where F^R and F^L are the flux components using the right and left constructed states respectively (i.e., $F^R = F(q^R_{i+1/2})$ and $F^L = F(q^L_{i+1/2})$). The local wave propagation speed $c_{i+1/2}$, is given by the maximum absolute eigenvalue of the Jacobian matrix of F between the cells i and $i+1$, i.e.,

$$c_{i+1/2} = \max(r(A_i), r(A_{i+1})) \tag{19}$$

where $r(A)$ represents the spectral radius of matrix A. For our case, the spectral radius may simply be determined as $r(A) = \max(|u|, |u-a|, |u+a|)$. Thus our final expression for the wave propagation speed in a cell can be written as

$$c_{i+1/2} = \max(|u|_i, |u-a|_i, |u+a|_i, |u|_{i+1}, |u-a|_{i+1}, |u+a|_{i+1}). \tag{20}$$

3.2.4. Roe Scheme

The Godunov theorem states the the exact values of the face fluxes can be computed by the following relation if the Jacobian matrix is constant [69]

$$F_{i+1/2} = \frac{1}{2}(F^R + F^L) - \frac{1}{2}R|\Lambda|L\left(q^R_{i+1/2} - q^L_{i+1/2}\right) \tag{21}$$

where $|\Lambda|$ is the diagonal matrix obtained after a similarity transformation of the Jacobian matrix A. For the Euler equations however, A is a function of the conserved variables (i.e., $A = A(q)$). To account for this, the Roe Riemann solver [70,71] estimates the interfacial fluxes in an approximate Godunov manner

$$F_{i+1/2} = \frac{1}{2}(F^R + F^L) - \frac{1}{2}\tilde{R}|\tilde{\Lambda}|\tilde{L}\left(q^R_{i+1/2} - q^L_{i+1/2}\right) \tag{22}$$

where the superscript tilde represents the Roe density weighted average between the left and right states. Our modified eigensystem can be constructed by using the following averaged values of the conserved variables instead

$$\tilde{u} = \frac{u_R\sqrt{\rho_R} + u_L\sqrt{\rho_L}}{\sqrt{\rho_R} + \sqrt{\rho_L}} \tag{23}$$

$$\tilde{v} = \frac{v_R\sqrt{\rho_R} + v_L\sqrt{\rho_L}}{\sqrt{\rho_R} + \sqrt{\rho_L}} \tag{24}$$

$$\tilde{H} = \frac{H_R\sqrt{\rho_R} + H_L\sqrt{\rho_L}}{\sqrt{\rho_R} + \sqrt{\rho_L}} \tag{25}$$

where the left and right states of the un-averaged conserved variables are available from the WENO-5 reconstruction described earlier. We note that the Roe averaged speed of sound is now a function of these averaged variables. We also utilize Harten's [72] entropy fix approach for expansion shocks by replacing Roe averaged eigenvalues with

$$|\tilde{\lambda_i}| = \begin{cases} |\tilde{\lambda}_i|, & \text{if } |\tilde{\lambda}_i| \geq 2\epsilon\tilde{a} \\ \tilde{\lambda}_i^2/(4\epsilon\tilde{a}), & \text{if } |\tilde{\lambda}_i| < 2\epsilon\tilde{a} \end{cases} \tag{26}$$

where $\tilde{a}$ is the Roe averaged speed of sound and $\tilde{\lambda}_i$ are the eigenvalues of the Roe averaged Jacobian matrix $\tilde{A}$. Here ϵ is a small positive number and is typically chosen as $\epsilon = 0.1$ in our simulations.

3.2.5. HLL Scheme

Harten et al. [73] assumed that the lower and upper bounds on the characteristic speeds can be used in the solution of the Riemann problem. These upper and lower bounds were given by [74–76]

$$S_L = \min(u_L, u_R) - \max(a_L, a_R) \tag{27}$$

$$S_R = \max(u_L, u_R) + \max(a_L, a_R) \tag{28}$$

where S_L and S_R are the lower and upper bounds on the left and right state characteristics speeds. The HLL solver thus takes the following form

$$F_{i+1/2} = \begin{cases} F^L, & \text{if } S_L \geq 0 \\ F^R, & \text{if } S_R \leq 0 \\ \frac{S_R F^L - S_L F^R + S_L S_R (q^R_{i+1/2} - q^L_{i+1/2})}{S_R - S_L}, & \text{otherwise.} \end{cases} \tag{29}$$

3.2.6. AUSM Scheme

The advection upstream splitting method (or AUSM) was proposed to provide an alternative approach to low-diffusion flux-splitting methods [77,78]. This approach is based on recognizing that the inviscid flux consists of a distinct convective flux and a pressure flux. A cell interface advection Mach number is defined to determine convective flow quantities. Our investigation uses the AUSM approach for its lower dissipative behavior through the following expression of the interface numerical flux

$$F_{i+1/2} = \frac{M_{i+1/2}}{2}\left[\begin{pmatrix}\rho a\\ \rho a u\\ \rho a v\\ \rho a H\end{pmatrix}^L + \begin{pmatrix}\rho a\\ \rho a u\\ \rho a v\\ \rho a H\end{pmatrix}^R\right] - \frac{|M_{i+1/2}|}{2}\left[\begin{pmatrix}\rho a\\ \rho a u\\ \rho a v\\ \rho a H\end{pmatrix}^R - \begin{pmatrix}\rho a\\ \rho a u\\ \rho a v\\ \rho a H\end{pmatrix}^L\right] + \begin{pmatrix}0\\ p_L^+ + p_R^-\\ 0\\ 0\end{pmatrix} \tag{30}$$

where

$$M_{i+1/2} = M_L^+ + M_R^- \tag{31}$$

in which the directional convective Mach number ($M = u/a$) is given as

$$M^\pm = \begin{cases} \pm\frac{1}{4}(M \pm 1)^2, & \text{if } |M| \leq 1 \\ \frac{1}{2}(M \pm |M|), & \text{otherwise.} \end{cases} \tag{32}$$

The following split formula is used for the pressure

$$p^\pm = \begin{cases} p\frac{1}{4}(M \pm 1)^2(2 \mp M), & \text{if } |M| \leq 1 \\ p\frac{1}{2}(M \pm |M|)/M, & \text{otherwise.} \end{cases} \tag{33}$$

We note that there are many variants of the classic AUSM solver [79–84], which are used to obtain more accurate and robust results in all-speed regimes.

3.3. Central Reconstruction Schemes

This approach relies on a traditional finite volume formulation for calculating the density, energy and velocities at the cell interfaces following which the cell face fluxes are calculated as a function of these reconstructed variables. The following 6-point stencil symmetric non-dissipative scheme is used for face reconstruction of the conserved quantity [85]

$$q_{i+1/2} = a(q_{i+1} + q_i) + b(q_{i+2} + q_{i-1}) + c(q_{i+3} + q_{i-2}) \tag{34}$$

where the stencil coefficients are given by

$$a = 37/60; \quad b = -2/15; \quad c = 1/60. \tag{35}$$

Once the relevant face quantities are determined from the nodal values, the fluxes may be calculated for use in Equation (7). The considerable simplicity of this approach is apparent. This simple central scheme may be combined with a classic relaxation filtering method as shown below to remove high wavenumber content and prevent floating point overflow. This can be shown to be an explicit large eddy simulation approach for conservation laws [44].

3.3.1. Relaxation Filtering

A low-pass spatial filter sweep is carried out using a seven point stencil (as given in [86]), wherein it is denoted as a standard relaxation filtering procedure if carried out a the end of time integration or as a modified relaxation filtering procedure when carried out at each substep of the TVDRK3 algorithm. The motivation of low-pass spatial filtering is to remove high frequency content from the solution field at every sweep and thus prevent any accumulation of noise (which might eventually lead to Gibbs oscillations). For the purpose of filtering, we use the following expression

$$\bar{q}_i = q_i - \sigma\left(f_0 q_i + f_1(q_{i+1} + q_{i-1}) + f_2(q_{i+2} + q_{i-2}) + f_3(q_{i+3} + q_{i-3})\right) \tag{36}$$

where

$$f_0 = 5/16; \quad f_1 = -15/64; \quad f_2 = 3/32; \quad f_3 = -1/64 \tag{37}$$

and $\sigma = [0, 1]$ is a parameter that controls filter dissipation strength with $\sigma = 0$ being completely non-dissipative and $\sigma = 1$ being most dissipative. The most dissipative $\sigma = 1$ case corresponds to a Gaussian-type filter with a full attenuation at the smallest grid cut-off scale. The choice of applying this filter sweep at each substep of the time integration or at its conclusion would appear to have ramifications on its success in preventing oscillations in the solutions for the conserved variable. This shall be examined in detail in the results section. As expected, the use of this explicit LES approach for the discrete hyperbolic equations leads to a significant reduction in computational expense. However, it may be prone to stability issues when used for strongly shocked flows (particularly for lower values of the free parameter controlling dissipation). It must be noted here that there is no restriction on the choice of filtering strategies when it comes to the relaxation filtering approach, any low-pass spatial filtering process (of course with a similarly shaped transfer function) would achieve similar results [87]. The transfer function of the filter chosen here is plotted in Figure 1 and can be represented as

$$T(\omega) = 1 - \sigma\left(f_0 + 2f_1\cos(\omega) + 2f_2\cos(2\omega) + 2f_3\cos(3\omega)\right) \tag{38}$$

where ω is the modified wavenumber. A value of $\sigma = 1$ gives us the 6th order binomial smoothing filter which is derived in [88]. Using a Taylor series expansion procedure in Equation (36), it is easy to show that the leading truncation error term in this linear filter becomes sixth order accurate

$$\bar{q}_i = q_i + \frac{\sigma}{64}h^6 q^{(VI)} \tag{39}$$

where h is considered our grid discretization size and the Roman numeral superscript on q denotes the 6th order differentiation. We note that Equation (36) is used in each direction sequentially (i.e., first in

y direction, the direction of domain decomposition, to eliminate any extra MPI data exchange cost due to the filtering processes).

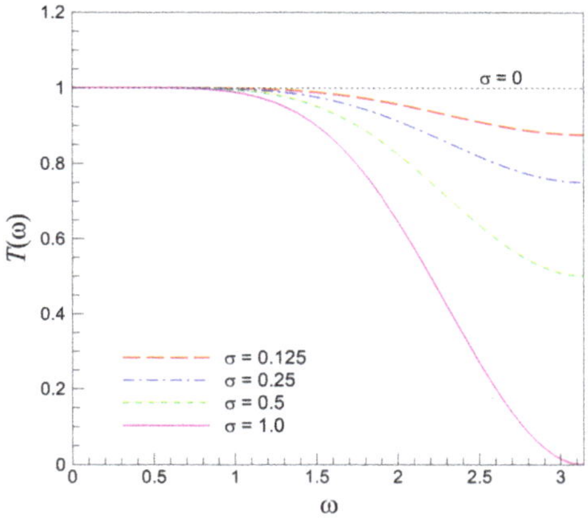

Figure 1. Transfer function for low-pass spatial filter used in standard and modified relaxation filtering. Notice reduction of area under curve with increasing values of σ implying more dissipation.

3.3.2. Shock Filtering

The shock filtering approach devised by Bogey et al. [89] relies on an adaptive filtering approach to remove noise from the solution field. The adaptive quantum of dissipation is calculated from the local solution to modify the free parameter (σ) as mentioned in the previous subsection. The basic kernel of the shock-capturing filtering approach is

$$\bar{q}_i = q_i - \left(\sigma^s_{i+1/2} D_{i+1/2} - \sigma^s_{i-1/2} D_{i-1/2}\right), \tag{40}$$

where

$$D_{i+1/2} = \sum_{j=1-n}^{n} c_j q_{i+j} \text{ and } D^n_{i-1/2} = \sum_{j=1-n}^{n} c_j q_{i+j-1}. \tag{41}$$

In this work, we choose $n = 1$ to give us the following simplified version of the above equation

$$D_{i+1/2} = c_0 q_i + c_1 q_{i+1} \text{ and } D_{i-1/2} = c_0 q_{i-1} + c_1 q_i \tag{42}$$

with $c_0 = 0.25$ and $c_1 = -c_0$. The filtering parameter σ^s is calculated adaptively through extracting the high-wavenumber components of the pressure field using a second-order filtering process given by

$$\tilde{p}_i = (-p_{i+1} + 2p_i - p_{i-1})/4. \tag{43}$$

The magnitude of the high-pass filtered pressure squared term is then calculated as

$$\tilde{p}_i^{2m} = \frac{1}{2}\left((\tilde{p}_i - \tilde{p}_{i+1})^2 + (\tilde{p}_i - \tilde{p}_{i-1})^2\right). \tag{44}$$

We can now devise a shock sensor through the ratio r given by

$$r_i = \frac{\tilde{p}_i^{2m}}{p_i^2} + \epsilon^s \tag{45}$$

where $\epsilon^s = 10^{-16}$ is needed to avoid numerical divergence. We note that it is possible to determine a shock detection through the local dilatation rather than the pressure (which is particularly useful for discerning turbulent fluctuations and shocks), but we limit the investigation to the pressure magnitude version of the shock sensor to align ourselves with the WENO-Riemann solver architecture which have primarily been developed for the purpose of shock capturing. Our filtering parameter may now be updated dynamically by

$$\sigma_i^s = \frac{1}{2}\left(1 - \frac{r_{th}}{r_i} + \left|1 - \frac{r_{th}}{r_i}\right|\right). \tag{46}$$

For $r_i \leq r_{th}$, the filtering magnitude $\sigma^s = 0$. Values of $r_i > r_{th}$ give us nonzero values of σ^s with $\sigma^s \to 1$ for $r_i \to +\infty$. The threshold parameter r_{th} is thus our free parameter for the addition of dissipation and has been set between values of 10^{-4} and 10^{-7}. Our cell face value for the filtering parameter σ^s can be approximated as

$$\sigma_{i+1/2}^s = \frac{1}{2}\left(\sigma_{i+1}^s + \sigma_i^s\right) \text{ and } \sigma_{i-1/2}^s = \frac{1}{2}\left(\sigma_i^s + \sigma_{i-1}^s\right). \tag{47}$$

We remark here that this nonlinear explicit filtering scheme is expected to be more dissipative than the standard (or modified) central relaxation filtering approach in the previous subsection since it employs a considerably lower order of reconstruction (2nd order).

4. Results

This section is devoted to a quantification of the performance of the various numerical approaches outlined above. We first define the initial conditions of the two test cases before outlining our MPI methodology followed by which the results of the different numerical solvers are explained in the form of density contours, kinetic energy spectra and total energy plots.

4.1. Problem Definitions

As mentioned previously, we utilize the DSL problem and the RSI test case for the evaluation of our numerical methods. The DSL problem consists of the simulation of an initially perturbed compressible shear layer in two dimensions. This test problem characterizes the solvers' ability to evolve a sine wave perturbation into two-dimensional turbulence. We expect to see the instability trigger small-scale vortical structures at the sharp density interface initially which eventually transition (through nonlinear interactions) to a completely turbulent field. We must note here that the turbulent field obtained after sufficient progress in the simulation are due to growth of the shear layer instabilities. Figure 2a describes the initial conditions for the DSL problem. Our computational domain is a square of unit side length with periodic boundary conditions in all directions. The initial conditions are specified as follows

$$\rho(x,y) = \begin{cases} 1.0, & \text{if } |y| \geq 0.25 \\ 2.0, & \text{if } |y| < 0.25 \end{cases} \tag{48}$$

$$u(x,y) = \begin{cases} 1.0, & \text{if } |y| \geq 0.25 \\ -1.0, & \text{if } |y| < 0.25 \end{cases} \tag{49}$$

$$v(x,y) = \lambda \sin(2\pi n x/L) \tag{50}$$

$$p(x,y) = 2.5. \tag{51}$$

One can observe that the vertical component of the velocity is perturbed using a single mode sine wave ($n = 2$, $L = 1.0$) and the amplitude of the perturbation is set at $\lambda = 0.01$. The DSL numerical experiments are solved to a final dimensionless time of $t = 5$. The second test case we have selected for this investigation is that of a shocked flow (i.e., the RSI test case) given by configuration 3 in [15] as shown in Figure 2b. Details about the configuration in question may also be found in [16]. This problem is designed with different initial conditions specified in each quadrant of our computational domain. These initial conditions may be varied to obtain a variety of combinations of shocks and contact discontinuities. The final time we consider for the RSI problem is $t = 0.5$ and the transmissive boundary conditions are used in both directions (e.g., $q_{b-k/2} = q_{b+k/2}$, where $k = 1, 2$, and 3 and the subscript b refers to the boundary edge). In both problems, the ratio of specific heats is considered to be constant $\gamma = 7/5$. The $\mathbb{R}^2$ simulation domain for all experiments is set $(x, y) \in [-0.5, 0.5] \times [-0.5, 0.5]$.

Both test cases are evaluated using three coarse grid resolutions given by $N^2 = 256^2$, 1024^2 and 4096^2 points with a $N^2 = 16,384^2$ solution obtained using the Roe Riemann solver as a reference high fidelity simulation (HFS). Density contours and kinetic energy spectra scaling are provided for the purpose of evaluation of the numerical schemes in the DSL test case. Classical k^{-3} scaling from KBL theory is also provided for the purpose of comparison [90–92]. Only density contours are provided for the RSI test case (for visually ascertaining shock capturing ability).

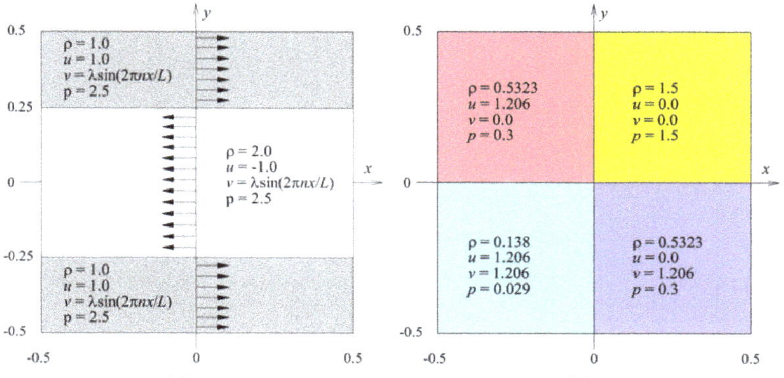

Figure 2. Representative initial conditions for the (**a**) dual shear layer (DSL) and (**b**) Riemann Shock Interaction (RSI) test cases. Note that we use periodic boundary conditions on all sides for the DSL problem while using open boundary conditions in all directions for the RSI test case.

4.2. MPI Methodology

For the purpose of parallelization, the OpenMPI [52] standard library is used to speed up our computation. Using the familiar principles of distributed multiprocessing, our physical domain is split up into several smaller computational domains which are then distributed to different computational units (or processes). Each process solves the governing equations on its own slice of the physical domain before communicating boundary data to its neighbors. This domain decomposition and data transfer can be observed in Figure 3. One can observe that there is only one direction in which data exchange is needed, since the physical domain is distributed into even slices with the same number of points in the x direction. It must be noted that in case of the DSL problem, data must also be exchanged between the top and bottom computational slices cut from the physical domain due its periodic boundary conditions. Data exchange is programmed to occur at the end of each time integration substep as well as during the explicit filtering process. This is because the calculation of derivatives and the filtering process requires data values from beyond the computational domain allocated to the process. This extra data belonging to the neighboring domains is stored in 'ghost' arrays for each computational slice. The ghost arrays must thus be updated concurrently with the solution through send and receive commands. The process of sending and receiving data from these ghost arrays is done using non-blocking send `MPI_Isend` and non-blocking receive `MPI_Irecv` commands. These non-blocking communication privileges imply that the function calls return before the communication is completed. This is done to ensure that there is no deadlock. A deadlock implies a concurrent use of synchronous send or receive buffers (which may only be used by one thread). The use of the aforementioned non-blocking commands alleviates this issue but also requires the use of the `MPI_wait` command to ensure all communication is complete before the code executes any further on each thread. The interested reader is referred to [93] for a up to date tutorial on the MPI standard.

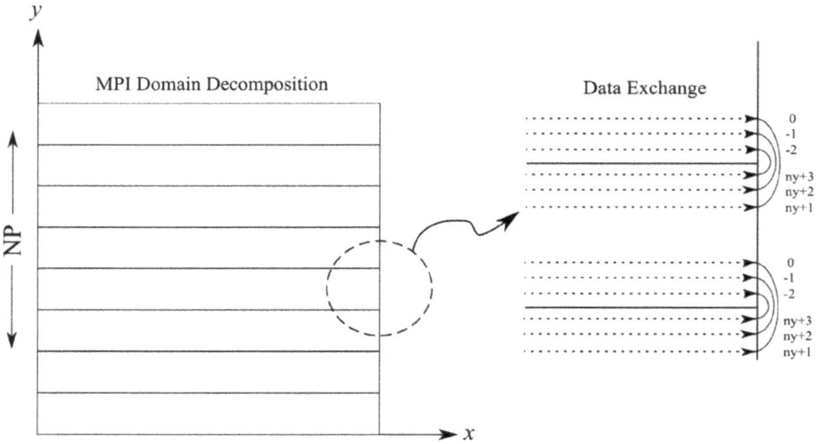

Figure 3. A schematic of our domain decomposition. Data is exchanged in the y direction prior to each time integration substep and during the explicit filtering process.

A computational performance assessment of our code is detailed in Figure 4 where both strong and weak scalings are described. In a weak scaling test, the problem size is proportionally scaled with the number of processes, so that the computational work per core remains essentially constant. Ideally, the relative runtime (the run time scaled by that on one process) should remain close to unity as the number of processes and the problem size are increased [94]. The weak scaling test demonstrates how the parallel overhead scales as a function of NP and provides some measure of how the communication complexity of an algorithm degrades as the process count increases. In a strong scaling test, a fixed problem size is considered and a varying number of processes are used to solve the problem. The parallel method should exhibit run times inversely proportional to the number of processes used, or alternatively, the speedup should vary linearly with NP. The strong scaling test will also show limiting behavior for a fixed problem size, since for a large enough NP, there will not be sufficient computational work available on each process to adequately overlap parallel communication overhead. Speedup (S) is defined as the ratio of execution time of the serial algorithm to execution time of the parallel algorithm with NP processes (i.e., $S = t_1/t_{NP}$).

One can observe a marked increase in relative runtime (and also a corresponding decrease in efficiency) as the order of the processes (NP) becomes closer to the order of the number of grid points in each direction. This is because the relative cost of data exchange becomes higher. From a computational perspective, a future improvement of the current implementation of our code would be to decompose the physical domain through a resolution dependent heuristic for optimal efficiency. We note, however, that for the purpose of this investigation, the speed up offered enabled us to obtain our reference HFS data (i.e., approximately 268 million degrees of freedom) in approximately 400 hours of simulation time (on 432 processes) for the DSL case and 50 h (on 288 processes) for the RSI case.

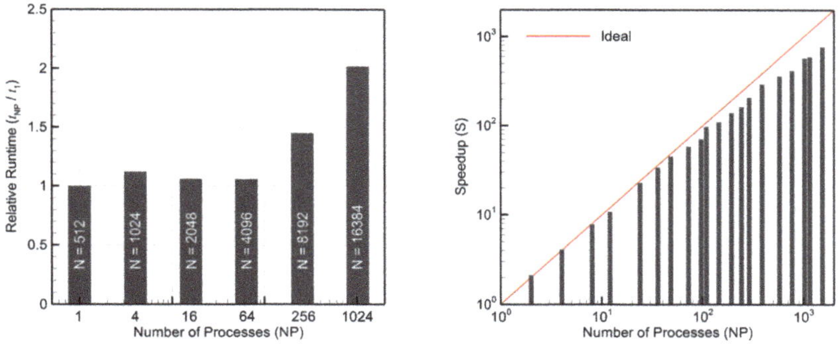

(**a**) Weak scaling

(**b**) Strong scaling

Figure 4. Message passing interface (MPI) performance indicators. (**a**) Weak scaling keeps the problem-per-node size constant while increasing the number of nodes. (**b**) Strong scaling spread the same size problem across more nodes.

4.3. Dual Shear Layer (DSL) Problem

In this subsection, we perform an exhaustive investigation into the performance of the different numerical techniques described previously on the DSL test case. Figure 5 shows two density contours (at $t = 5$) for the Roe and HLL Riemann solvers at the highest resolution (i.e., corresponding to $N^2 = 16,384^2$). The high degrees of freedom leads to the capture of several small scale structures in both solvers. We remark here that a considerable loss in correlation can be observed between the two solver solutions manifesting itself in a phase difference. This is due to the different amount of dissipation imparted to the system by each Riemann solver - the different perturbations cause a different evolution of the flow field (a hallmark of deterministic chaos). We remark here that the research in numerical scheme development should ideally be aligned to the goal of improved predictions of both statistical quantities such as kinetic energy spectra and of phase. Figure 5 also describes the total energy evolution of the system. It must be noted that an ideal evolution of the Euler equations would conserve energy for all time but the dissipation due to our numerical methods eventually reduce the total kinetic energy of the system. Once the instability transitions to a turbulent field, one can see a loss of phase between the HLL and Roe solvers from the total energy evolution plot. The total kinetic energy of the solution field is defined as the spatial average of the instantaneous kinetic energy at all points and our averaging operator is given by

$$E(t) = \frac{1}{A} \int \int E(x, y, t) dx dy \tag{52}$$

where A is the area of the physical domain and $E(x, y, t)$ is the instantaneous kinetic energy per unit mass at a particular point in the solution field (i.e., $E = 1/2(u^2 + v^2)$). Indeed, we compute the integral

$$E(t) = \frac{1}{N_x N_y} \sum_{i=1}^{N_x} \sum_{j=1}^{N_y} \frac{1}{2} \left(u_{i,j}^2(t) + v_{i,j}^2(t) \right) \tag{53}$$

using the cell centered values of velocity components.

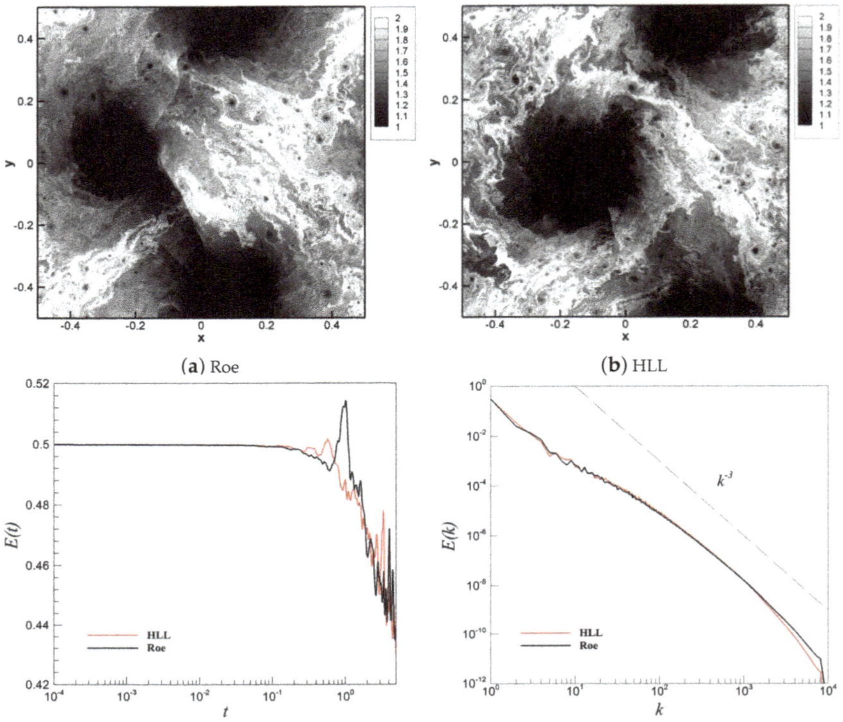

(**a**) Roe (**b**) HLL

(**c**) Total kinetic energy (**d**) Kinetic energy spectra ($t = 5$)

Figure 5. High fidelity ($N^2 = 16,384^2$) density contours at $t = 5$ for the Roe (**a**) and HLL (**b**) Riemann solvers. Accordingly, the evolution of the total kinetic energy content (**c**) as well as the kinetic energy spectral scaling (**d**) of the flow are shown for both cases. Note the phase difference in the solutions once it transitions to turbulence.

Figure 5 also details the difference between the HLL and the Roe solvers in terms of final time kinetic energy spectra. It is clear that the flow fields (for these HFS) have shown the k^{-3} scaling from KBL theory in the inertial range. The existence of this scaling has also been observed in other benchmark two-dimensional configurations (e.g., a single-mode Richtmyer-Meshkov instability problem [95]) using the same underlying governing equations. From the comparison of the HLL and Roe Riemann solvers in terms of spectral scaling, one can visually represent the difference in the dissipative nature of both solvers. We observe, as expected, that both approaches represent the critical inertial and integral length scales with the same degree of accuracy. The kinetic energy spectra that we have represented here can be calculated using the following expression in wavenumber space [96]

$$\hat{E}(\mathbf{k},t) = \frac{1}{2}|\hat{\mathbf{u}}(\mathbf{k},t)|^2 \tag{54}$$

where $\hat{\mathbf{u}}(\mathbf{k},t)$ is the Fourier transform of the velocity vector in the wavenumber space. Equation (54) can be also rewritten in terms of velocity components

$$\hat{E}(\mathbf{k},t) = \frac{1}{2}\left(|\hat{u}(\mathbf{k},t)|^2 + |\hat{v}(\mathbf{k},t)|^2\right) \tag{55}$$

where we compute velocity components $\hat{u}(\mathbf{k},t)$ and $\hat{v}(\mathbf{k},t)$ using a fast Fourier transform algorithm [97]. Finally, the spectra can be calculated by angle averaging in the following manner

$$E(k,t) = \sum_{k-\frac{1}{2} \leq |\acute{\mathbf{k}}| < k+\frac{1}{2}} \hat{E}(\acute{\mathbf{k}},t) \tag{56}$$

where $k = |\mathbf{k}| = \sqrt{k_x^2 + k_y^2}$.

Figure 6 shows the time evolution of the solution field in the form of three snapshots of the density contours at $t = 1, 3$ and 5 for the Roe Riemann solver. We have also included the transition to turbulence for our reference HFS. These snapshots are also provided for the coarser grids and show a considerable reduction in the small scale details with increasing coarseness. For the coarsest resolution of $N^2 = 256^2$ one can see that there are very few small scale details and only large scale structures are preserved. The absence of smaller structures with increasing coarsening also gives evidence of spatial filtering being performed implicitly by the ILES approach. We can also observe the presence of acoustic waves in the central belt of high density in the earlier stages of the instability.

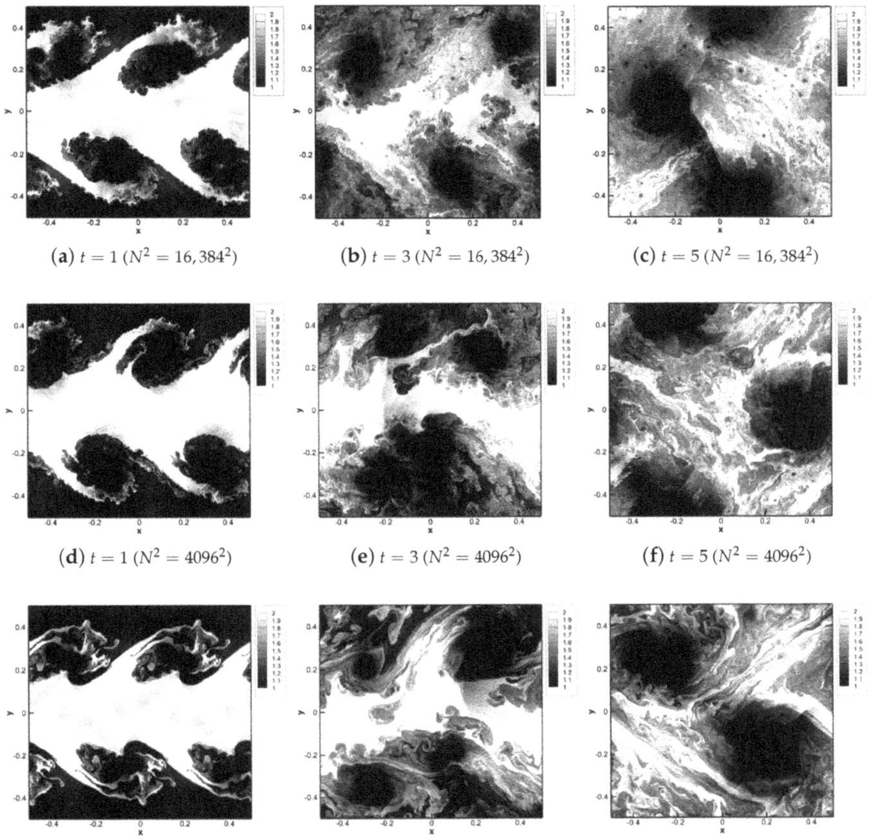

(**a**) $t = 1$ ($N^2 = 16,384^2$) (**b**) $t = 3$ ($N^2 = 16,384^2$) (**c**) $t = 5$ ($N^2 = 16,384^2$)

(**d**) $t = 1$ ($N^2 = 4096^2$) (**e**) $t = 3$ ($N^2 = 4096^2$) (**f**) $t = 5$ ($N^2 = 4096^2$)

(**g**) $t = 1$ ($N^2 = 1024^2$) (**h**) $t = 3$ ($N^2 = 1024^2$) (**i**) $t = 5$ ($N^2 = 1024^2$)

Figure 6. *Cont.*

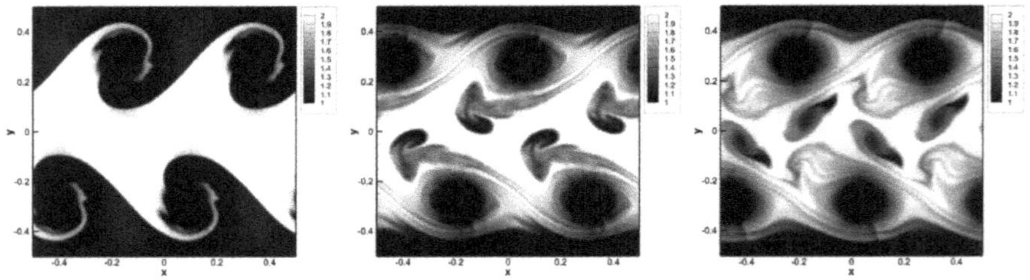

(**j**) $t = 1$ ($N^2 = 256^2$) (**k**) $t = 3$ ($N^2 = 256^2$) (**l**) $t = 5$ ($N^2 = 256^2$)

Figure 6. DSL density contours with time obtained through the Roe solver. High fidelity data with $N^2 = 16,384^2$ (**a–c**) and coarse grid simulations at $N^2 = 4096^2$ (**d–f**), $N^2 = 1024^2$ (**g–i**) and $N^2 = 256^2$ (**j–l**). Note absence of smaller structures in the coarse grid runs implying spatial filtering being performed implicitly.

Similar trends are seen in Figure 7 which details the coarse grid simulations for the HLL Riemann solver. The slightly higher dissipation of this approach is apparent with a reduced number of smaller structures for a similar resolution when compared to the Roe approach (a visual examination of the $N^2 = 1024^2$ case confirms this). This can also be seen in Figure 8 where the Rusanov solver is used for the construction of interfacial fluxes. Both the Rusanov and HLL Riemann solvers are considered amongst the more dissipative approaches for solving a system of hyperbolic equations. The Rusanov approach happens to be the most dissipative solver we have investigated here. This becomes clear when the kinetic energy spectra for each solver are compared (later). Figure 9 shows the performance of the AUSM solver (which is the least dissipative Riemann solver in this investigation) and it can be seen that it triggers a turbulent phase (due to a reduced dissipation of numerical noise) even in the coarsest of runs. Spectral scaling confirms this lack of dissipation.

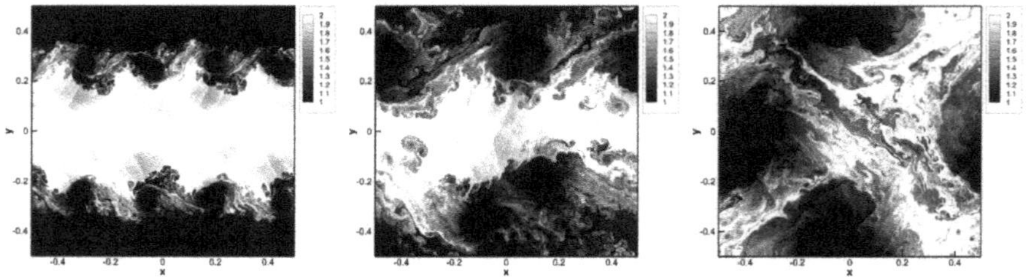

(**a**) $t = 1$ ($N^2 = 4096^2$) (**b**) $t = 3$ ($N^2 = 4096^2$) (**c**) $t = 5$ ($N^2 = 4096^2$)

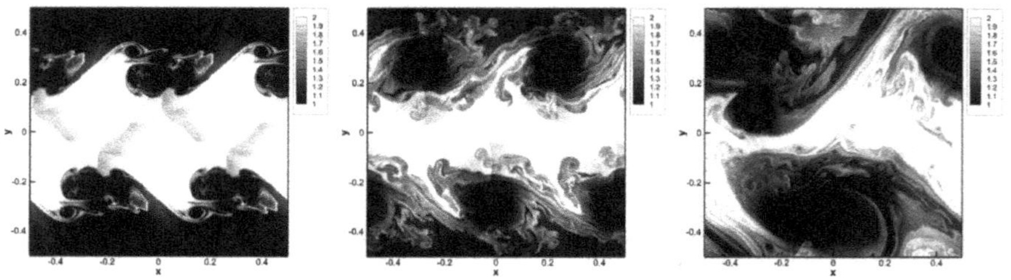

(**d**) $t = 1$ ($N^2 = 1024^2$) (**e**) $t = 3$ ($N^2 = 1024^2$) (**f**) $t = 5$ ($N^2 = 1024^2$)

Figure 7. *Cont.*

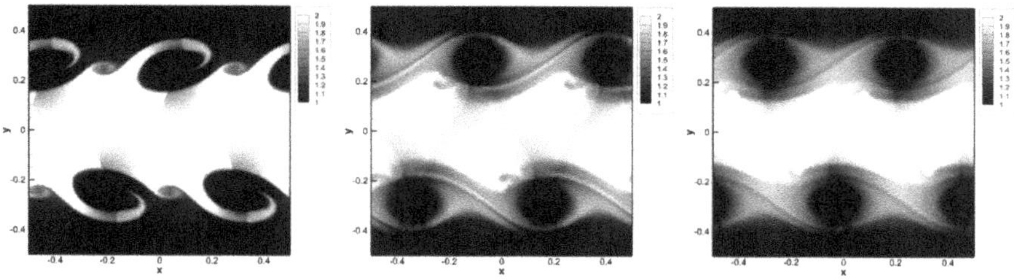

(g) $t = 1$ ($N^2 = 256^2$) **(h)** $t = 3$ ($N^2 = 256^2$) **(i)** $t = 5$ ($N^2 = 256^2$)

Figure 7. DSL density contours with time obtained through the HLL Riemann solver with coarse grid resolutions of $N^2 = 4096^2$ (**a–c**), $N^2 = 1024^2$ (**d–f**), $N^2 = 256^2$ (**g–i**).

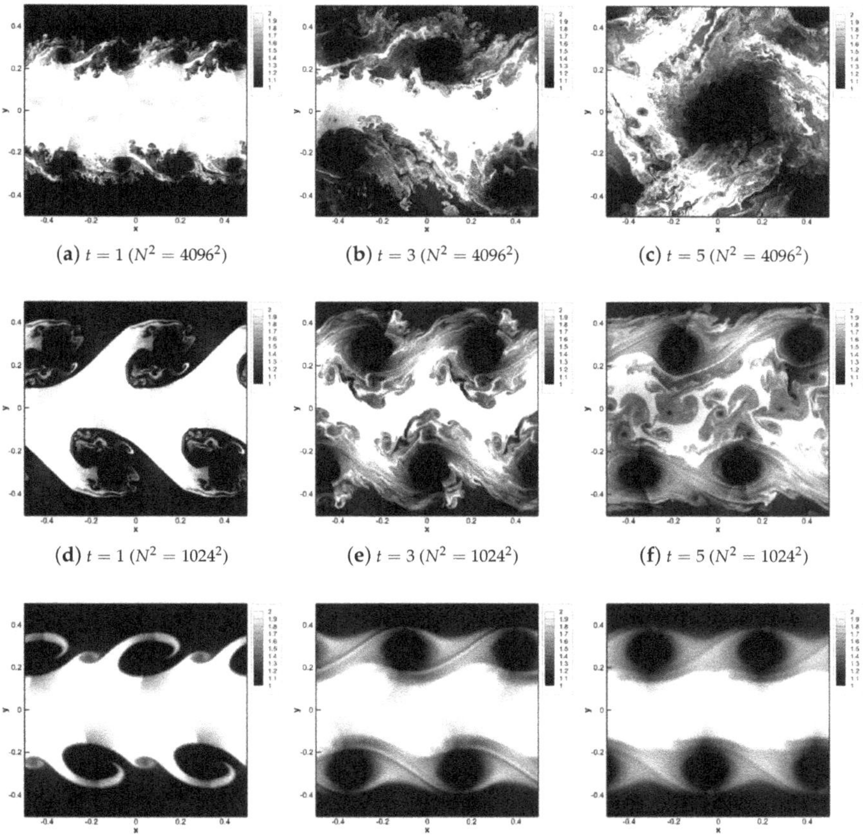

(a) $t = 1$ ($N^2 = 4096^2$) **(b)** $t = 3$ ($N^2 = 4096^2$) **(c)** $t = 5$ ($N^2 = 4096^2$)

(d) $t = 1$ ($N^2 = 1024^2$) **(e)** $t = 3$ ($N^2 = 1024^2$) **(f)** $t = 5$ ($N^2 = 1024^2$)

(g) $t = 1$ ($N^2 = 256^2$) **(h)** $t = 3$ ($N^2 = 256^2$) **(i)** $t = 5$ ($N^2 = 256^2$)

Figure 8. DSL density contours with time obtained through the Rusanov Riemann solver with coarse grid resolutions of $N^2 = 4096^2$ (**a–c**), $N^2 = 1024^2$ (**d–f**), $N^2 = 256^2$ (**g–i**).

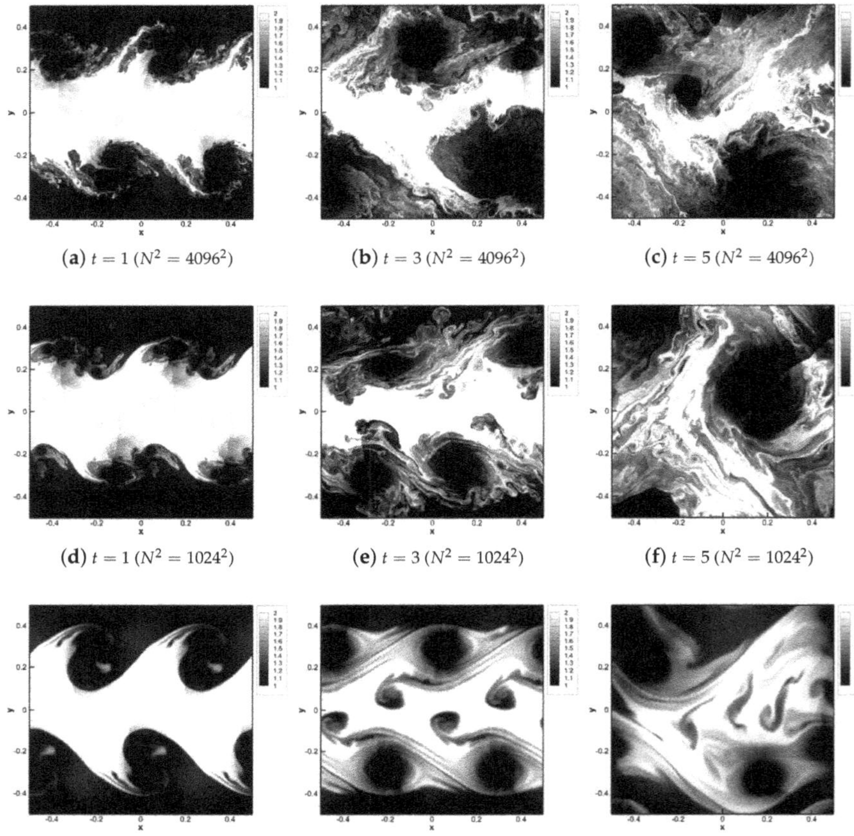

(**a**) $t = 1$ ($N^2 = 4096^2$) (**b**) $t = 3$ ($N^2 = 4096^2$) (**c**) $t = 5$ ($N^2 = 4096^2$)

(**d**) $t = 1$ ($N^2 = 1024^2$) (**e**) $t = 3$ ($N^2 = 1024^2$) (**f**) $t = 5$ ($N^2 = 1024^2$)

(**g**) $t = 1$ ($N^2 = 256^2$) (**h**) $t = 3$ ($N^2 = 256^2$) (**i**) $t = 5$ ($N^2 = 256^2$)

Figure 9. DSL density contours with time obtained through the AUSM Riemann solver with coarse grid resolutions of $N^2 = 4096^2$ (**a–c**), $N^2 = 1024^2$ (**d–f**), $N^2 = 256^2$ (**g–i**).

The DSL problem was also simulated using the central scheme for constructing interfacial fluxes and subsequent low-pass spatial filtering (dubbed the relaxation filtering approach). This approach produced accurate solutions for the DSL problem without the use of any upwinding as seen in Figure 10 for a free filtering parameter value of $\sigma = 0.5$. It can also be observed that small scale structure capture is similar to the performance of the Roe solver. This is an important observation—the proposed relaxation filtering approach now allows us to control the dissipation in an active manner through its filtering parameter. This can be observed through density contours for a lower value of $\sigma = 0.25$ as shown in Figure 11 where much more smaller structures are captured and for the highest value of $\sigma = 1.0$ where larger scales are preserved at the expense of the smallest ones (in Figure 12). This potentially allows for experimentation with different classes of low-pass spatial filters (with different transfer functions) in order to obtain the desired amount of scale capture. We also note the considerable computational ease of implementation for this framework and reduced computational expense in comparison to the relatively involved nature of the WENO reconstructions and the Riemann solvers. We make a note of caution though that the choice of the filtering parameter and the associated filter also play a role in the stability of the numerical scheme [98]. It is entirely possible that a very low quantity of dissipation may lead to the emergence of nonphysical Gibbs oscillations and possible floating point overflow (i.e., $\sigma = 0$ refers to a perfectly central scheme without any numerical dissipation). In this regard, the Riemann solvers used with the WENO-5 reconstruction are decidedly superior with their conditional upwinding using nonlinear weights in their stencils. One possible approach to improve the stability of this approach is to implement the low-pass spatial filtering sweep at the end of each time integration substep. In theory, this would improve the stability of the relaxation filtering approach. We find that the relaxation filtering done at multiple substeps performs just as well as the regular relaxation filtering approach and it shall also be seen that this approach does not represent a considerable computational overhead in comparison to the regular relaxation filtering approach. The density contours of this modified implementation of the relaxation filtering approach are shown in Figure 13. To the authors' experience, the relaxation filtering approach with the filtering strength of $\sigma = 0.5$ yields accurate results on both fine and coarse grids in practice, although it doesn't fulfil full attenuation at the cut-off grid scale (i.e., see Figure 1).

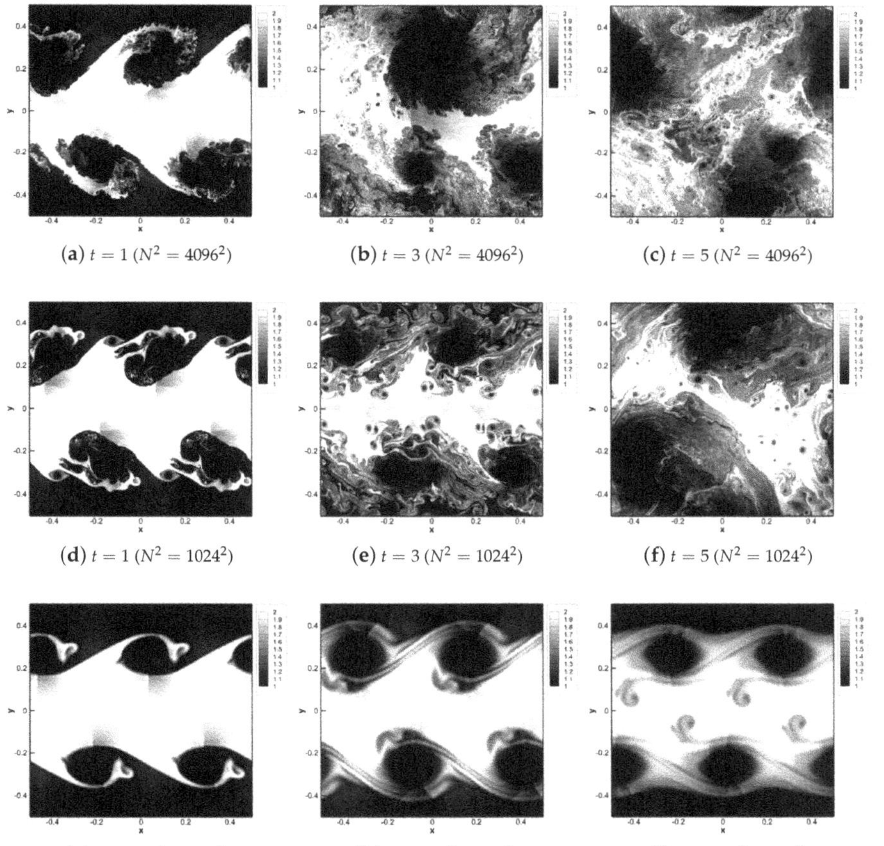

(**a**) $t = 1$ ($N^2 = 4096^2$) (**b**) $t = 3$ ($N^2 = 4096^2$) (**c**) $t = 5$ ($N^2 = 4096^2$)

(**d**) $t = 1$ ($N^2 = 1024^2$) (**e**) $t = 3$ ($N^2 = 1024^2$) (**f**) $t = 5$ ($N^2 = 1024^2$)

(**g**) $t = 1$ ($N^2 = 256^2$) (**h**) $t = 3$ ($N^2 = 256^2$) (**i**) $t = 5$ ($N^2 = 256^2$)

Figure 10. DSL density contours with time obtained through the standard relaxation filtering approach (using $\sigma = 0.5$) with coarse grid resolutions of $N^2 = 4096^2$ (**a–c**), $N^2 = 1024^2$ (**d–f**), $N^2 = 256^2$ (**g–i**).

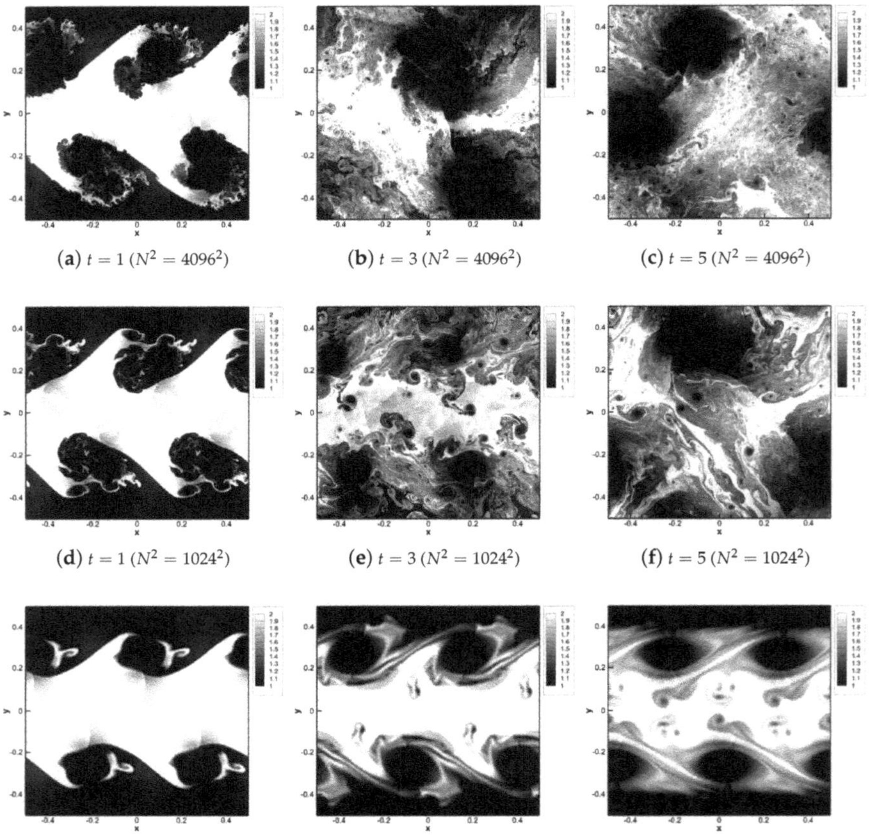

(**a**) $t = 1$ ($N^2 = 4096^2$) (**b**) $t = 3$ ($N^2 = 4096^2$) (**c**) $t = 5$ ($N^2 = 4096^2$)

(**d**) $t = 1$ ($N^2 = 1024^2$) (**e**) $t = 3$ ($N^2 = 1024^2$) (**f**) $t = 5$ ($N^2 = 1024^2$)

(**g**) $t = 1$ ($N^2 = 256^2$) (**h**) $t = 3$ ($N^2 = 256^2$) (**i**) $t = 5$ ($N^2 = 256^2$)

Figure 11. DSL density contours with time obtained through the standard relaxation filtering approach with (using $\sigma = 0.25$) with coarse grid resolutions of $N^2 = 4096^2$ (**a–c**), $N^2 = 1024^2$ (**d–f**), $N^2 = 256^2$ (**g–i**).

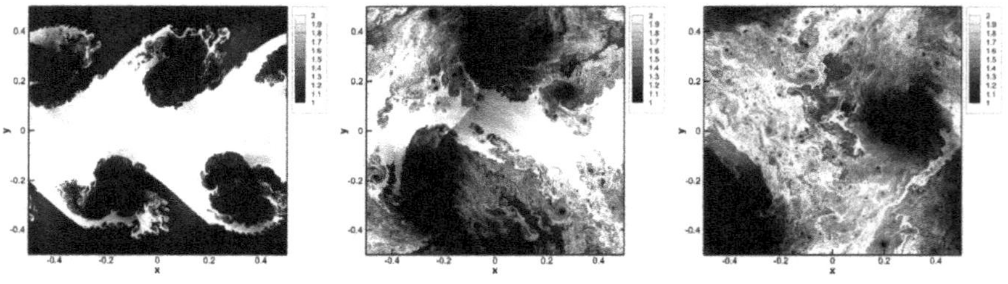

(**a**) $t = 1$ ($N^2 = 4096^2$) (**b**) $t = 3$ ($N^2 = 4096^2$) (**c**) $t = 5$ ($N^2 = 4096^2$)

Figure 12. *Cont.*

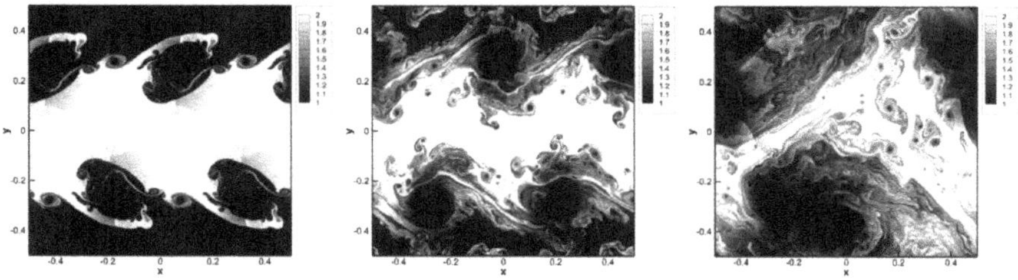

(d) $t = 1$ ($N^2 = 1024^2$) **(e)** $t = 3$ ($N^2 = 1024^2$) **(f)** $t = 5$ ($N^2 = 1024^2$)

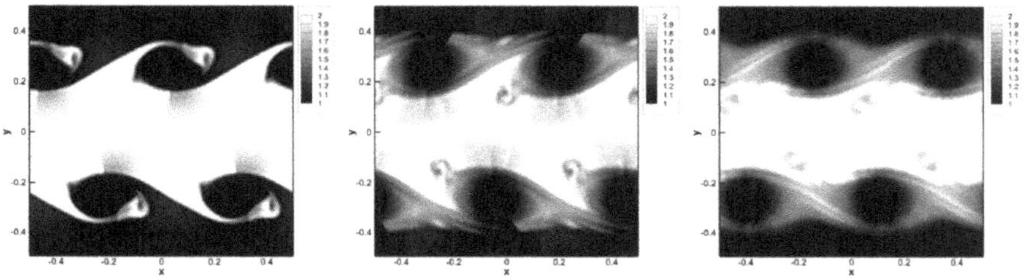

(g) $t = 1$ ($N^2 = 256^2$) **(h)** $t = 3$ ($N^2 = 256^2$) **(i)** $t = 5$ ($N^2 = 256^2$)

Figure 12. DSL density contours with time obtained through the standard relaxation filtering approach (using $\sigma = 1.0$) with coarse grid resolutions of $N^2 = 4096^2$ (**a**–**c**), $N^2 = 1024^2$ (**d**–**f**), $N^2 = 256^2$ (**g**–**i**).

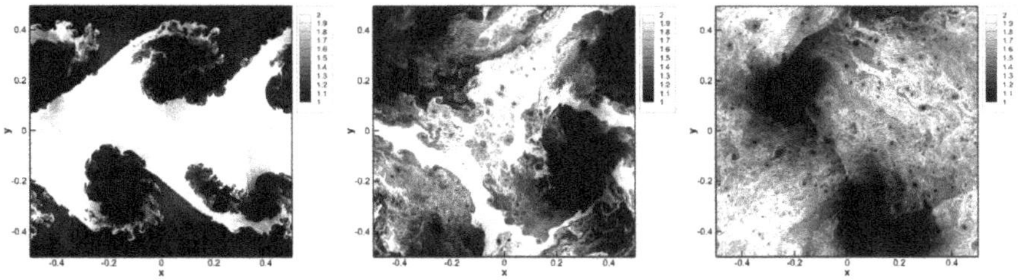

(a) $t = 1$ ($N^2 = 4096^2$) **(b)** $t = 3$ ($N^2 = 4096^2$) **(c)** $t = 5$ ($N^2 = 4096^2$)

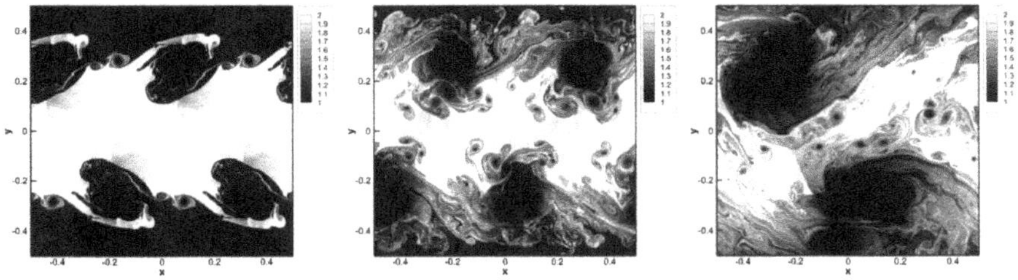

(d) $t = 1$ ($N^2 = 1024^2$) **(e)** $t = 3$ ($N^2 = 1024^2$) **(f)** $t = 5$ ($N^2 = 1024^2$)

Figure 13. *Cont.*

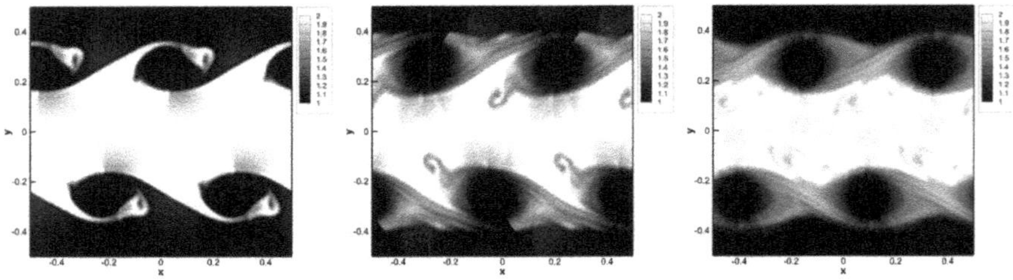

(g) $t = 1$ ($N^2 = 256^2$) **(h)** $t = 3$ ($N^2 = 256^2$) **(i)** $t = 5$ ($N^2 = 256^2$)

Figure 13. DSL density contours with time obtained through the modified relaxation filtering approach (using $\sigma = 1.0$) with coarse grid resolutions of $N^2 = 4096^2$ (**a–c**), $N^2 = 1024^2$ (**d–f**), $N^2 = 256^2$ (**g–i**).

The adaptive implementation of the relaxation filtering approach (called the shock filtering approach here) was also implemented to solve the DSL problem as shown in Figure 14. Here, as mentioned previously, the control parameter for dissipation is the shock threshold parameter r_{th} which controls the sensitivity of the shock sensor. At this high value of $r_{th} = 10^{-5}$, it is seen that the shock sensor provides the least amount of dissipation and results in turbulence being triggered even for the coarsest runs. In fact, on closer examination, spurious numerical oscillations can also be seen in the density contours. It is also observed that the larger vortices in the coarse grained simulations are less well defined. This fact will be confirmed later in the kinetic energy spectra comparisons where we shall also observe a deviation from the expected inertial range scaling. On reducing the magnitude of r_{th} we increase the dissipation of the shock filtering scheme as shown in Figures 15 and 16. It is seen that for coarser runs, the increased dissipation manages to damp the initial perturbation to the flow field and prevent any sort of transition to turbulence (as seen in the case with $r_{th} = 10^{-7}$.

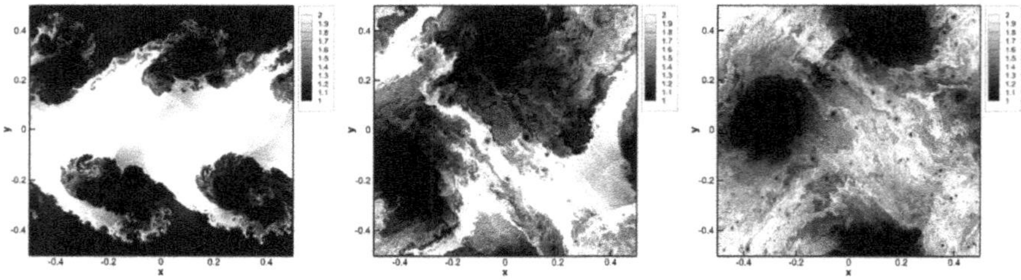

(a) $t = 1$ ($N^2 = 4096^2$) **(b)** $t = 3$ ($N^2 = 4096^2$) **(c)** $t = 5$ ($N^2 = 4096^2$)

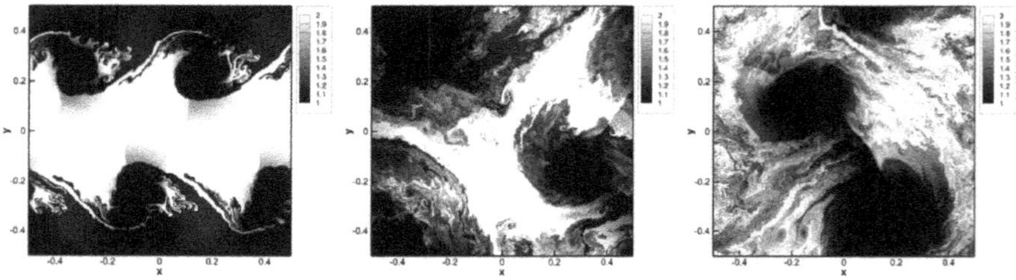

(d) $t = 1$ ($N^2 = 1024^2$) **(e)** $t = 3$ ($N^2 = 1024^2$) **(f)** $t = 5$ ($N^2 = 1024^2$)

Figure 14. *Cont.*

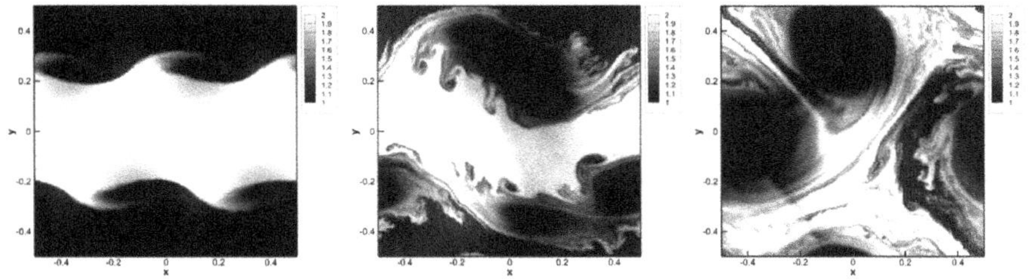

(g) $t = 1$ ($N^2 = 256^2$) **(h)** $t = 3$ ($N^2 = 256^2$) **(i)** $t = 5$ ($N^2 = 256^2$)

Figure 14. DSL density contours with time obtained through the shock filtering approach (using $r_{th} = 10^{-5}$) with coarse grid resolutions of $N^2 = 4096^2$ (**a–c**), $N^2 = 1024^2$ (**d–f**), $N^2 = 256^2$ (**g–i**).

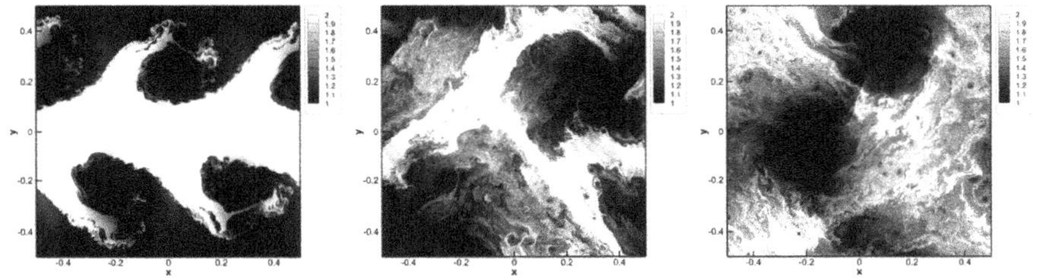

(a) $t = 1$ ($N^2 = 4096^2$) **(b)** $t = 3$ ($N^2 = 4096^2$) **(c)** $t = 5$ ($N^2 = 4096^2$)

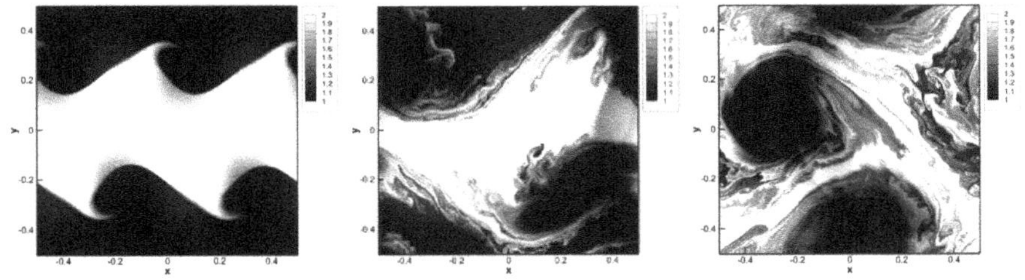

(d) $t = 1$ ($N^2 = 1024^2$) **(e)** $t = 3$ ($N^2 = 1024^2$) **(f)** $t = 5$ ($N^2 = 1024^2$)

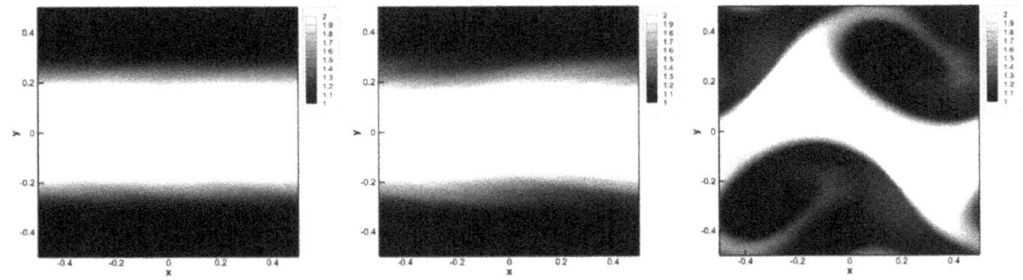

(g) $t = 1$ ($N^2 = 256^2$) **(h)** $t = 3$ ($N^2 = 256^2$) **(i)** $t = 5$ ($N^2 = 256^2$)

Figure 15. DSL density contours with time obtained through the shock filtering approach (using $r_{th} = 10^{-6}$) with coarse grid resolutions of $N^2 = 4096^2$ (**a–c**), $N^2 = 1024^2$ (**d–f**), $N^2 = 256^2$ (**g–i**).

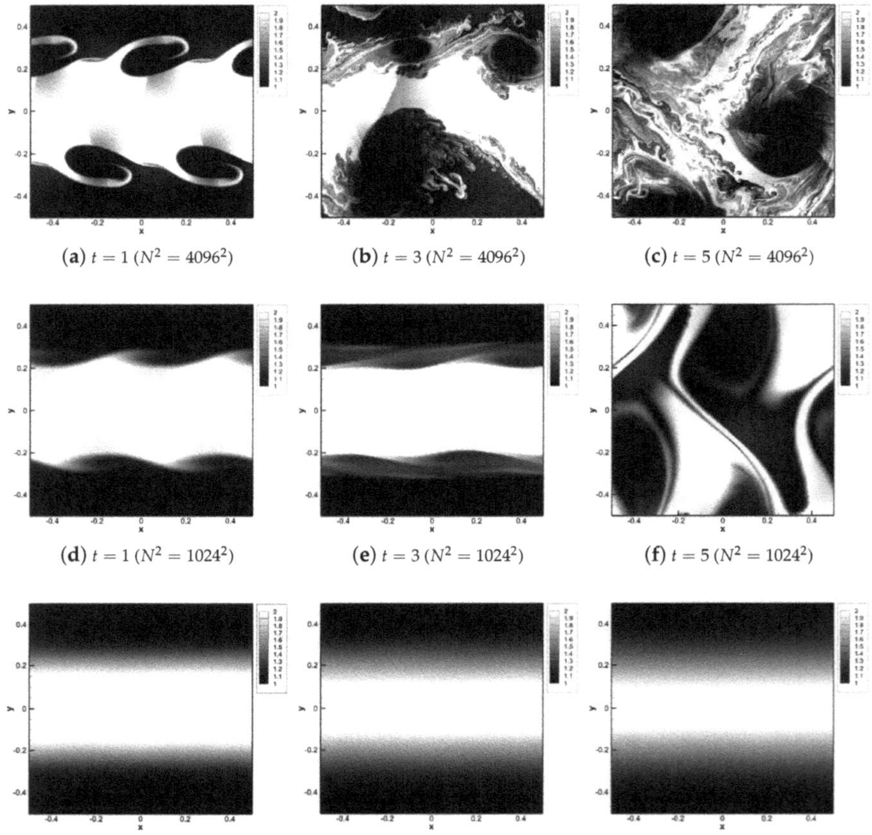

(a) $t = 1$ ($N^2 = 4096^2$) (b) $t = 3$ ($N^2 = 4096^2$) (c) $t = 5$ ($N^2 = 4096^2$)

(d) $t = 1$ ($N^2 = 1024^2$) (e) $t = 3$ ($N^2 = 1024^2$) (f) $t = 5$ ($N^2 = 1024^2$)

(g) $t = 1$ ($N^2 = 256^2$) (h) $t = 3$ ($N^2 = 256^2$) (i) $t = 5$ ($N^2 = 256^2$)

Figure 16. DSL density contours with time obtained through the shock filtering approach (using $r_{th} = 10^{-7}$) with coarse grid resolutions of $N^2 = 4096^2$ (**a–c**), $N^2 = 1024^2$ (**d–f**), $N^2 = 256^2$ (**g–i**).

Figure 17 shows the performance of the aforementioned numerical schemes quantified in a statistical sense through kinetic energy spectra scaling at different resolutions for the four different WENO-Riemann combinations. Plots showing the total energy in the flow are also reported. It is generally seen that the flow transitions to turbulence somewhere around $t = 1$. For the purpose of comparison, we utilize HFS statistical data from the Roe Riemann solver (with $N^2 = 16,384^2$). We also include the expected theoretical scaling given by KBL theory for kinetic energy cascades in two-dimensional turbulence (i.e., $E(k) \propto k^{-3}$). This figure details the dissipative behavior of the ILES techniques for three different coarse grid simulations. It can be seen that the integral length scales are captured well by each WENO-Riemann combination whereas the difference in dissipative behavior becomes apparent at the higher wavenumbers in the cascade. Our previous observations about the dissipation of small scale structures by the different Riemann solvers in the ILES approach are confirmed where it can be seen that the AUSM solver is seen to be the least dissipative in comparison to the other approaches. The Rusanov solver is seen to be the most dissipative. The Roe and HLL approach are generally seen to be in between the Rusanov and AUSM approaches. This is also confirmed through the total energy content plots which show the Rusanov and HLL solvers reducing the total energy content of the flow at the same rate until turbulence ensues.

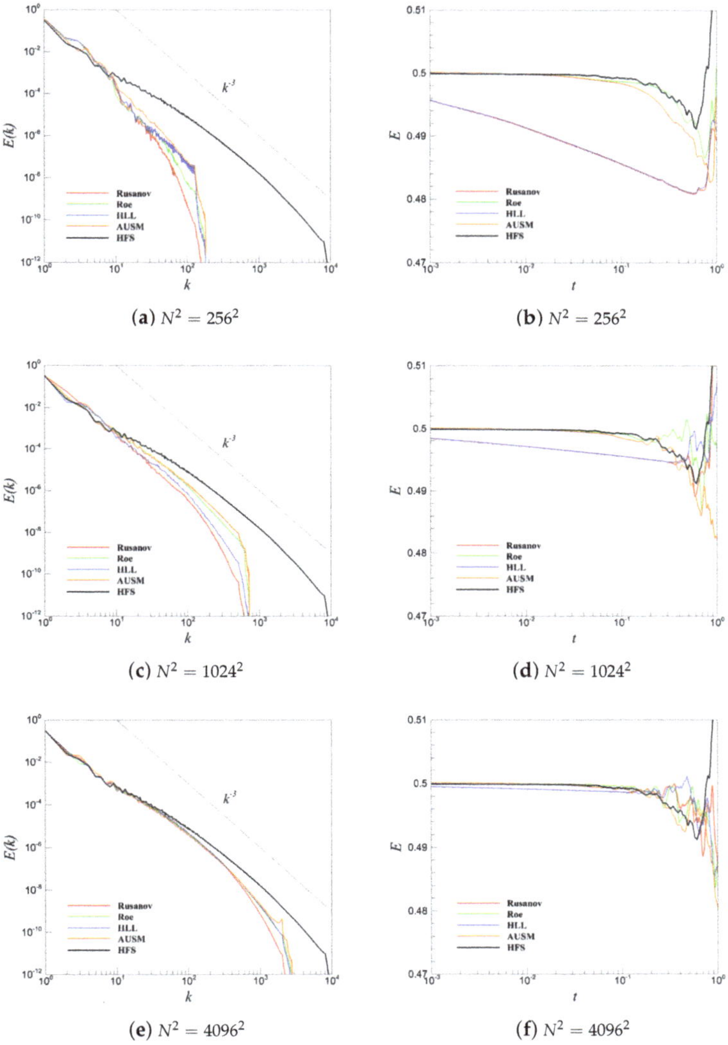

(**a**) $N^2 = 256^2$ (**b**) $N^2 = 256^2$

(**c**) $N^2 = 1024^2$ (**d**) $N^2 = 1024^2$

(**e**) $N^2 = 4096^2$ (**f**) $N^2 = 4096^2$

Figure 17. Kinetic energy spectra (**a**,**c**,**e**) for the DSL problem obtained using four different weighted essentially non-oscillatory (WENO)-Riemann solver combinations at time $t = 5$. High fidelity kinetic energy spectra from a Roe simulation at $N^2 = 16,384^2$ data points included as reference. Total energy of the flow (**b**,**d**,**f**) also plotted from $t = 0$ to $t = 1$.

We detail the dissipative behavior of the standard relaxation filtering approach (i.e., relaxation filtering at the end of time integration) in Figure 18. The effect of the filtering parameter σ is clearly seen with higher values of σ proving more dissipative. We note that using a different family of filters would perhaps lead to a greater control over the dissipative characteristics of this approach. The interested reader is directed to [99] for a comparison of the effects of different filters on large eddy simulation type approaches. In general, a very good approximation of the HFS kinetic energy spectra is obtained by this method. The integral length scales are also captured relatively well. The total energy content plots also confirm the dissipative behavior of the explicit filter with larger values of σ generally reducing the total energy content at a faster place. A curious observation here is that the relaxation filtering approach destabilizes the flow at a comparatively similar point in time for all coarse resolutions. This could imply a resolution independent dissipation phenomenon.

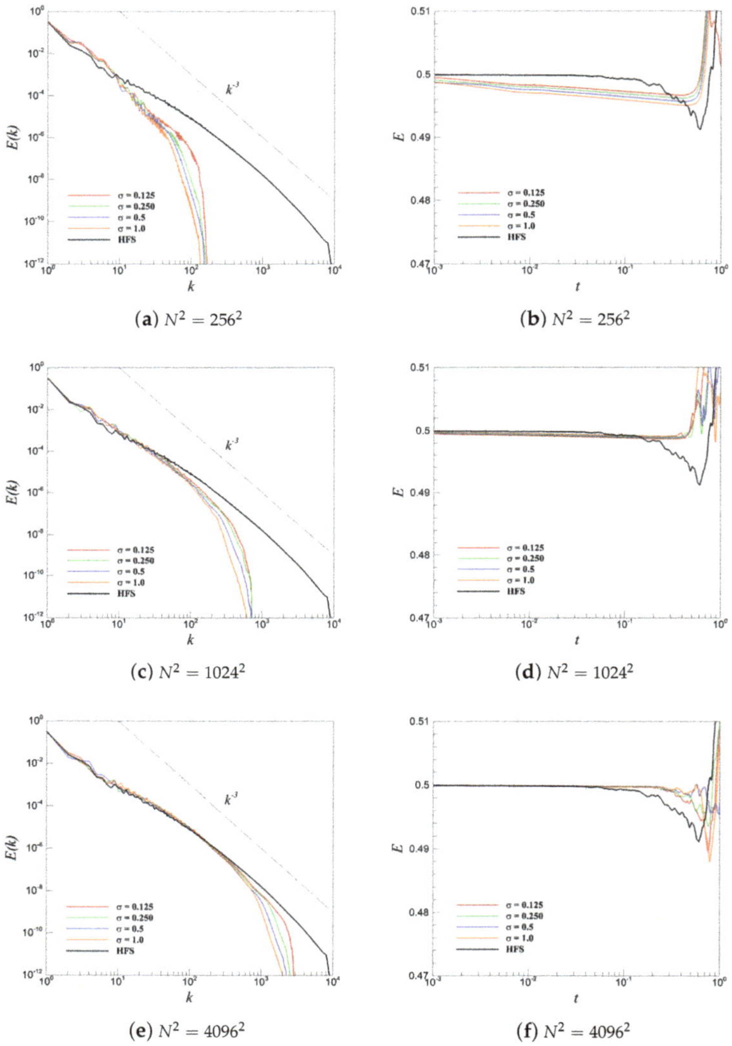

(**a**) $N^2 = 256^2$ (**b**) $N^2 = 256^2$

(**c**) $N^2 = 1024^2$ (**d**) $N^2 = 1024^2$

(**e**) $N^2 = 4096^2$ (**f**) $N^2 = 4096^2$

Figure 18. Kinetic energy spectra (**a**,**c**,**e**) for the DSL problem obtained using the standard relaxation filtering approach at time $t = 5$. High fidelity kinetic energy spectra from a Roe simulation at $N^2 = 16,384^2$ data points included as reference. Total energy of the flow (**b**,**d**,**f**) also plotted from $t = 0$ to $t = 1$.

Figure 19 shows the performance of the shock relaxation filter for the DSL problem. It can immediately be noticed that the integral length scales are captured incorrectly at the coarsest resolutions by this approach. The highest value of the shock threshold ($r_{th} = 10^{-4}$) fails to return a solution (due to floating point overflow) for the $N^2 = 4096^2$ resolution but albeit with limited success in terms of scaling capture for the other two resolutions. This points to a resolution dependent dissipation mechanism which would imply a need to vary the shock sensor threshold manually. Increasing the resolution of our simulations leads to a better performance at the integral length scales by this approach but a considerable proneness to aliasing error is seen at the cut-off wavenumbers (particularly for higher values of the shock threshold parameter). A comparatively larger amount of inertial range dissipation is also seen in the form of a downward shifting of the coarse grained kinetic energy spectra. This can be attributed to the lower order filtering procedure when compared to the higher order stencil used in the standard relaxation filtering approach. The total energy content plots also outline the same trends with far more rapid fall off in the total energy of the solution field in comparison to the ILES and standard relaxation filtering method. We clarify that the $r_{th} = 10^{-4}$ simulation eventually led to a floating point overflow for the $N^2 = 4096^2$ case (as seen in the final time kinetic energy spectra plot). We tentatively conclude that the proposed relaxation filtering mechanism performs in a better manner than the adaptive shock capturing approach for this investigation.

Figure 20 shows the effect of the WENO-5 flux reconstruction parameter p on both versions of our nonlinear weight calculator (i.e., WENO-JS and WENO-Z) for the coarse resolution of $N^2 = 1024^2$. The kinetic energy spectra and total energy plots shown here are for the Roe Riemann solver. An inspection of both these quantities proves inconclusive about the effect of p when it comes to the trends in dissipation. On examining Figure 21, we can see once again that the choice of p does not affect the kinetic energy spectra at the final time of the simulation. However, the total energy content plots show that lower values of p preserve the total energy of the flow before it becomes turbulent. Figure 22 shows the effect of the choice of the flux reconstruction (i.e., WENO-JS or WENO-Z) for the ILES approach. It is seen that the choice of the nonlinear weight calculation assumes more importance when utilizing a more dissipative solver (such as the HLL or Rusanov Riemann solver) where the WENO-Z reconstruction proves to be less dissipative than the WENO-JS approach. On the contrary, for the Roe and AUSM solvers (which are less dissipative), both methods perform in the same manner.

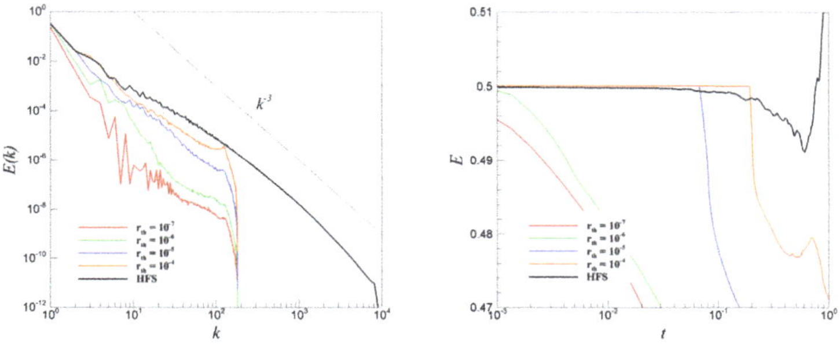

(**a**) $N^2 = 256^2$ (**b**) $N^2 = 256^2$

Figure 19. *Cont.*

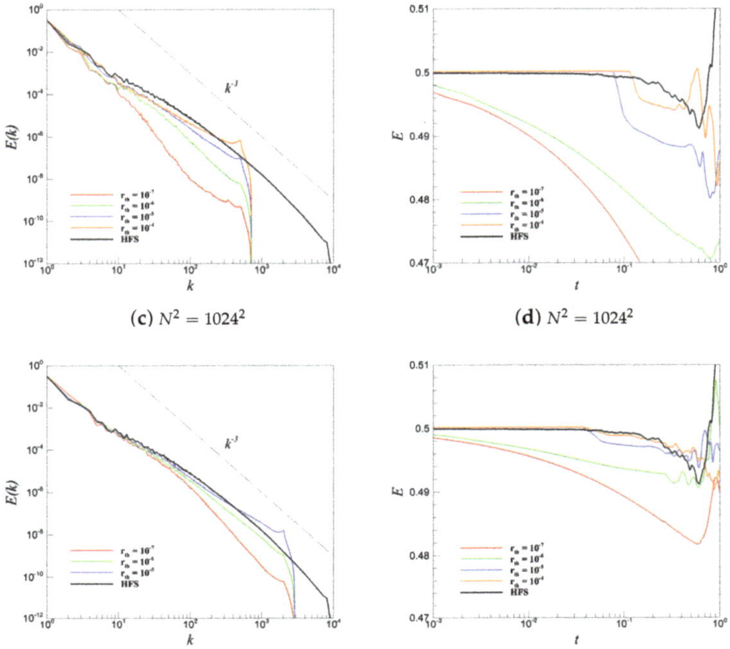

(**c**) $N^2 = 1024^2$

(**d**) $N^2 = 1024^2$

(**e**) $N^2 = 4096^2$

(**f**) $N^2 = 4096^2$

Figure 19. Kinetic energy spectra (**a**,**c**,**e**) for the DSL problem obtained using the shock relaxation filtering approach at time $t = 5$. High fidelity kinetic energy spectra from a Roe simulation at $N^2 = 16,384^2$ data points included as reference. Total energy of the flow (**b**,**d**,**f**) also plotted from $t = 0$ to $t = 1$.

A final examination of the DSL problem is made with respect to the implementation of the relaxation filtering approach. Kinetic energy spectra from the normal implementation (where the relaxation filtering sweep is carried out at the end of each time integration) and the modified implementation (with filtering at the end of each timestep) are compared in Figure 23 for four different values of the filtering parameter σ at the resolution of $N^2 = 1024^2$. We observe that both approaches add the same amount of dissipation for integral length scale capture of features and show slight differences in dissipative behavior at the highest wavenumbers. We note that spurious oscillations can be arrested with much more certainty in the modified implementation of the relaxation filtering kernel. The relative dissipative behavior of both implementations remains relatively constant over the scale of σ values.

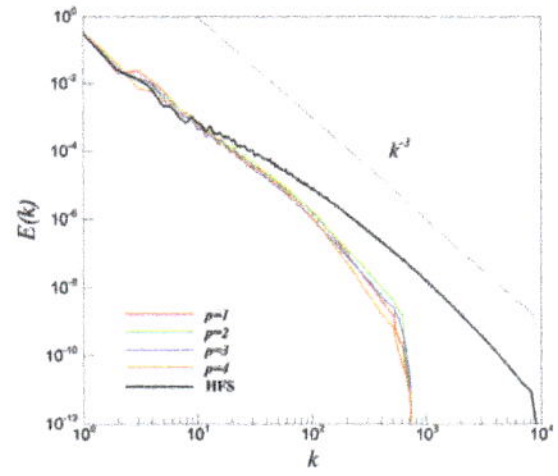

(**a**) WENO-JS

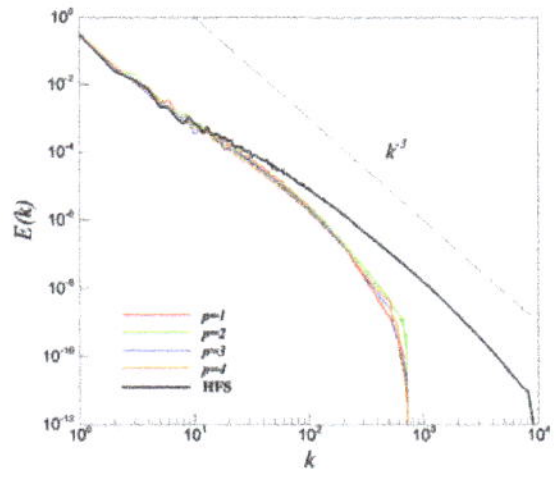

(**b**) WENO-Z

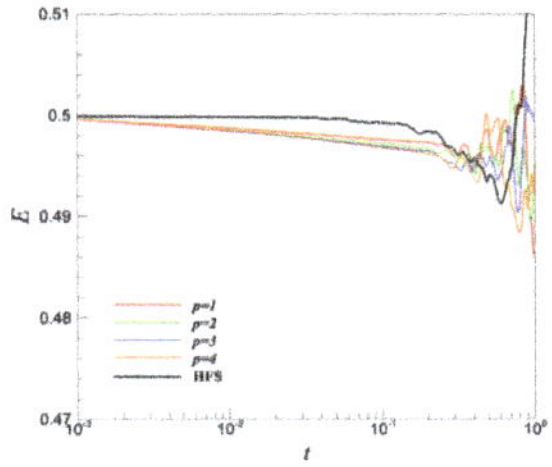

(**c**) WENO-JS

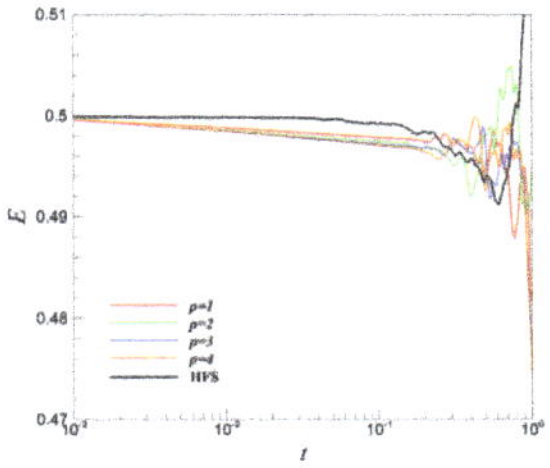

(**d**) WENO-Z

Figure 20. Effect of p in the smoothness indicator on dissipative behavior for the DSL problem. Sensitivity is checked with both the WENO-JS and WENO-Z implementations for the Roe solver at $N^2 = 1024^2$ with kinetic energy spectra (**a**,**b**) and total energy plots (**c**,**d**).

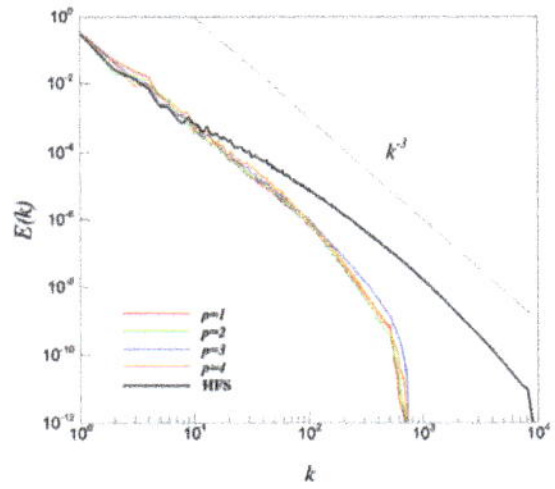

(**a**) WENO-JS

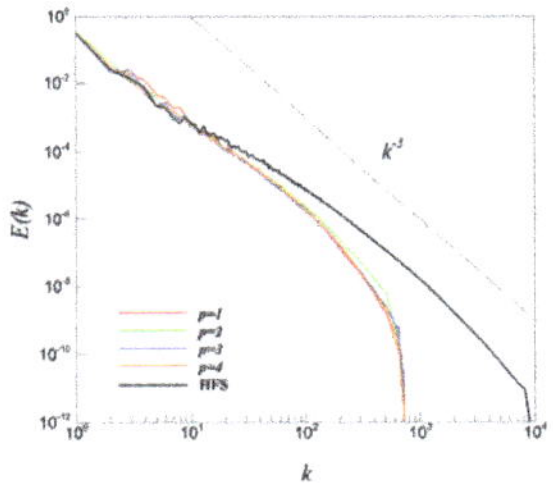

(**b**) WENO-Z

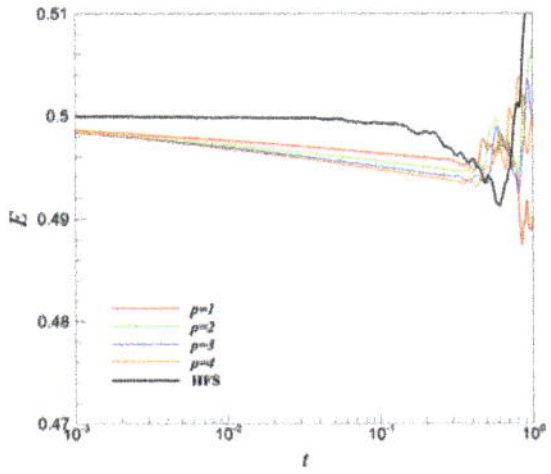

(**c**) WENO-JS

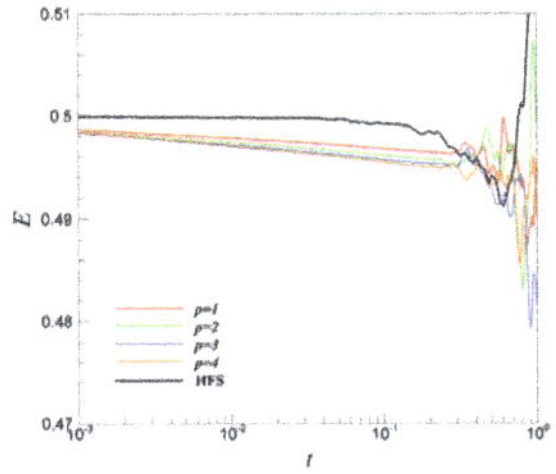

(**d**) WENO-Z

Figure 21. Effect of p in the smoothness indicator on dissipative behavior for the DSL problem. Sensitivity is checked with both the WENO-JS and WENO-Z implementations for the HLL solver at $N^2 = 1024^2$ with kinetic energy spectra (**a**,**b**) and total energy plots (**c**,**d**).

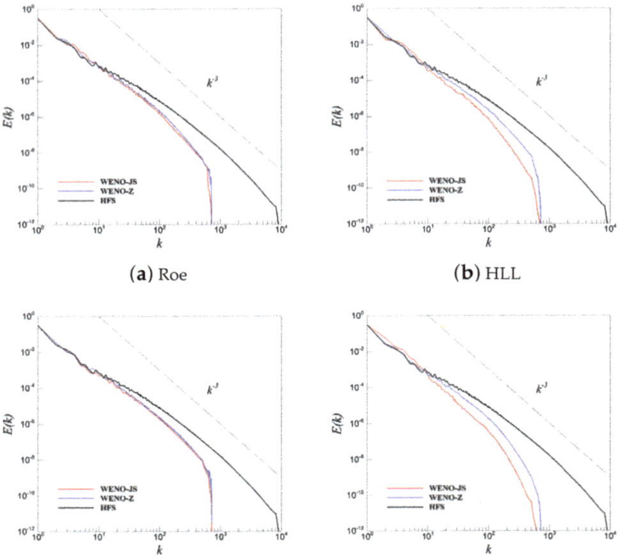

(**a**) Roe (**b**) HLL

(**c**) AUSM (**d**) Rusanov

Figure 22. A comparison of the WENO-JS and WENO-Z scheme through effect on DSL spectra at a resolution of $N^2 = 1024^2$. (**a**) Roe; (**b**) HLL; (**c**) AUSM; (**d**) Rusanov.

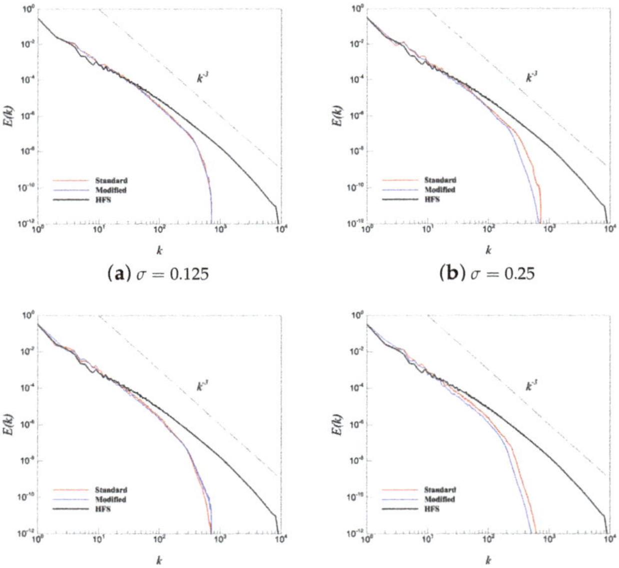

(**a**) $\sigma = 0.125$ (**b**) $\sigma = 0.25$

(**c**) $\sigma = 0.5$ (**d**) $\sigma = 1.0$

Figure 23. A comparison of the standard and modified implementations of the central filtering scheme for a resolution of $N^2 = 1024^2$ on the DSL problem. (**a**) $\sigma = 0.125$; (**b**) $\sigma = 0.25$; (**c**) $\sigma = 0.5$; (**d**) $\sigma = 1.0$.

4.4. Riemann Shock Interaction (RSI) Problem

The RSI test case is also studied for the shock capturing ability of the previously described numerical schemes. Density contours at $t = 0.5$ are displayed for each of the different numerical approaches for the purpose of determining their relative differences in dissipation and shock capturing ability. Figure 24 shows the HFS density contours obtained using the four Riemann solvers combined with WENO flux reconstructions. At this fine resolution, it is seen that all four WENO-Riemann combinations are equally able to capture the smallest relevant length scales in the flow. The shock evolution of this configuration is generally expected to be symmetric about the center line of the shock plume and the same is noticed in the Roe, HLL and Rusanov solvers. The AUSM Riemann solver, however, shows a slight skew in the formation of the plume which may be associated with low dissipation inherent to the scheme. It is possible that this low dissipation coupled with the Kelvin-Helmholtz vortices being formed during the evolution of the shock lead to the observed skewing. Figure 25 shows the results of the coarse grid resolution runs for this test case using the 4 Riemann solvers, it is apparent from the visual examination of the density contours shown here that the AUSM solver remains the least dissipative of all the Riemann approaches studied here and shows the skewing of its shock even at the relatively coarse grid run of $N^2 = 4096^2$. A successive reduction in the presence of small scale features can also be seen with increasing coarseness in our simulations. The coarsest runs also display non-physical striations (commonly known as shock oscillations) in the density contours which are due insufficient dissipation at that level of grid size. These striations are most pronounced for the Roe and AUSM solvers as one might expect. These shock oscillations are commonly visible in the contour plots when their lower frequency modes remain undamped. A characteristics based reconstruction procedure (i.e., interpolating the characteristic variables) would reduce these oscillations, albeit with added expense.

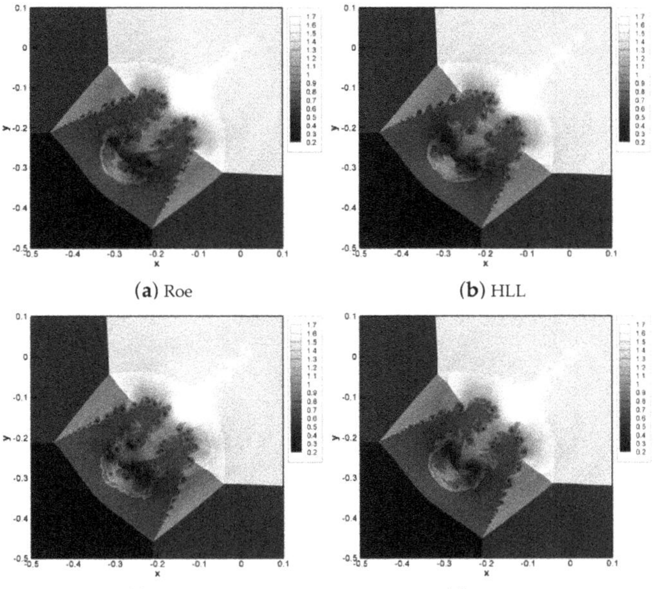

(**a**) Roe (**b**) HLL (**c**) AUSM (**d**) Rusanov

Figure 24. A comparison of the effect of different WENO-Riemann solver combinations on density contours at $t = 0.5$ for the RSI test case at a resolution of $N^2 = 16,384^2$. (**a**) Roe; (**b**) HLL; (**c**) AUSM; (**d**) Rusanov.

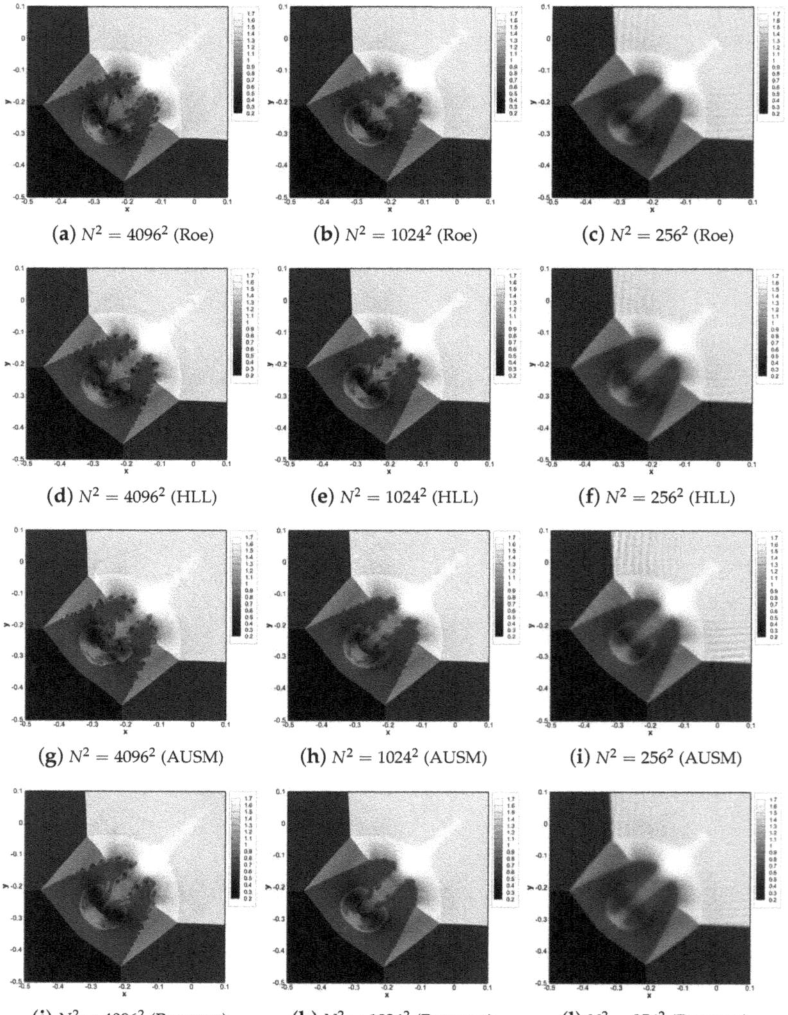

(**a**) $N^2 = 4096^2$ (Roe) (**b**) $N^2 = 1024^2$ (Roe) (**c**) $N^2 = 256^2$ (Roe)

(**d**) $N^2 = 4096^2$ (HLL) (**e**) $N^2 = 1024^2$ (HLL) (**f**) $N^2 = 256^2$ (HLL)

(**g**) $N^2 = 4096^2$ (AUSM) (**h**) $N^2 = 1024^2$ (AUSM) (**i**) $N^2 = 256^2$ (AUSM)

(**j**) $N^2 = 4096^2$ (Rusanov) (**k**) $N^2 = 1024^2$ (Rusanov) (**l**) $N^2 = 256^2$ (Rusanov)

Figure 25. RSI test case density contours at $t = 0.5$ for different coarse grid resolutions and WENO-Riemann solver combination ((**a–c**)—Roe, (**d–f**)—HLL, (**g–i**)—AUSM, (**j–l**)—Rusanov). Notice differences in dissipative behavior from presence of vortical structures.

Figure 26 shows the performance of the standard relaxation filtering employed at the end of the time integration procedure (using a filter parameter of $\sigma = 0.5$) for three coarse resolutions as well as performance of shock relaxation filter for three different r_{th} values. It can be seen that the standard relaxation filtering approach leads to a good result for the density field and may be considered comparable in quality with the shock capturing WENO-Riemann schemes. The shock filtering approach is seen to be adept at capturing the most sharp gradients in the field but higher higher values of r_{th} lead to a loss in the symmetry of the shock plume with the evidence of a skewing (for example in subfigure (g)) even without small scale capture. The highest value of the shock capturing threshold parameter used $r_{th} = 10^{-5}$ displays a considerable skewing of the shock plume with much less small scale capture as compared to the standard relaxation filtering kernel with $\sigma = 0.5$. It must be observed here that the striations observed in the ILES runs do not appear to be visible using this methodology. This is due to the low-pass spatial filtering characteristics of the explicit filtering kernel. The higher frequency components of these oscillations however may be present in the solution field as they are usually present to a certain extent in each computational realization even on very high fidelity simulations.

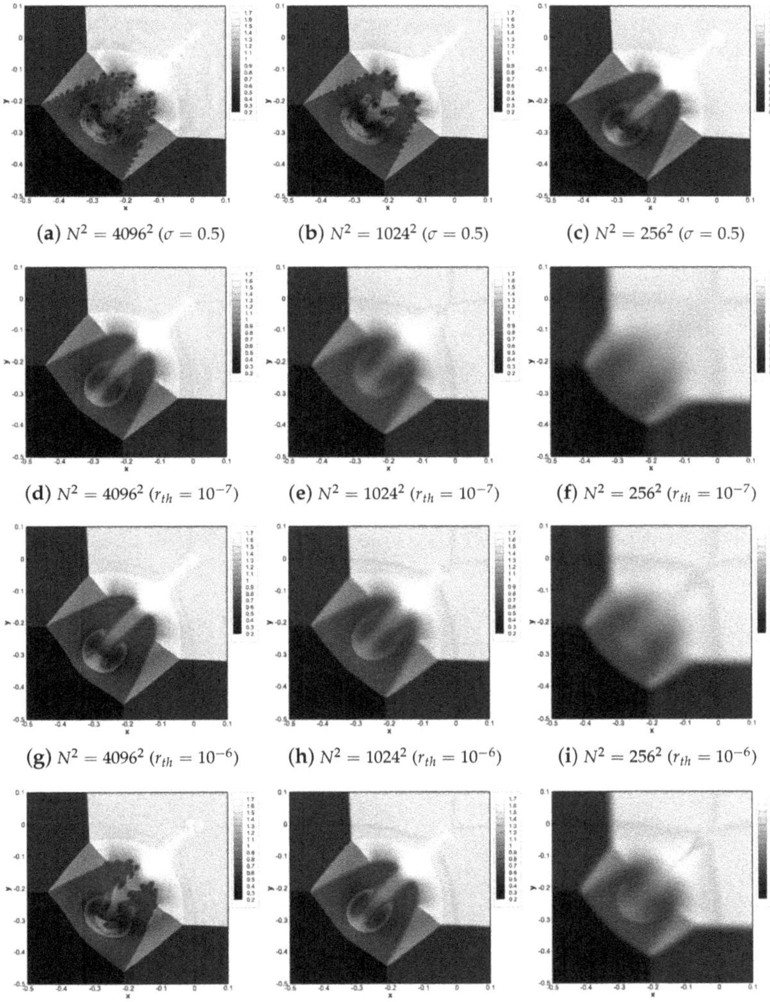

(a) $N^2 = 4096^2$ ($\sigma = 0.5$) **(b)** $N^2 = 1024^2$ ($\sigma = 0.5$) **(c)** $N^2 = 256^2$ ($\sigma = 0.5$)

(d) $N^2 = 4096^2$ ($r_{th} = 10^{-7}$) **(e)** $N^2 = 1024^2$ ($r_{th} = 10^{-7}$) **(f)** $N^2 = 256^2$ ($r_{th} = 10^{-7}$)

(g) $N^2 = 4096^2$ ($r_{th} = 10^{-6}$) **(h)** $N^2 = 1024^2$ ($r_{th} = 10^{-6}$) **(i)** $N^2 = 256^2$ ($r_{th} = 10^{-6}$)

(j) $N^2 = 4096^2$ ($r_{th} = 10^{-5}$) **(k)** $N^2 = 1024^2$ ($r_{th} = 10^{-5}$) **(l)** $N^2 = 256^2$ ($r_{th} = 10^{-5}$)

Figure 26. RSI test case density contours at $t = 0.5$ for different coarse grid resolutions and explicit filtering dissipative approaches ((**a–c**) — standard relaxation filter with $\sigma = 0.5$, (**d–f**) — shock filter with $r_{th} = 10^{-7}$, (**g–i**) — shock filter with $r_{th} = 10^{-6}$, (**j–l**) — shock filter with $r_{th} = 10^{-5}$).

Figures 27 and 28 compare the WENO-JS and WENO-Z approaches to cell face flux reconstruction for the RSI test case for two different values of p. From the density contours displayed at a coarse resolution of $N^2 = 1024^2$ and 4096^2, it is difficult to conclude about the dissipative behavior of either reconstruction mechanisms for this particular case. However, the faint appearance of striations for the WENO-JS case with $p = 2$ seems to imply that this particular configuration of the reconstruction mechanism is less dissipative in comparison to the others. Finally, Figure 29 shows a one to one comparison of the instantaneous density contours for the relaxation filtering approach in its standard (i.e., low-pass spatial filtering at the end of the time integration) and modified (i.e., low-pass spatial filtering at each substep of the time integration) approach with the filtering strength of $\sigma = 0.5$. A visual examination does not show any large deviation in the dissipative behavior of both implementations of the relaxation filtering approach with the absence of any major skewing of the shock plume. However an examination of the $N^2 = 1024^2$ instance does indicate a more dissipative behavior of the modified approach. We note however, the difference in the dissipative behavior of either of these implementations is minimal (as shown in the double shear layer test case through the examination of kinetic energy spectra). The added number of filter sweeps acts as a preventive methodology for shock discontinuities leading to Gibbs phenomena.

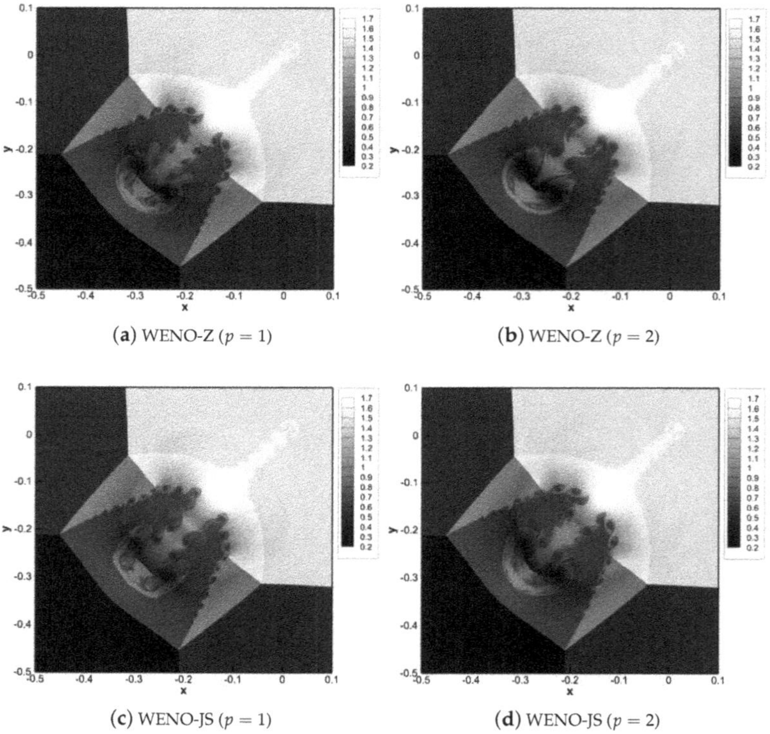

(**a**) WENO-Z ($p = 1$)

(**b**) WENO-Z ($p = 2$)

(**c**) WENO-JS ($p = 1$)

(**d**) WENO-JS ($p = 2$)

Figure 27. A comparison of the (**c**,**d**) WENO-JS and (**a**,**b**) WENO-Z scheme (for different p values) through effect on density contours for the RSI test case at a resolution of $N^2 = 1024^2$ for the Roe solver.

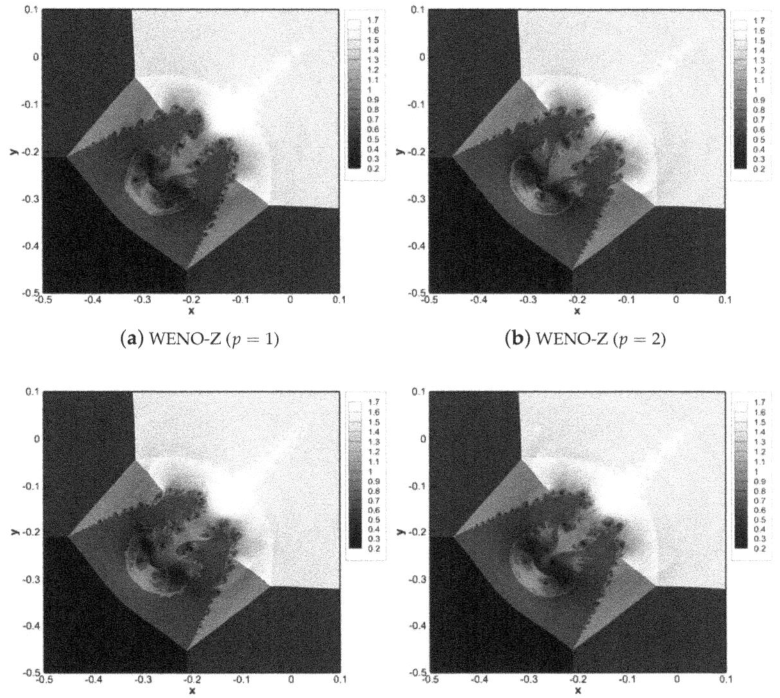

(**a**) WENO-Z ($p = 1$) (**b**) WENO-Z ($p = 2$)

(**c**) WENO-JS ($p = 1$) (**d**) WENO-JS ($p = 2$)

Figure 28. A comparison of the (**c**,**d**) WENO-JS and (**a**,**b**) WENO-Z scheme (for different p values) through effect on density contours for the RSI test case at a resolution of $N^2 = 4096^2$ for the Roe solver.

Table 1 shows a compilation of computational expense for the RSI test case for the three coarsest resolutions each using a different number of processes. We have used 12 process for $N^2 = 256^2$, 48 processes for $N^2 = 1024^2$ and 192 processes for $N^2 = 4096^2$. A broad message from this table can be obtained from the final column which tabulates CPU times for $N^2 = 4096^2$ case where the relaxation and shock filtering approaches are seen to be around 4 times faster (on average) than the WENO-Riemann solver ILES method. This adds value to the conclusions stated above, wherein the standard relaxation filtering approach is seen to perform as well as the ILES approach but at a much lower computational expense. We must also note that the modified standard relaxation approach does not add a significant amount of overhead to the standard relaxation filtering methodology. We must clarify that the tabulated times at the lower resolutions are not indicative due to the shorter duration of simulations and the unequal load distribution on the shared HPC cluster. Indeed, the Rusanov solver requires the least computational work load due to its simplicity. It does not involve any conditional statements or characteristic transformations. One possible reason for this inhomogeneous performance of compute clusters could be inhomogeneous hardware equipment (i.e., uneven CPU or memory equipment of the nodes). Another reason could be software related interactions. We report here our assessments only for a particular simulation at a certain time. Running the same executable on a different occasion with a different set of compute nodes would yield results in a slightly different wall-clock time. Therefore, their assessments will be valid only in a statistical sense. However, we would like to highlight that the central scheme, with relaxation filtering, offers a competitive approach to ILES and is much more computationally efficient than WENO-based schemes.

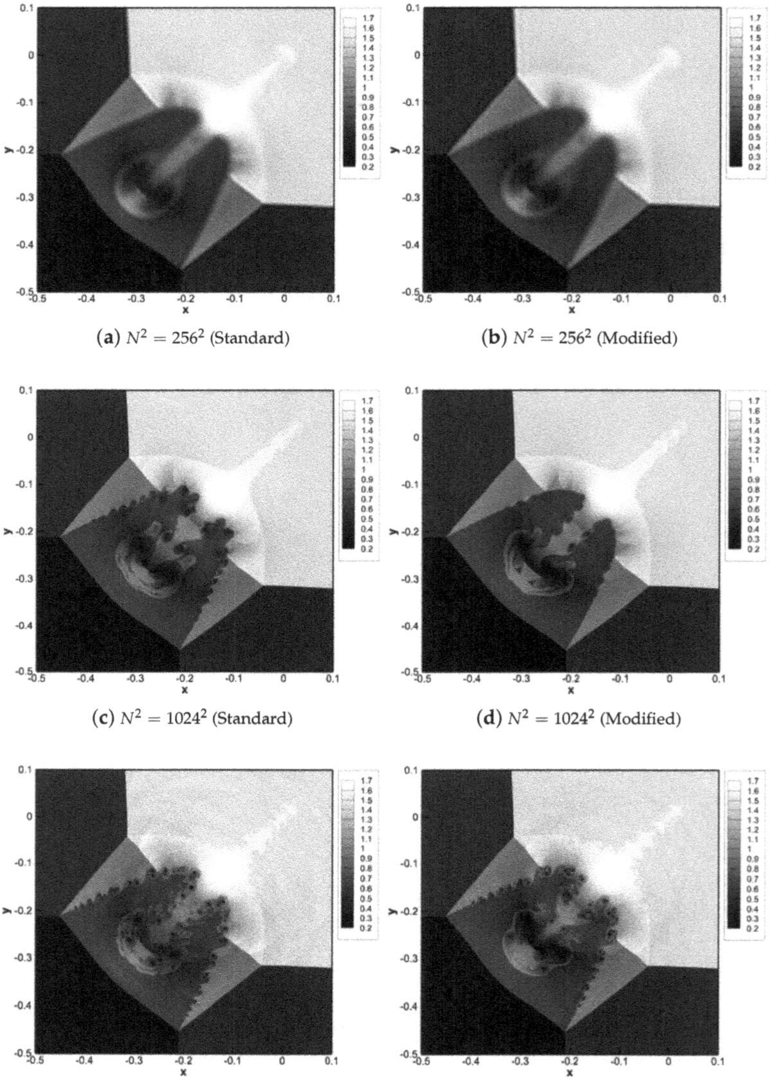

(**a**) $N^2 = 256^2$ (Standard) (**b**) $N^2 = 256^2$ (Modified)

(**c**) $N^2 = 1024^2$ (Standard) (**d**) $N^2 = 1024^2$ (Modified)

(**e**) $N^2 = 4096^2$ (Standard) (**f**) $N^2 = 4096^2$ (Modified)

Figure 29. A comparison of the different implementations of the central scheme with the relaxation filtering procedure with $\sigma = 0.5$ (i.e., standard (**a**,**c**,**e**) or modified (**b**,**d**,**f**)) for the RSI test case.

Table 1. CPU times for different solvers for the RSI test case. 12 processes were used for $N^2 = 256^2$, 48 processes were used for $N^2 = 1024^2$ and 192 processes were used for $N^2 = 4096^2$.

	CPU Time (in Hours)		
Solver	$N^2 = 256^2$	$N^2 = 1024^2$	$N^2 = 4096^2$
Rusanov	4.05×10^{-3}	2.02×10^{-2}	0.90
Roe	5.29×10^{-3}	6.14×10^{-2}	1.39
HLL	6.75×10^{-3}	5.57×10^{-2}	1.02
AUSM	3.68×10^{-3}	7.79×10^{-2}	1.15
Relaxation filtering (standard)	1.73×10^{-3}	2.18×10^{-2}	0.32
Relaxation filtering (modified)	2.21×10^{-3}	2.88×10^{-2}	0.38
Shock filtering	2.28×10^{-3}	1.86×10^{-2}	0.35

5. Conclusions

This study presents an in depth analysis of both implicit large eddy simulation (ILES) methods and explicit filtering approaches for the purpose of simulating compressible hydrodynamic turbulence by solving the two-dimensional Euler equations. We start by outlining the mathematical formulation of our governing equations followed by which we introduce the two major computational approaches we have used: (i) upwind-biased ILES formulations and (ii) central non-dissipative reconstructions equipped with explicit filtering procedures. To test the fidelity of the numerical methods being examined here, we use the dual shear layer (DSL) test case as well as a Riemann shock interaction (RSI) problem to evaluate their kinetic energy spectra resolving and shock capturing capabilities. Both test cases are simple representations for the physics of many real world phenomena and are well suited for benchmarking the advantages and disadvantages of our chosen numerical schemes. Evaluations of the different numerical schemes are done through the comparison of statistical data such as the kinetic energy spectra and total energy as well as through qualitative analyses of the instantaneous flow structures. A final comparison of the computational expense of the different numerical approaches is also reported.

In order to examine the general nature of the upwind-biased based numerical schemes, we utilize four different WENO-Riemann solver combinations (i.e., the Roe, HLL, AUSM and Rusanov solvers) with slightly different characteristics in terms of dissipative behavior and fine scale structure capture ability. The upwinding required for using these Riemann solvers in their implicit LES formulation is obtained through nonlinear biased interfacial flux constructions using the fifth order accurate WENO approach. Two versions of the smoothness indicators (the WENO-JS and WENO-Z formulations) are used for the reconstruction and the parameter p in these smoothness indicators is also varied for a sensitivity analysis. These ILES solvers (with different flux reconstruction and p configurations) are used to simulate coarse grained solution fields with $N^2 = 256^2, 1024^2$, and 4096^2 resolutions for the DSL and RSI problems. These coarse grained runs are compared to high fidelity simulation data obtained through the simulation of the test cases at $N^2 = 16,384^2$ using the Roe Riemann solver. The theoretical scaling from KBL theory is also provided for the purpose of comparison. Predictably, it is observed that the Roe and AUSM solvers are less dissipative as compared to the HLL and Rusanov solvers and that all implementations of the Riemann solvers in the ILES framework are successful in capturing the integral length scales (along with a fair length of the inertial range in the turbulent cascade). Simulations which utilize different values of p are seen to remain more or less invariant in terms of kinetic energy spectra scaling. It is noted, however, that the choice of the smoothness indicator (i.e., whether WENO-JS or WENO-Z) has important ramifications on the dissipative nature of the more dissipative HLL and Rusanov Riemann solvers for which the WENO-JS scheme is seen to be more dissipative. The low-dissipation of the AUSM Riemann solver also causes a skewing of the shock plume generated in the RSI test case when compared to the other Roe, Rusanov and HLL approaches. In the RSI case, nonphysical striations (i.e., shock oscillations) are also observed at the coarsest simulation of $N^2 = 256^2$ for the WENO-Riemann solver combinations. These striations are most pronounced, once again, for the less dissipative AUSM approach. In our future studies, we plan to examine the results of the WENO interpolations done using the characteristics variables and in terms of their effects on these oscillations.

The DSL and RSI test cases are also simulated using the relaxation filtering approach where a low-pass spatial filtering sweep is applied to the solution field. This sweep may be applied at the end of each timestep (which we call the standard relaxation filter) or at each substep of the time integration process (denoted the modified relaxation filter). The dissipative nature of the filter is controlled by the free parameter σ which varies between 0 and 1 with the former implying no filtering and the latter denoting highest dissipation. We can remark here that the $\sigma = 1$ case corresponds to a standard Gaussian shaped filter. A shock capturing version of the aforementioned filter is also used where the filtering strengths are calculated adaptively through a heuristic based on the local solution field. The model parameter which controls dissipation is a shock sensor r_{th} which is varied from 10^{-7} to 10^{-4} with lower values being more dissipative. It is seen that the standard relaxation filtering approach performs in a manner similar to the ILES approach with excellent integral length scale capture as well as improved inertial range scale capture. It is also shown that a significant decrease in computational expense is also obtained. The standard and modified relaxation filters are also shown to be more or less similar to each other in dissipative behavior. The modified relaxation filter is implemented and tested primarily due to the fact that the added number of filter sweeps at each timestep would be less susceptible to Gibbs oscillations. In addition to the standard and modified relaxation filtering approach, the shock filtering methodology is tested on the DSL and RSI test cases for different values of its shock threshold parameter where it is observed that integral length scales are captured poorly at the coarsest resolutions. It is seen that this filtering technique is prone to over-dissipating in the inertial range (which shows itself as a downward shift of the averaged kinetic energy cascade). A pile up of kinetic energy at the cutoff wavenumber is also seen for the shock threshold values tested in this work. Although a proper choice of r_{th} may result in an accurate prediction, we conclude that the determination of the optimum value for the threshold parameter r_{th} is not that trivial and depends on the resolution as well. The explicit filters were also applied to the RSI test case where it was seen that the low frequency shock oscillations visible in the $N^2 = 256^2$ test case were dissipated effectively by the choice of our free parameters. However, it was noted that the shock filtering approach caused a distinct skewing of the shock plume at higher values of r_{th}. Once again, the performance of the standard and modified relaxation filters was comparable to the ILES approach.

We conclude by revisiting the questions posed in the introduction and may summarize our findings as follows:

1. It can be stated with a certain degree of confidence that the central schemes implemented with relaxation filters perform in a comparable manner to the WENO-Riemann solver combinations tested here and that the standard and modified relaxation filters represent a viable alternative to the ILES framework for the test cases investigated here. Not only are they seen to capture small scale features and their nonlinear interactions accurately, they also appear to perform well when it comes to capturing shocks. However, WENO based ILES schemes are attractive, because they do not require any ad-hoc tuning or filtering parameters.
2. We note a significant reduction in computational complexity due to the relative simplicity of the numerical algorithm for the relaxation filtering framework. In addition, the computational expense of this framework is seen to be much lower in comparison to the different WENO based ILES solvers. We find that the relaxation filtering (or *explicit filtering*) approach is approximately four times faster than the WENO reconstruction based ILES approaches.
3. The free modeling parameters associated with the proposed numerical framework perform as expected with a changing transfer function shape corresponding to a modification of the dissipative behavior of the numerical method. This also allows us to actively control the dissipation in the numerical method through model free parameters σ and r_{th} for the standard and shock relaxation filtering approach respectively.

4. We have performed a thorough investigation of the WENO-Riemann solver based ILES schemes where it is verified that the AUSM Riemann solver is the least dissipative whereas the Rusanov solver is the most. The flux reconstruction scheme does have an effect on the dissipative behavior of the ILES methodology when it comes to the expressions for the calculation of the smoothness indicators (i.e., WENO-JS and WENO-Z), but for the dissipative Riemann solvers (Rusanov and HLL) only. The WENO-JS scheme is seen to be more dissipative than the WENO-Z scheme for these two solvers and more or less identical for the less dissipative AUSM and Roe solvers. It is seen that p does not have a major effect on the dissipative effect of the scheme for this investigation. However, less dissipative results are obtained for lower p values.

Our closing comments are made on the fact that the standard or modified relaxation filtering framework could possibly represent a viable tool for the efficient computation of hyperbolic conservation law problems. The main conclusion is that the central scheme, with relaxation filtering, offers a competitive approach to ILES and is much more computationally efficient than WENO-based schemes. We aim to perform further investigations for other benchmark flows to lend more strength to this conclusion.

Acknowledgments: The computing for this project was performed by using resources from the High Performance Computing Center (HPCC) at Oklahoma State University.

Author Contributions: Omer San conceived and designed the study; Romit Maulik and Omer San performed the simulations, analyzed the data, made helpful observations and wrote the paper.

Conflicts of Interest: The authors declare no conflict of interest.

References

1. Titarev, V.; Toro, E. WENO schemes based on upwind and centred TVD fluxes. *Comput. Fluids* **2005**, *34*, 705–720.
2. Shu, C.W. High order weighted essentially nonoscillatory schemes for convection dominated problems. *SIAM Rev.* **2009**, *51*, 82–126.
3. Pirozzoli, S. Numerical methods for high-speed flows. *Annu. Rev. Fluid Mech.* **2011**, *43*, 163–194.
4. Martín, M.P.; Taylor, E.M.; Wu, M.; Weirs, V.G. A bandwidth-optimized WENO scheme for the effective direct numerical simulation of compressible turbulence. *J. Comput. Phys.* **2006**, *220*, 270–289.
5. Grinstein, F.F.; Margolin, L.G.; Rider, W.J. *Implicit Large Eddy Simulation: Computing Turbulent Fluid Dynamics*; Cambridge University Press: Cambridge, UK, 2007.
6. Karaca, M.; Lardjane, N.; Fedioun, I. Implicit Large Eddy Simulation of high-speed non-reacting and reacting air/H_2 jets with a 5th order WENO scheme. *Comput. Fluids* **2012**, *62*, 25–44.
7. Zhao, S.; Lardjane, N.; Fedioun, I. Comparison of improved finite-difference WENO schemes for the implicit large eddy simulation of turbulent non-reacting and reacting high-speed shear flows. *Comput. Fluids* **2014**, *95*, 74–87.
8. Hickel, S.; Egerer, C.P.; Larsson, J. Subgrid-scale modeling for implicit large eddy simulation of compressible flows and shock-turbulence interaction. *Phys. Fluids* **2014**, *26*, 106101.
9. Lund, T. The use of explicit filters in large eddy simulation. *Comput. Math. Appl.* **2003**, *46*, 603–616.
10. Visbal, M.R.; Gaitonde, D.V. On the use of higher-order finite-difference schemes on curvilinear and deforming meshes. *J. Comput. Phys.* **2002**, *181*, 155–185.
11. Visbal, M.R.; Rizzetta, D. Large-eddy simulation on curvilinear grids using compact differencing and filtering schemes. *J. Fluids Eng.* **2002**, *124*, 836–847.
12. Thomson, W. Hydrokinetic solutions and observations. *Philos. Mag. Ser.* **1871**, *42*, 362–377.
13. Hwang, K.J.; Goldstein, M.; Kuznetsova, M.; Wang, Y.; Viñas, A.; Sibeck, D. The first in situ observation of Kelvin-Helmholtz waves at high-latitude magnetopause during strongly dawnward interplanetary magnetic field conditions. *J. Geophys. Res. Space Phys.* **2012**, *117*, A08233.
14. San, O.; Kara, K. Evaluation of Riemann flux solvers for WENO reconstruction schemes: Kelvin-Helmholtz instability. *Comput. Fluids* **2015**, *117*, 24–41.

15. Lax, P.D.; Liu, X.D. Solution of two-dimensional Riemann problems of gas dynamics by positive schemes. *SIAM J. Sci. Comput.* **1998**, *19*, 319–340.
16. San, O.; Kara, K. Numerical assessments of high-order accurate shock capturing schemes: Kelvin-Helmholtz type vortical structures in high-resolutions. *Comput. Fluids* **2014**, *89*, 254–276.
17. Li, Z.; Zhang, Y.; Chen, H. A low dissipation numerical scheme for Implicit Large Eddy Simulation. *Comput. Fluids* **2015**, *117*, 233–246.
18. Fu, L.; Hu, X.Y.; Adams, N.A. A family of high-order targeted ENO schemes for compressible-fluid simulations. *J. Comput. Phys.* **2016**, *305*, 333–359.
19. Drikakis, D.; Hahn, M.; Mosedale, A.; Thornber, B. Large eddy simulation using high-resolution and high-order methods. *Philos. Trans. R. Soc. Lond. A Math. Phys. Eng. Sci.* **2009**, *367*, 2985–2997.
20. Tsoutsanis, P.; Titarev, V.A.; Drikakis, D. WENO schemes on arbitrary mixed-element unstructured meshes in three space dimensions. *J. Comput. Phys.* **2011**, *230*, 1585–1601.
21. Margolin, L.; Rider, W.; Grinstein, F. Modeling turbulent flow with implicit LES. *J. Turbul.* **2006**, *7*, N15.
22. Zhu, H.; Fu, S.; Shi, L.; Wang, Z. Implicit large-eddy simulation for the high-order flux reconstruction method. *AIAA J.* **2016**, *54*, 2721–2733.
23. Shu, C.W.; Osher, S. Efficient implementation of essentially non-oscillatory shock-capturing schemes. *J. Comput. Phys.* **1988**, *77*, 439–471.
24. Shu, C.W.; Osher, S. Efficient implementation of essentially non-oscillatory shock-capturing schemes, II. *J. Comput. Phys.* **1989**, *83*, 32–78.
25. Zhu, J.; Qiu, J. A new fifth order finite difference WENO scheme for solving hyperbolic conservation laws. *J. Comput. Phys.* **2016**, *318*, 110–121.
26. Domaradzki, J.A.; Xiao, Z.; Smolarkiewicz, P.K. Effective eddy viscosities in implicit large eddy simulations of turbulent flows. *Phys. Fluids* **2003**, *15*, 3890–3893.
27. Qiu, J.; Shu, C.W. On the construction, comparison, and local characteristic decomposition for high-order central WENO schemes. *J. Comput. Phys.* **2002**, *183*, 187–209.
28. Jia, F.; Gao, Z.; Don, W.S. A spectral study on the dissipation and dispersion of the WENO schemes. *J. Sci. Comput.* **2015**, *63*, 49–77.
29. Tsoutsanis, P.; Antoniadis, A.F.; Drikakis, D. WENO schemes on arbitrary unstructured meshes for laminar, transitional and turbulent flows. *J. Comput. Phys.* **2014**, *256*, 254–276.
30. Hu, C.; Shu, C.W. Weighted essentially non-oscillatory schemes on triangular meshes. *J. Comput. Phys.* **1999**, *150*, 97–127.
31. Jiang, G.S.; Wu, C.c. A high-order WENO finite difference scheme for the equations of ideal magnetohydrodynamics. *J. Comput. Phys.* **1999**, *150*, 561–594.
32. Li, G.; Xing, Y. High order finite volume WENO schemes for the Euler equations under gravitational fields. *J. Comput. Phys.* **2016**, *316*, 145–163.
33. Latini, M.; Schilling, O.; Don, W.S. Effects of WENO flux reconstruction order and spatial resolution on reshocked two-dimensional Richtmyer-Meshkov instability. *J. Comput. Phys.* **2007**, *221*, 805–836.
34. Schilling, O.; Latini, M. High-order WENO simulations of three-dimensional reshocked Richtmyer-Meshkov instability to late times: dynamics, dependence on initial conditions, and comparisons to experimental data. *Acta Math. Sci.* **2010**, *30*, 595–620.
35. Thornber, B.; Mosedale, A.; Drikakis, D. On the implicit large eddy simulations of homogeneous decaying turbulence. *J. Comput. Phys.* **2007**, *226*, 1902–1929.
36. Boris, J.; Grinstein, F.; Oran, E.; Kolbe, R. New insights into large eddy simulation. *Fluid Dyn. Res.* **1992**, *10*, 199–228.
37. Hickel, S.; Adams, N.A.; Domaradzki, J.A. An adaptive local deconvolution method for implicit LES. *J. Comput. Phys.* **2006**, *213*, 413–436.
38. Watanabe, T.; Sakai, Y.; Nagata, K.; Ito, Y.; Hayase, T. Implicit large eddy simulation of a scalar mixing layer in fractal grid turbulence. *Phys. Scr.* **2016**, *91*, 074007.
39. Egerer, C.P.; Schmidt, S.J.; Hickel, S.; Adams, N.A. Efficient implicit LES method for the simulation of turbulent cavitating flows. *J. Comput. Phys.* **2016**, *316*, 453–469.
40. Domaradzki, J.; Adams, N. Direct modelling of subgrid scales of turbulence in large eddy simulations. *J. Turbul.* **2002**, *3*, N24.

41. Denaro, F.M. What does Finite Volume-based implicit filtering really resolve in Large-Eddy Simulations? *J. Comput. Phys.* **2011**, *230*, 3849–3883.
42. Ducros, F.; Ferrand, V.; Nicoud, F.; Weber, C.; Darracq, D.; Gacherieu, C.; Poinsot, T. Large-eddy simulation of the shock/turbulence interaction. *J. Comput. Phys.* **1999**, *152*, 517–549.
43. Mittal, R.; Moin, P. Suitability of upwind-biased finite difference schemes for large-eddy simulation of turbulent flows. *AIAA J.* **1997**, *35*, 1415–1417.
44. Mathew, J.; Lechner, R.; Foysi, H.; Sesterhenn, J.; Friedrich, R. An explicit filtering method for large eddy simulation of compressible flows. *Phys. Fluids* **2003**, *15*, 2279–2289.
45. Mathew, J.; Foysi, H.; Friedrich, R. A new approach to LES based on explicit filtering. *Int. J. Heat Fluid Flow* **2006**, *27*, 594–602.
46. Kremer, F.; Bogey, C. Large-eddy simulation of turbulent channel flow using relaxation filtering: Resolution requirement and Reynolds number effects. *Comput. Fluids* **2015**, *116*, 17–28.
47. Fauconnier, D.; Bogey, C.; Dick, E. On the performance of relaxation filtering for large-eddy simulation. *J. Turbul.* **2013**, *14*, 22–49.
48. Bull, J.R.; Jameson, A. Explicit filtering and exact reconstruction of the sub-filter stresses in large eddy simulation. *J. Comput. Phys.* **2016**, *306*, 117–136.
49. Bogey, C.; Bailly, C. A family of low dispersive and low dissipative explicit schemes for flow and noise computations. *J. Comput. Phys.* **2004**, *194*, 194–214.
50. Berland, J.; Lafon, P.; Daude, F.; Crouzet, F.; Bogey, C.; Bailly, C. Filter shape dependence and effective scale separation in large-eddy simulations based on relaxation filtering. *Comput. Fluids* **2011**, *47*, 65–74.
51. Bose, S.T.; Moin, P.; You, D. Grid-independent large-eddy simulation using explicit filtering. *Phys. Fluids* **2010**, *22*, 105103.
52. Gropp, W.; Lusk, E.; Skjellum, A. *Using MPI: Portable Parallel Programming with the Message-Passing Interface*; MIT Press: Cambridge, MA, USA, 1999; Volume 1.
53. Laney, C.B. *Computational Gasdynamics*; Cambridge University Press: Cambridge, UK, 1998.
54. Pletcher, R.H.; Tannehill, J.C.; Anderson, D. *Computational Fluid Mechanics and Heat Transfer*; CRC Press: New York, NY, USA, 2012.
55. Van Der Burg, J.; Kuerten, J.G.M.; Zandbergen, P.J. Improved shock-capturing of Jameson's scheme for the Euler equations. *Int. J. Numer. Methods Fluids* **1992**, *15*, 649–671.
56. Parent, B. Positivity-preserving high-resolution schemes for systems of conservation laws. *J. Comput. Phys.* **2012**, *231*, 173–189.
57. Gottlieb, S.; Shu, C.W. Total variation diminishing Runge-Kutta schemes. *Math. Comput. Am. Math. Soc.* **1998**, *67*, 73–85.
58. Liu, X.D.; Osher, S.; Chan, T. Weighted essentially non-oscillatory schemes. *J. Comput. Phys.* **1994**, *115*, 200–212.
59. Jiang, G.S.; Shu, C.W. Efficient implementation of weighted ENO schemes. *J. Comput. Phys.* **1996**, *126*, 202–228.
60. Borges, R.; Carmona, M.; Costa, B.; Don, W.S. An improved weighted essentially non-oscillatory scheme for hyperbolic conservation laws. *J. Comput. Phys.* **2008**, *227*, 3191–3211.
61. Henrick, A.K.; Aslam, T.D.; Powers, J.M. Mapped weighted essentially non-oscillatory schemes: achieving optimal order near critical points. *J. Comput. Phys.* **2005**, *207*, 542–567.
62. Ha, Y.; Kim, C.H.; Lee, Y.J.; Yoon, J. An improved weighted essentially non-oscillatory scheme with a new smoothness indicator. *J. Comput. Phys.* **2013**, *232*, 68–86.
63. Kim, C.H.; Ha, Y.; Yoon, J. Modified non-linear weights for fifth-order weighted essentially non-oscillatory schemes. *J. Sci. Comput.* **2016**, *67*, 299–323.
64. Huang, C. WENO scheme with new smoothness indicator for Hamilton-Jacobi equation. *Appl. Math. Comput.* **2016**, *290*, 21–32.
65. Balsara, D.S.; Shu, C.W. Monotonicity preserving weighted essentially non-oscillatory schemes with increasingly high order of accuracy. *J. Comput. Phys.* **2000**, *160*, 405–452.
66. Balsara, D.S.; Garain, S.; Shu, C.W. An efficient class of WENO schemes with adaptive order. *J. Comput. Phys.* **2016**, *326*, 780–804.
67. Aràndiga, F.; Baeza, A.; Belda, A.; Mulet, P. Analysis of WENO schemes for full and global accuracy. *SIAM J. Numer. Anal.* **2011**, *49*, 893–915.

68. Rusanov, V. The calculation of the interaction of non-stationary shock waves with barriers. *Zhurnal Vychislitel'noi Matematiki i Matematicheskoi Fiziki* **1961**, *1*, 267–279.
69. Godunov, S.K. A difference method for numerical calculation of discontinuous solutions of the equations of hydrodynamics. *Matematicheskii Sbornik* **1959**, *89*, 271–306.
70. Roe, P.L. Approximate Riemann solvers, parameter vectors, and difference schemes. *J. Comput. Phys.* **1981**, *43*, 357–372.
71. Li, X.S.; Li, X.L. All-speed Roe scheme for the large eddy simulation of homogeneous decaying turbulence. *Int. J. Comput. Fluid Dyn.* **2016**, *30*, 69–78.
72. Harten, A. High resolution schemes for hyperbolic conservation laws. *J. Comput. Phys.* **1983**, *49*, 357–393.
73. Harten, A.; Lax, P.D.; Van Leer, B. On upstream differencing and Godunov-type schemes for hyperbolic conservation laws. In *Upwind and High-Resolution Schemes*; Springer: Berlin/Heildelberg, Germany, 1997; pp. 53–79.
74. Toro, E.F. *Riemann Solvers and Numerical Methods for Fluid Dynamics: A Practical Introduction*; Springer: Berlin/Heildelberg, Germany, 2013.
75. Davis, S. Simplified second-order Godunov-type methods. *SIAM J. Sci. Stat. Comput.* **1988**, *9*, 445–473.
76. Trangenstein, J.A. *Numerical Solution of Hyperbolic Partial Differential Equations*; Cambridge University Press: Cambridge, UK, 2009.
77. Liou, M.S.; Steffen, C.J. A new flux splitting scheme. *J. Comput. Phys.* **1993**, *107*, 23–39.
78. Kundu, A.; De, S.; Thangadurai, M.; Dora, C.; Das, D. Numerical visualization of shock tube-generated vortex–wall interaction using a fifth-order upwind scheme. *J. Vis.* **2016**, *19*, 667–678.
79. Edwards, J.R.; Liou, M.S. Low-diffusion flux-splitting methods for flows at all speeds. *AIAA J.* **1998**, *36*, 1610–1617.
80. Liou, M.S. A sequel to AUSM: AUSM+. *J. Comput. Phys.* **1996**, *129*, 364–382.
81. Wada, Y.; Liou, M.S. An accurate and robust flux splitting scheme for shock and contact discontinuities. *SIAM J. Sci. Comput.* **1997**, *18*, 633–657.
82. Kim, K.H.; Lee, J.H.; Rho, O.H. An improvement of AUSM schemes by introducing the pressure-based weight functions. *Comput. Fluids* **1998**, *27*, 311–346.
83. Kim, K.H.; Kim, C.; Rho, O.H. Methods for the accurate computations of hypersonic flows: I. AUSMPW+ scheme. *J. Comput. Phys.* **2001**, *174*, 38–80.
84. Liou, M.S. A sequel to AUSM, Part II: AUSM+-up for all speeds. *J. Comput. Phys.* **2006**, *214*, 137–170.
85. Hyman, J.M.; Knapp, R.J.; Scovel, J.C. High order finite volume approximations of differential operators on nonuniform grids. *Phys. D Nonlinear Phenom.* **1992**, *60*, 112–138.
86. Ricot, D.; Marié, S.; Sagaut, P.; Bailly, C. Lattice Boltzmann method with selective viscosity filter. *J. Comput. Phys.* **2009**, *228*, 4478–4490.
87. San, O. Analysis of low-pass filters for approximate deconvolution closure modelling in one-dimensional decaying Burgers turbulence. *Int. J. Comput. Fluid Dyn.* **2016**, *30*, 20–37.
88. Jahne, B. *Digital Image Processing: Concepts, Algorithms, and Scientific Aplications*; Springer: Berlin/Heildelberg, Germany, 1997.
89. Bogey, C.; De Cacqueray, N.; Bailly, C. A shock-capturing methodology based on adaptative spatial filtering for high-order non-linear computations. *J. Comput. Phys.* **2009**, *228*, 1447–1465.
90. Kraichnan, R.H. Inertial ranges in two-dimensional turbulence. *Phys. Fluids* **1967**, *10*, 1417–1423.
91. Batchelor, G. Computation of the energy spectrum in homogeneous two-dimensional turbulence. *Phys. Fluids* **1969**, *12*, 233–239.
92. Leith, C. Atmospheric predictability and two-dimensional turbulence. *J. Atmos. Sci.* **1971**, *28*, 145–161.
93. Gropp, W. *Tutorial on MPI: The Message-Passing Interface*; Mathematics and Computer Science Division Argonne National Laboratory: Argonne, IL, USA, 2009.
94. Paolucci, S.; Zikoski, Z.J.; Grenga, T. WAMR: An adaptive wavelet method for the simulation of compressible reacting flow. Part II. The parallel algorithm. *J. Comput. Phys.* **2014**, *272*, 842–864.
95. Schilling, O.; Latini, M.; Don, W.S. Physics of reshock and mixing in single-mode Richtmyer-Meshkov instability. *Phys. Rev. E* **2007**, *76*, 026319.
96. Kida, S.; Murakami, Y.; Ohkitani, K.; Yamada, M. Energy and flatness spectra in a forced turbulence. *J. Phys. Soc. Jpn.* **1990**, *59*, 4323–4330.

97. Press, W.H.; Teukolsky, S.A.; Vetterling, W.T.; Flannery, B.P. *Numerical Recipes in FORTRAN*; Cambridge University Press: Cambridge, UK, 1992.
98. Shu, C.W.; Don, W.S.; Gottlieb, D.; Schilling, O.; Jameson, L. Numerical convergence study of nearly incompressible, inviscid Taylor-Green vortex flow. *J. Sci. Comput.* **2005**, *24*, 1–27.
99. San, O.; Staples, A.E.; Iliescu, T. A posteriori analysis of low-pass spatial filters for approximate deconvolution large eddy simulations of homogeneous incompressible flows. *Int. J. Comput. Fluid Dyn.* **2015**, *29*, 40–66.

fluids

MDPI

Article

Evolutionary Optimization of Colebrook's Turbulent Flow Friction Approximations

Dejan Brkić [1,*] and Žarko Ćojbašić [2,*]

1 European Commission, Joint Research Centre, 21027 Ispra VA, Italy
2 Faculty of Mechanical Engineering, University of Niš, 18000 Niš, Serbia
* Correspondence: dejanbrkic0611@gmail.com (D.B.); zcojba@ni.ac.rs (Ž.Ć.); Tel.: +381-64-254-3668 (D.B.)

Academic Editor: William Layton
Received: 1 March 2017; Accepted: 1 April 2017; Published: 6 April 2017

Abstract: This paper presents evolutionary optimization of explicit approximations of the empirical Colebrook's equation that is used for the calculation of the turbulent friction factor (λ), i.e., for the calculation of turbulent hydraulic resistance in hydraulically smooth and rough pipes including the transient zone between them. The empirical Colebrook's equation relates the unknown flow friction factor (λ) with the known Reynolds number (R) and the known relative roughness of the inner pipe surface (ε/D). It is implicit in the unknown friction factor (λ). The implicit Colebrook's equation cannot be rearranged to derive the friction factor (λ) directly, and therefore, it can be solved only iteratively [λ = f(λ, R, ε/D)] or using its explicit approximations [λ ≈ f(R, ε/D)], which introduce certain error compared with the iterative solution. The optimization of explicit approximations of Colebrook's equation is performed with the aim to improve their accuracy, and the proposed optimization strategy is demonstrated on a large number of explicit approximations published up to date where numerical values of the parameters in various existing approximations are changed (optimized) using genetic algorithms to reduce maximal relative error. After that improvement, the computational burden stays unchanged while the accuracy of approximations increases in some of the cases very significantly.

Keywords: Colebrook equation; Colebrook–White; Moody diagram; turbulent flow; hydraulic resistance; Darcy friction; pipes; genetic algorithms; optimization techniques; error analysis

1. Introduction

In this paper, more accurate explicit approximations of Colebrook's equation are presented. The Colebrook Equation (1) relates hydraulic flow friction (λ) through the Reynolds number (R) and the relative roughness (ε/D) of the inner pipe surface, but in an implicit way; λ = f(λ, R, ε/D) [1–18]. On the other hand, to express flow friction (λ) in an explicit way, a number of approximations can be used; λ ≈ f(R, ε/D) [19–44]. Such approximations carry a certain error compared with the iterative solution of the original equation. Increased accuracy of the approximations is achieved using genetic algorithms where the numerical values of the parameters in various existing approximations are changed (optimized) with the goal to reduce error [45–52].

The Colebrook equation is empirical, and hence, its accuracy can be disputed; it is still accepted in engineering practice as sufficiently accurate. It is still widely used in petroleum, mining, mechanical, civil and in all branches of engineering that deal with fluid flow.

Hydraulic resistance: Hydraulic resistance in general depends on the flow rate [53–59]. To make things even more complex, hydraulic resistance is usually expressed through the flow friction factor, such as Darcy's (λ), where further pressure drop and flow rate are correlated with the well-known formula by Darcy and Weisbach. In the non-linear Darcy–Weisbach law for pipe flow, Darcy's friction factor (λ) is variable and always depends on flow. This assumption stands also if Fanning's friction is in use, since its physical meaning is equal to Darcy's friction (λ) with the difference that Fanning used the hydraulic radius while Darcy used the diameter to define the friction factor. Darcy's friction factor, known also under the names of Moody or Darcy–Weisbach, is four-times greater than Fanning's friction factor. The Fanning friction factor is more commonly used by chemical engineers and those following the British convention.

Colebrook equation: To be more complex, as already said, the widely-used empirical and nonlinear Colebrook's Equation (1) for calculation of Darcy's friction factor (λ) is iterative, i.e., implicit in the fluid flow friction factor since the unknown friction factor appears on the both sides of the equation [$\lambda_0 = f(\lambda_0, R, \varepsilon/D)$] [2]. This unknown friction factor (λ) cannot be extracted to be on the left side of the equal sign analytically, i.e., with no use of some kind of mathematical simplifications. Better to say, it can be expressed explicitly only if approximate calculus takes place.

$$\frac{1}{\sqrt{\lambda_0}} = -2 \cdot \log_{10}\left(\frac{2.51}{R \cdot \sqrt{\lambda_0}} + \frac{\varepsilon}{3.71 \cdot D}\right) \tag{1}$$

λ_0 denotes the high precision iterative solution of Colebrook's equation, which is treated here as accurate; R denotes the Reynolds number; while ε/D denotes the relative roughness of the inner pipe surfaces. All three mentioned values are dimensionless.

The Colebrook equation is also known as the Colebrook–White equation or simply the CW equation [1,2]. This equation is valuable for the determination of hydraulic resistances for the turbulent regime in smooth and rough pipes including the turbulent zone between them, but it is not valid for the laminar regime. It describes a monotonic change in the friction factor (λ) during the turbulent flow in commercial pipes from smooth to fully rough. Moody's and Rouse's charts [3,4] represent the plots of the Colebrook equation over a very wide range of the Reynolds number (R from 2320 to 10^8) and relative roughness values (ε/D from 0 to 0.05). Besides some of its shortcomings [54], today, Colebrook's equation is accepted as the informal standard of accuracy for the calculation of the hydraulic friction factor (λ).

Accuracy: As already noted, the Colebrook equation is empirical, and therefore, its accuracy can be disputed; the equal sign "=" in "$\lambda_0 = f(\lambda_0, R, \varepsilon/D)$", i.e., in Equation (1), instead of the approximately equal sign "$\approx$", can be treated as accurate only conditionally [48]. In this paper, the iterative solution of Colebrook's equation (λ_0) after enough iterations is treated as accurate and is used for comparison as the standard of accuracy where the accuracy of the friction factor (λ) calculated using the shown approximations will be compared with it.

Lambert W-function: The Colebrook equation can be rearranged in explicit form only approximately [$\lambda \approx f(R, \varepsilon/D)$], where the approach with the Lambert W-function can be treated as a partial exemption from this rule [6–8,60–62], but also, further evaluation of the Lambert W-function function is approximate.

Looped network of pipes: The use of the accurate explicit approximations should be prioritized over the use of the iterative solution in the calculation of looped networks of pipes, since in that way, double iterative procedures, one for the Colebrook equation and one for the solution of the whole looped system of pipes, can be avoided [63–67].

Goal of the study: The goal is to increase the accuracy of the already available explicit approximation of Colebrook's equation. This is accomplished using genetic algorithms.

2. Genetic Algorithm Optimization Technique

Methodology: Genetic algorithms are one of the evolutionary computational intelligence techniques [45,46], inspired by Darwin's theory of biological evolution. Genetic algorithms provide solutions using randomly-generated strings (chromosomes) for different types of problems, searching the most suitable among chromosomes that make the population in the potential space of solutions. Genetic optimization is an alternative to the traditional optimal search approaches, which make it difficult to find the global optimum for nonlinear and multimodal optimization problems. Thus, genetic algorithms have been successful for example in solving combinatorial problems, control applications of parameter identification and control structure design, as well as in many other areas [47–52].

The used optimization approach: Here, the approach with genetic algorithms is implemented to optimize the parameters of the available approximations of the Colebrook equation for the hydraulic friction factor determination in order to improve their accuracy, at the same time retaining the previous complexity and computational burden of approximations. Small letters in the equations throughout the paper correspond to the numerical values before, while capital letters to the numerical values after optimization through genetic algorithms, as is picturesquely presented in Figure 1.

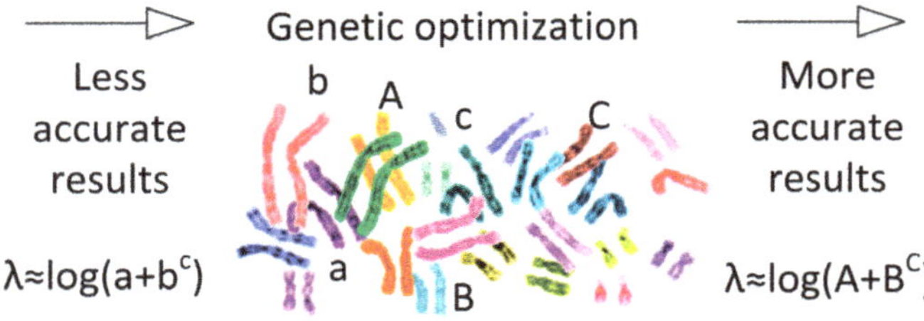

Figure 1. Picturesquely shown optimization using genetic algorithms.

Genetic algorithms are very powerful tools for optimization. Samadianfard [47] uses genetic programming, a sort of genetic algorithm, to develop his own explicit approximations to the Colebrook equation. Furthermore, genetic algorithms can be used together with some other techniques of artificial intelligence, such as neural networks [50–52].

Real coded genetic algorithms are used in this paper. The real coded genetic algorithms use the optimization designed cost function that minimizes maximal relative error, δ_{max}, as follow (2):

$$\left.\begin{array}{l} fitness = \max\limits_{\substack{i \\ i \in [1,n]}} (\delta); \quad \delta = \left|\frac{\lambda-\lambda_0}{\lambda_0}\right| \cdot 100\% \end{array}\right\} \quad (2)$$

In (2), δ denotes relative (percentage) error, λ_0 denotes the high precision iterative solution of Colebrook's equation, which is treated as accurate here, λ denotes the hydraulic friction factor solution calculated by each approximation considered and n denotes number of pairs of λ_0 and λ used for optimization (in our case, n = 90,000).

The fitness function is evaluated in a large number of 90 thousand points uniformly distributed in domains of the Reynolds number (R) and the relative roughness (ε/D). These domains correspond to those from the Moody diagram [3]; i.e., $2320 < Re < 10^8$, $10^6 < \varepsilon/D < 0.05$, where roughness ε usually for PVC and plastic pipes is 0.0015–0.007 mm, copper, lead, brass, aluminum (new) 0.001–0.002 mm and for steel commercial pipe 0.045–0.09 mm. The subjects of genetic optimization are coefficients in approximations, i.e., numeric coefficients in each approximation are changed by genetic algorithms in order to minimize the fitness function (2). In that way, approximations are changed in order to match the accuracy of the iterative solution of Colebrook's equation as much as possible. Simultaneous optimization of all coefficients in each approximation is attempted, while the range of values of parameters in which optimal solutions are searched is always in the arbitrary neighborhood of the initial values. Here, we chose to present the results obtained with the fitness function (2) in order to reduce the maximal error of each approximation as much as possible (assuming that the reduction of the maximal relative error is of the highest importance for the practical use of approximations). Genetic algorithms' performance depends on their parameter values, so genetic algorithm parameters were carefully selected by conducting numerous experiments. In the implemented algorithm, a real-coded population of 100 individuals, an elitism of 10 individuals and a scattered crossover function are used. All of the members are subjected to adaptive feasible mutation except for the elite. The individuals are randomly selected by the Roulette method. Optimization with genetic algorithms was carried out in MATLAB R2010a by MathWorks (Natick, MA, USA). The practical domain of the Reynolds number (R) and relative roughness of inner pipe surface (ε/D) are covered by a mesh of n = 90,000 points for this optimization. In these 90 thousand points, the iterative solution of the implicitly-given Colebrook equation, λ_0, and the non-iterative solution for every single observed approximation, λ, are calculated. The optimization of every single approximation lasts several hours. All evaluations of error were performed in MATLAB, with further confirmations in MS Excel to maintain full comparability with the study of Brkić [10] (for the use of iterative calculus in MS Excel Ver. 2007, see Brkić [11]; in Brkić and Tanasković [68], MS Excel is also used for other extensive, but non-iterative calculations). The mesh in MS Excel over the practical domain of the Reynolds number (R) and relative roughness of the inner pipe surface (ε/D) consists of n = 740 uniformly-distributed points.

Alternative optimization approaches: The main goal of the optimization in our case is to reduce the maximal error (δ_{max}) of the every single observed approximation. This means that sometimes, the average (mean) relative error in the practical range of the Reynolds number (R) and the relative roughness of inner pipe surface (ε/D) increases compared to the model of the observed approximation with initial, non-optimized values of the parameters. Of course, using genetic algorithm optimization with the function defined to reduce maximal error, this error is reduced more or less efficiently, which at the same time does not mean that average error is necessarily increased or decreased. Although the minimization of average error is not set as a goal by Equation (2), it can be reduced also during the optimization. Instead of the here already shown fitness function Equation (2), it can be redefined to simultaneously reduce average and maximal error Equation (3). In that way, both errors, i.e., maximal relative error and average (mean) relative error, can be reduced simultaneously for sure. This requires more one-off computational efforts compared with the approach in which only one type of error is reduced; in our case, maximal relative error, δ_{max}, while the fitness function is defined by Equation (2). In Equation (3), the first term reduces average (mean) relative error, δ_{avr}; the second term reduces maximal error δ; while weights k_1 and k_2 can be used to signify one of the terms and reduce the influence of other. In that case, a compromise between the reduction of the maximal and average relative error is obtained.

$$\left.\begin{aligned} fitness = k_1 \cdot (\delta_{avr}) + k_2 \cdot \max_{\substack{i \\ i\in[1,n]}} (\delta); \quad \delta_{avr} = \frac{1}{n}\sum_{i=1}^{n}\left(\left|\frac{\lambda-\lambda_0}{\lambda_0}\right| \cdot 100\%\right)_i \\ \delta = \left|\frac{\lambda-\lambda_0}{\lambda_0}\right| \cdot 100\% \end{aligned}\right\} \quad (3)$$

Furthermore, the fitness function can be set to reduce simultaneously the mean square error δ_{MSE} and maximal relative error δ, as in Equation (4). As already noted for Equation (3), the ratio between weight coefficients k_3 and k_4 determines the influence of the mean square error δ_{MSE} and the maximal relative error δ in the optimization. According to many different criteria, the values of coefficients in the existing explicit approximation to the Colebrook equation can be used. Using many computational resources, all three errors shown in our paper can be simultaneously reduced Equation (5), but such a procedure seems to be quite elusive.

$$\left.\begin{aligned} fitness = k_3 \cdot (\delta_{MSE}) + k_4 \cdot \max_{\substack{i \\ i\in[1,n]}} (\delta); \quad \delta_{MSE} = \frac{1}{n}\sum_{i=1}^{n}(\lambda-\lambda_0)_i^2 \\ \delta = \left|\frac{\lambda-\lambda_0}{\lambda_0}\right| \cdot 100\% \end{aligned}\right\} \quad (4)$$

$$\left.\begin{array}{ll} fitness = k_4 \cdot \underset{\substack{i \\ i\in[1,n]}}{\max}(\delta) + k_1\cdot(\delta_{avr}) + k_3\cdot(\delta_{MSE}); & \delta = \left|\frac{\lambda-\lambda_0}{\lambda_0}\right|\cdot 100\% \\ \delta_{avr} = \frac{1}{n}\sum\limits_{i=1}^{n}\left(\left|\frac{\lambda-\lambda_0}{\lambda_0}\right|\cdot 100\%\right)_i; & \delta_{MSE} = \frac{1}{n}\sum\limits_{i=1}^{n}(\lambda-\lambda_0)_i^2 \end{array}\right\} \quad (5)$$

3. Explicit Approximations of Colebrook's Equation

Colebrook's Equation (1) [2] suffers from being implicit in the unknown friction factor (λ). It requires an iterative solution where convergence to the final accuracy of the observed approximation typically requires less than seven iterations. As Brkić [10] proposed, here is used even a few thousand iterations to be sure that a sufficient value of accuracy for the friction factor, λ_0, is reached.

As already stated, the implicit Colebrook's equation cannot be rearranged to derive the friction factor directly in one step, while iterative calculus can cause a problem in the simulation of flow in a pipe system in which it may be necessary to evaluate the friction factor hundreds or thousands of times. This is the main reason for attempting to develop a relationship that is a reasonable and an as accurate as possible approximation for the Colebrook equation, but which is explicit in the friction factor. These approximations are used for the calculation of the friction factor (λ), which is compared with the very accurate solution (λ_0) calculated using the iterative procedure.

In this paper, 25 approximations are optimized: Brkić [19,20], Fang et al. [21], Ghanbari et al. [22], Papaevangelou et al. [23], Avci and Karagoz [24], Buzzelli [25], Sonnad and Goudar [26], Romeo et al. [27], Manadilli [28], Chen [29], Serghides [30], Haaland [31], Zigrang and Sylvester [32], Barr [33], Round [34], Shacham (available from [35]), Chen [36], Swamee and Jain [37], Eck [38], Wood [39] and Moody [40]. Ćojbašić and Brkić [42] already optimized the numerical values of the parameters by Romeo et al. [27] and by Serghides [30].

The accuracy of existing approximations of Colebrook's equation was thoroughly checked by many researchers [10–16]. Yıldırım [14] conducted a comprehensive analysis of existing correlations for single-phase friction, but he used Techdig 2.0 software to read the date from the Moody diagram, which caused remarkable reading error. One must be always aware that the Moody diagram [3] was constructed using Colebrook's equation [2] and not opposite. After all, the main conclusion of all papers [10–16] is that the relative error, δ, is non-uniformly distributed over the domain of the Reynolds number (R) and the relative roughness (ε/D).

The relative error δ is defined in Equations (2)–(4) of this paper, the average (mean) relative error δ_{avr} in Equation (3) and the mean square error δ_{MSE} in Equation (4). All three types of error are used in further text for the estimation of the accuracy of the examined explicit approximations of the Colebrook equation, but the accent is on the minimization of the maximal relative error, δ_{max}.

Using the shown genetic algorithm optimization technique, the values of existing parameters of the explicit approximations are improved compared to the iterative solution of Colebrook's equation. This means that the error of approximations decreases while the computational burden stays unchanged. In this section, new parameters are shown, and the reduction of the maximal relative error, δ_{max}, is estimated. The relative errors of the approximations shown in further text of this paper are calculated as $\delta = [(\lambda - \lambda_0)/\lambda_0]\cdot 100\%$, where λ is the Darcy friction factor calculated using the observed approximation, while λ_0 is the iterative solution of Colebrook's equation, which can be used as accurate after enough iterations (here, set to the maximal available number of iterations in MS Excel, which is 32,767, as explained in [10]).

Each of the 25 observed approximations is supplied with three diagrams; the first is the distribution of the relative error over the practical domain of applicability in engineering practice; the second is the same as the first, but with the relative error distribution after optimization; and the third is a comparative diagram. For the first two mentioned diagrams, the entire practical domain of the Reynolds number (R) and the relative roughness of the inner pipe surface (ε/D) is covered with a 740 point-mesh (diagrams produced in MS Excel). For the first two figures with approximations, same pace of error is used for non-optimized and for the optimized approximation, to provide a more easy comparison with the exceptions of Approximation (Appr.) 10; Equation (15), Appr. 11; Equation (16) and Appr. 14; Equation (19), where the optimization was extremely successfully performed. The mesh of 740 points is formed in MS Excel using 20 values of the relative roughness (ε/D) (shown in the related figures) and using 37 values of the Reynolds number (R); from 10^4–10^5 with a pace of 10^4, from 10^5–10^6 with a pace 10^5, from 10^6–10^7 with a pace of 10^6 and from 10^7–10^8 with a pace of 10^7.

According to Winning and Coole [16], using the value of mean square error δ_{MSE} defined in (4), all approximations can be classified into four groups (the very small error is lower than 10^{-11}; small is between 10^{-11} and 10^{-8}; medium is between 10^{-8} and 5×10^{-6}; and large is above 5×10^{-6}). This criterion is also used in further evaluation.

Regarding accuracy, it should be noted that the inner roughness of the pipe, ε, cannot be determined easily [17], so the physical interpretation of the relative roughness of the inner pipe surface (ε/D) is not the subject of this study.

For genetic algorithm optimization, MATLAB 2010a by MathWorks is used. For this purpose, a mesh of 90 thousand points over the entire practical domain of the Reynolds number (R) and the relative roughness of the inner pipe surface (ε/D) is generated. For these 90 thousand pairs of Reynolds number (R) and the relative roughness of inner pipe surface (ε/D), the friction factor (λ_0) is very accurately calculated to be used as a pattern during the procedure of optimization. Although genetic optimization allows for virtually all coefficients in each and every formula considered to be optimized in the search for the best result, which can be considered as an important advantage of the presented technique, the selection of the coefficients included in the optimization for each formula was the object of careful consideration. Not only the inclusion of more coefficients significantly widens the search space to be covered, it appears also that some approximations are highly sensitive to the changes of some coefficients. Therefore, subsets of coefficients to be optimized within each formula were selected on the basis of multiple trials, approaches considered by other authors and our own vast experience. The number of digits in each optimized coefficient was also limited as a tradeoff of the desired punctuality of approximation and the practical usability of the formula.

The efficiency of computing in the computer environment stays unchanged between non-optimized and related optimized approximations, since the model of the approximation stays unchanged; i.e., the number of logarithmic and power expressions stays unchanged [9,18]. Only the change of integer power to non-integer power in some approximation can increase the computational burden, but even then not significantly.

In the following Figures 2–26, symbols and zones with green and red color represent: the $\Delta\delta$-decreased level of maximal relative error δ_{max}; (1) zone of increased relative error δ (red); (2) zone of decreased relative error δ (green).

Brkić approximation (Appr. 1): Relevant parameters and errors related to the approximation by Brkić [19] after and before optimization (6) (Appr. 1) are given in Figure 2.

$$\left.\begin{matrix} \frac{1}{\sqrt{\lambda}} \approx -2 \cdot \log_{10}\left(\frac{2.18 \cdot a_1}{R} + \frac{1}{3.71} \cdot \frac{\varepsilon}{D}\right) \\ a_1 \approx \ln \frac{R}{1.816 \cdot \ln\left(\frac{1.1 \cdot R}{\ln(1+1.1 \cdot R)}\right)} \end{matrix}\right\} \rightarrow \left.\begin{matrix} \frac{1}{\sqrt{\lambda}} \approx -2.013 \cdot \log_{10}\left(\frac{2.261 \cdot A_1}{R} + \frac{1}{3.71} \cdot \frac{\varepsilon}{D}\right) \\ A_2 \approx \ln \frac{R}{2.479 \cdot \ln\left(\frac{1.1 \cdot R}{\ln(1+1.1 \cdot R)}\right)} \end{matrix}\right\} \quad (6)$$

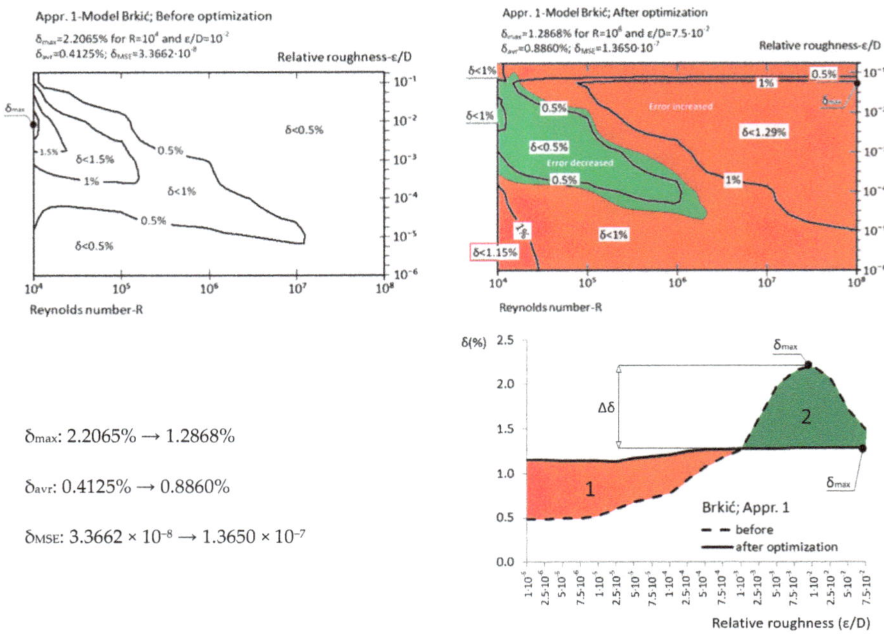

δ_{max}: 2.2065% → 1.2868%

δ_{avr}: 0.4125% → 0.8860%

δ_{MSE}: 3.3662 × 10^{-8} → 1.3650 × 10^{-7}

Figure 2. Relative error before and after optimization; Brkić (Appr. 1; Equation (6)).

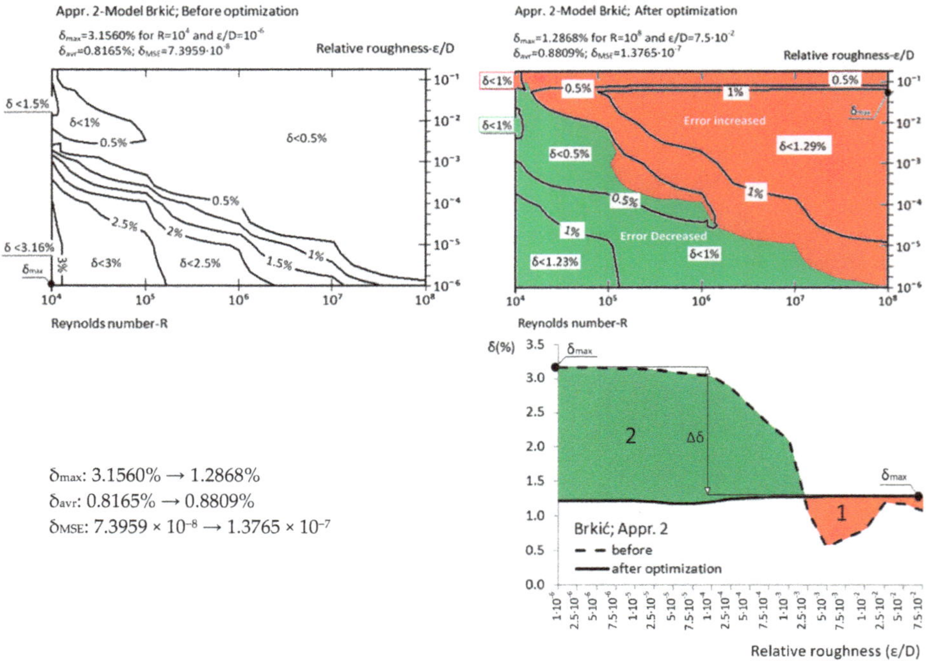

δ_{max}: 3.1560% → 1.2868%
δ_{avr}: 0.8165% → 0.8809%
δ_{MSE}: 7.3959 × 10^{-8} → 1.3765 × 10^{-7}

Figure 3. Relative error before and after optimization; Brkić (Appr. 2; Equation (7)).

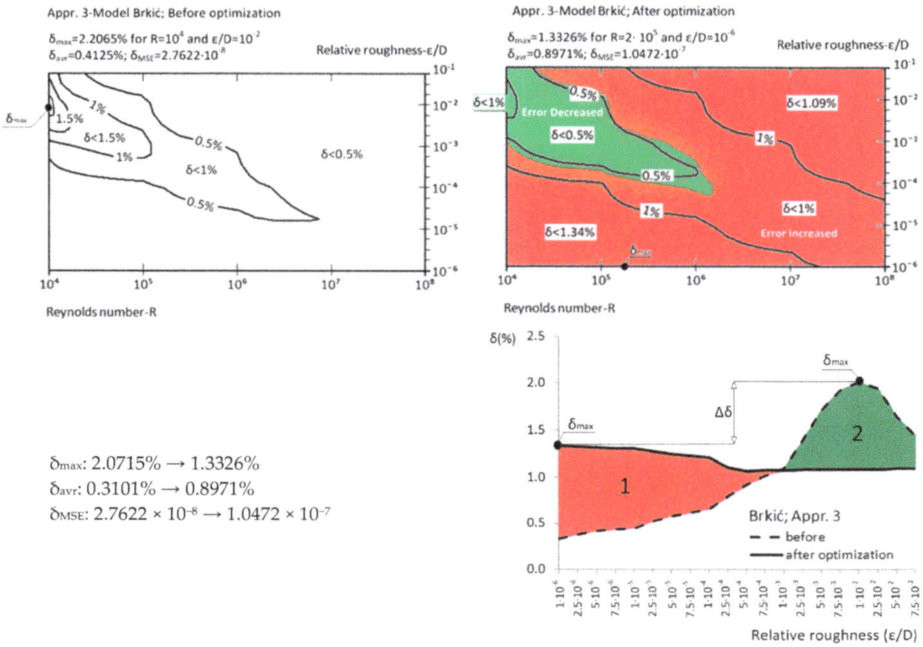

δ_{max}: 2.0715% → 1.3326%
δ_{avr}: 0.3101% → 0.8971%
δ_{MSE}: 2.7622 × 10^{-8} → 1.0472 × 10^{-7}

Figure 4. Relative error before and after optimization; Brkić (Appr. 3; Equation (8)).

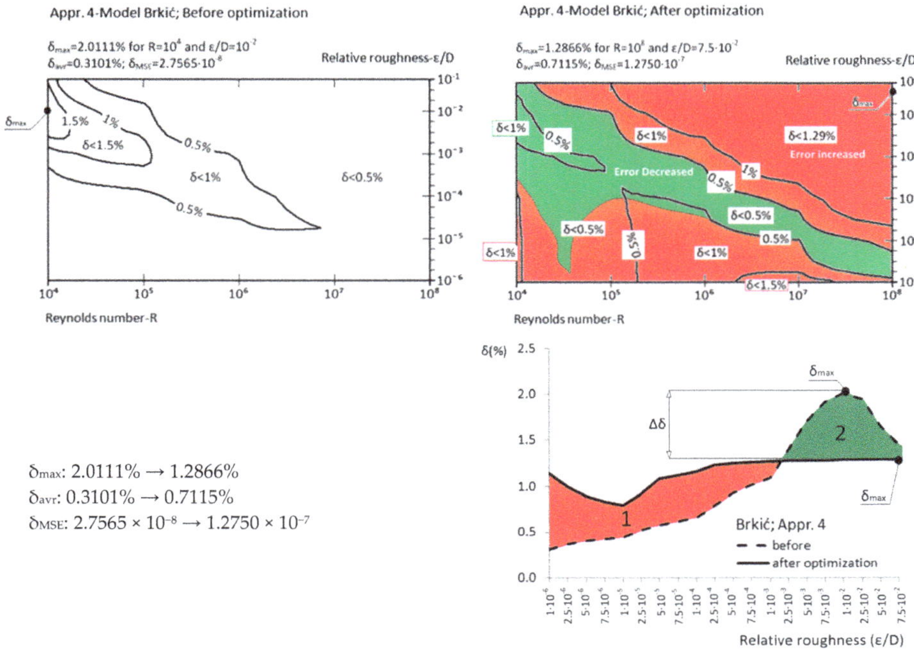

δ_{max}: 2.0111% → 1.2866%
δ_{avr}: 0.3101% → 0.7115%
δ_{MSE}: 2.7565 × 10^{-8} → 1.2750 × 10^{-7}

Figure 5. Relative error before and after optimization; Brkić (Appr. 4; Equation (9)).

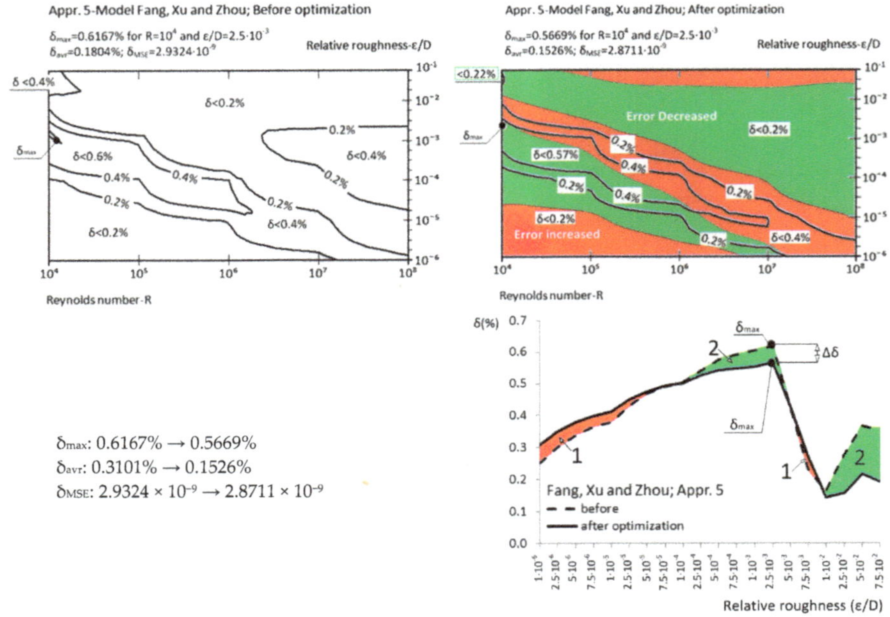

Figure 6. Relative error before and after optimization; Fang et al. (Appr. 5; Equation (10)).

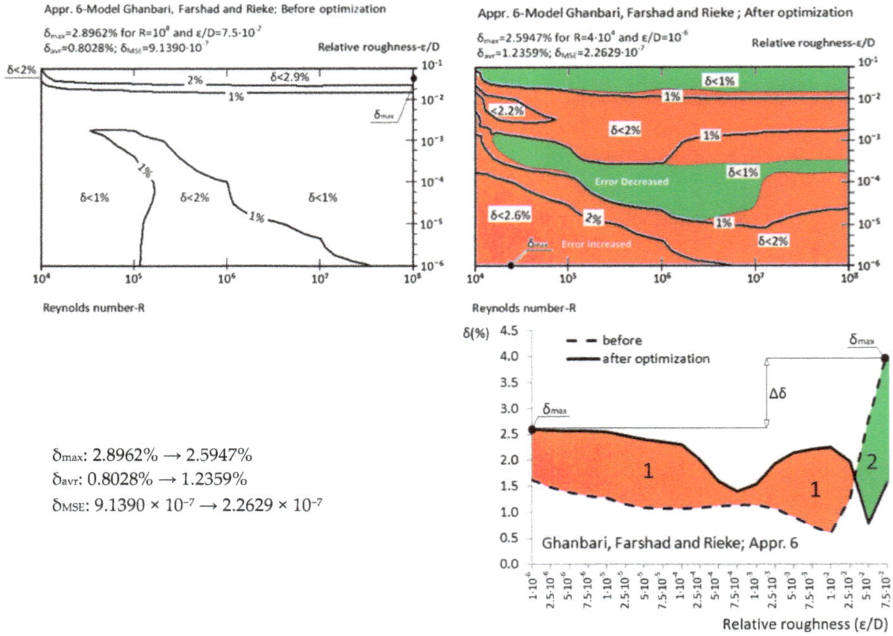

Figure 7. Relative error before and after optimization; Ghanbari et al. (Appr. 6; Equation (11)).

δ_{max}: 0.8248% $\rightarrow$ 0.7312%

δ_{avr}: 0.2001% $\rightarrow$ 0.2974%

δ_{MSE}: 1.2984 $\times$ 10^{-8} $\rightarrow$ 1.5319 $\times$ 10^{-8}

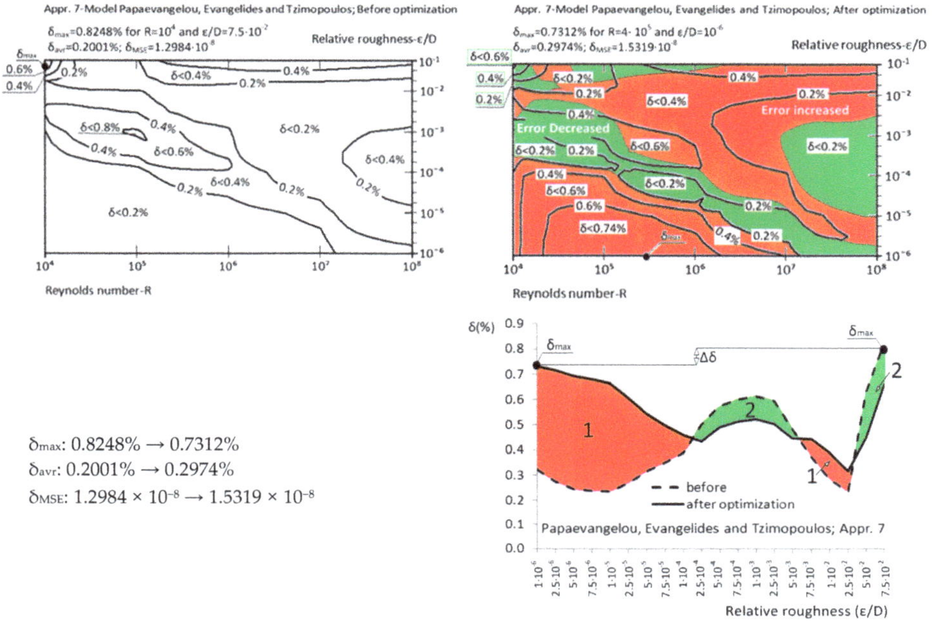

Figure 8. Relative error before and after optimization; Papaevangelou et al. (Appr. 7; Equation (12)).

δ_{max}: 4.7858% $\rightarrow$ 3.1259%

δ_{avr}: 1.2521% $\rightarrow$ 1.8650%

δ_{MSE}: 1.1611 $\times$ 10^{-6} $\rightarrow$ 3.1516 $\times$ 10^{-7}

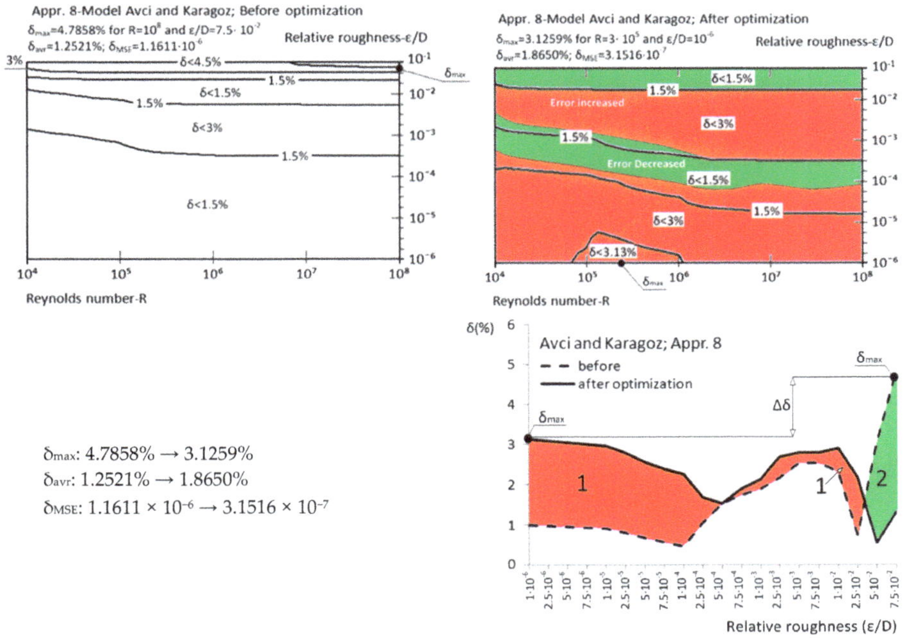

Figure 9. Relative error before and after optimization; Avci and Karagoz (Appr. 8; Equation (13)).

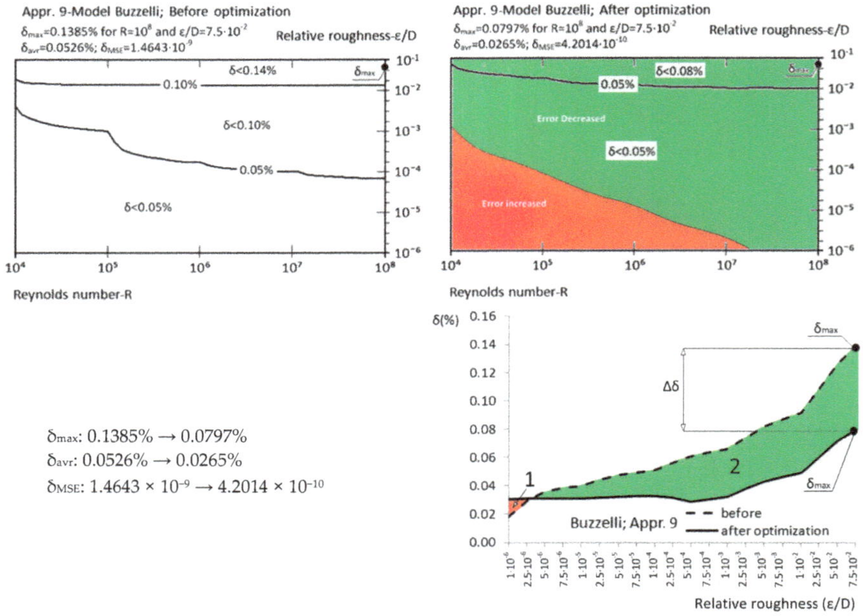

Figure 10. Relative error before and after optimization Buzzelli (Appr. 9; Equation (14)).

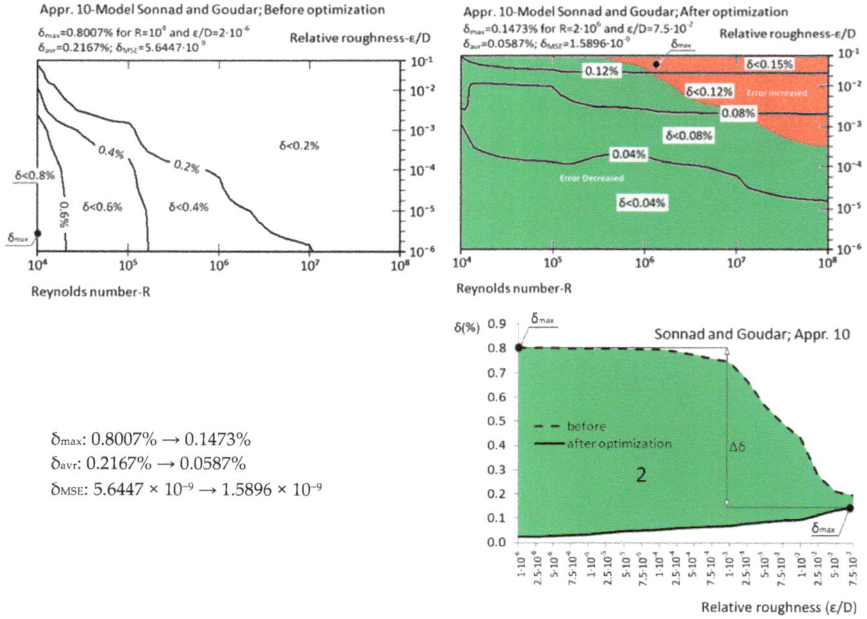

Figure 11. Relative error of Sonnad and Goudar (Appr. 10; Equation (15)) before and after optimization; optimized by Vatankhah and Kouchakzadeh [43,44].

δ_{max}: 0.1345% → 0.0083%
δ_{avr}: 0.0544% → 0.0037%
δ_{MSE}: 3.4379×10^{-10} → 4.3087×10^{-12}

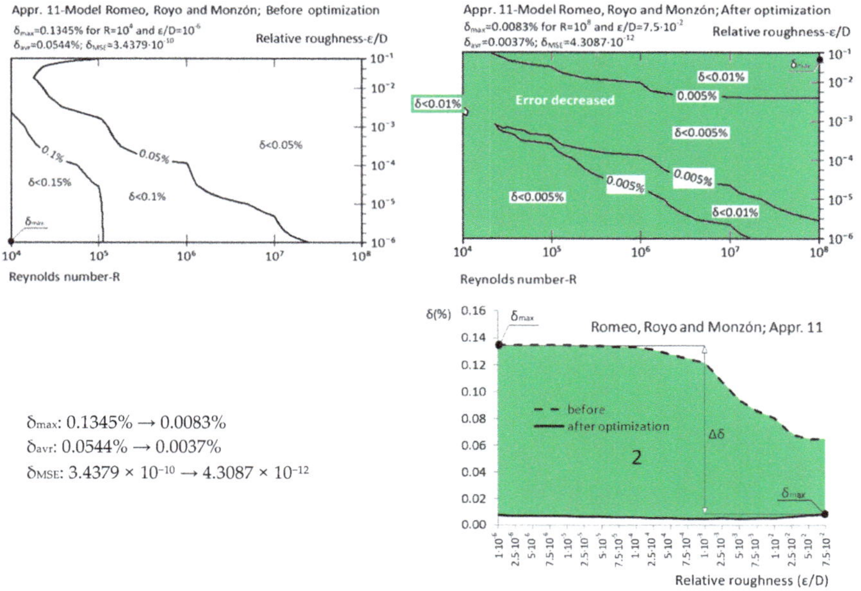

Figure 12. Relative error of Romeo, Royo and Monzón (Appr. 11; Equation (16)) before and after optimization; optimized by Ćojbašić and Brkić [42].

δ_{max}: 2.0651% → 1.5018%
δ_{avr}: 0.3716% → 0.5956%
δ_{MSE}: 3.4483×10^{-8} → 7.2942×10^{-8}

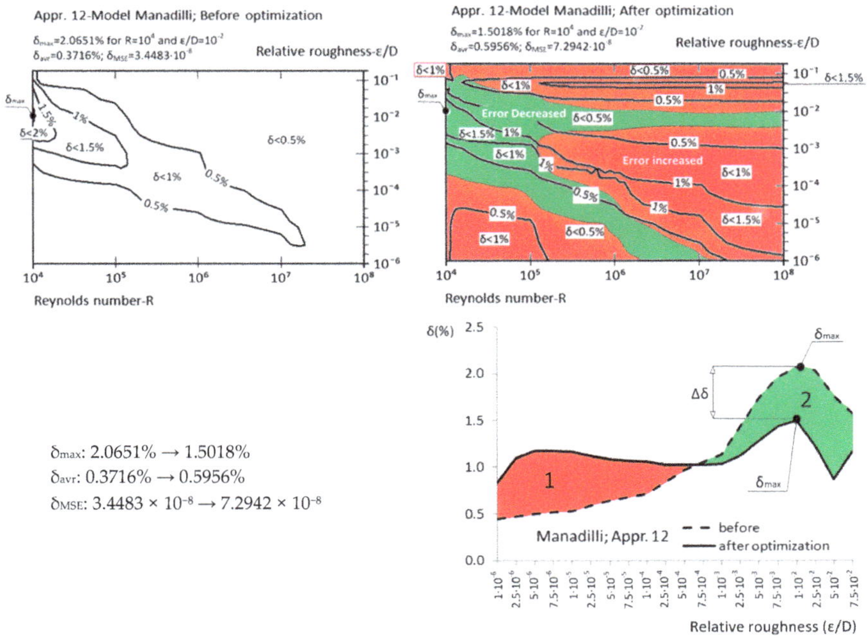

Figure 13. Relative error before and after optimization; Manadilli (Appr. 12; Equation (17)).

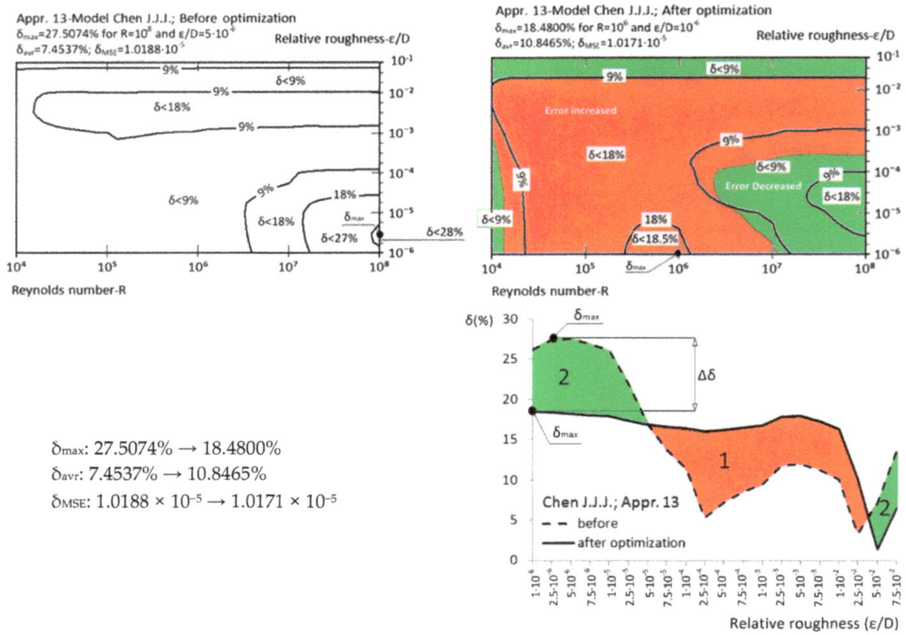

Figure 14. Relative error before and after optimization; Chen [29] (Appr. 13; Equation (18)).

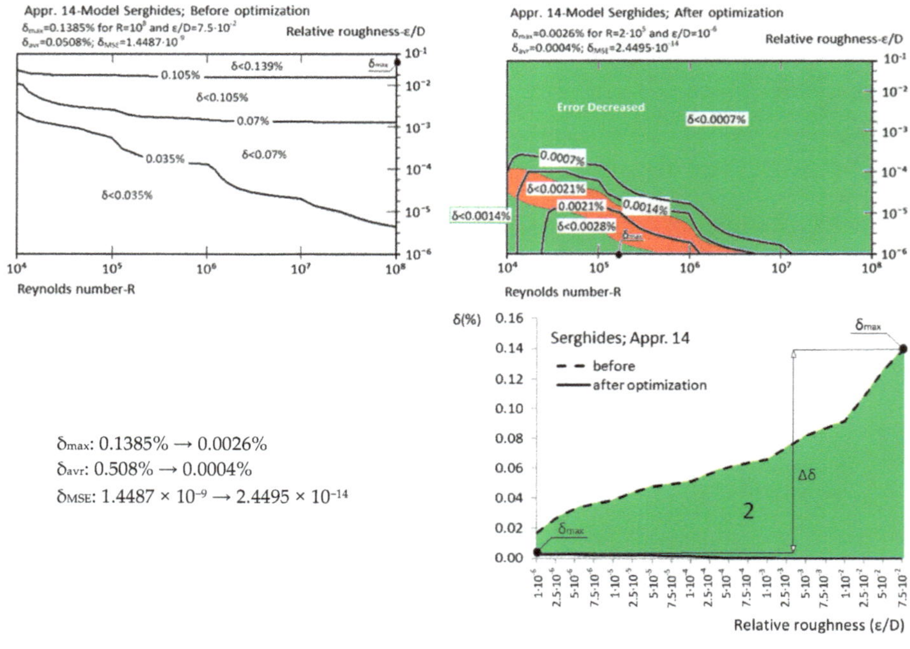

Figure 15. Relative error of Serghides (Appr. 14; Equation (19)) before and after optimization; optimized by Ćojbašić and Brkić [42].

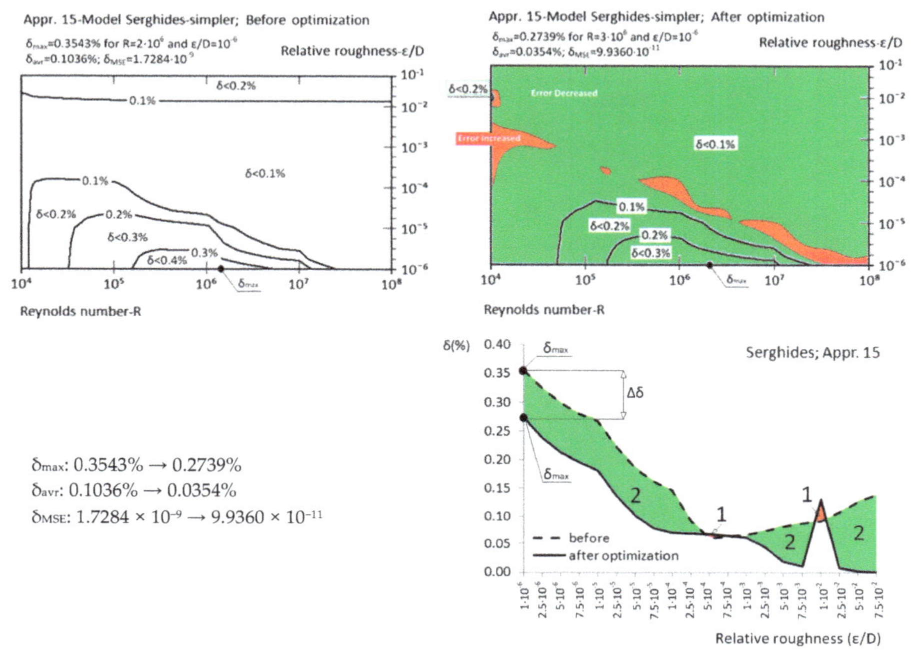

Figure 16. Relative error of simpler version before and after optimization; Serghides (Appr. 15; Equation (20)).

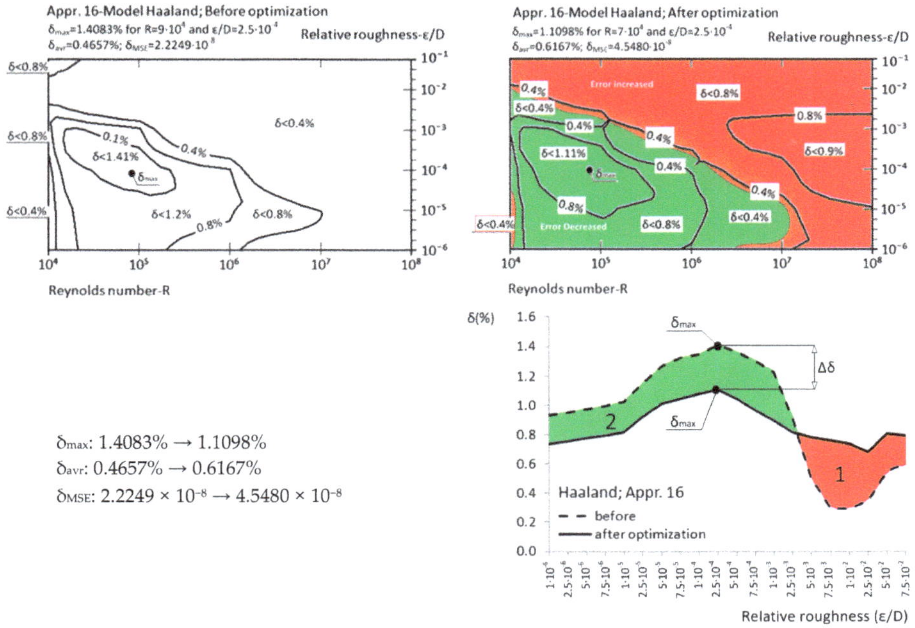

Figure 17. Relative error before and after optimization; Haaland (Appr. 16; Equation (21)).

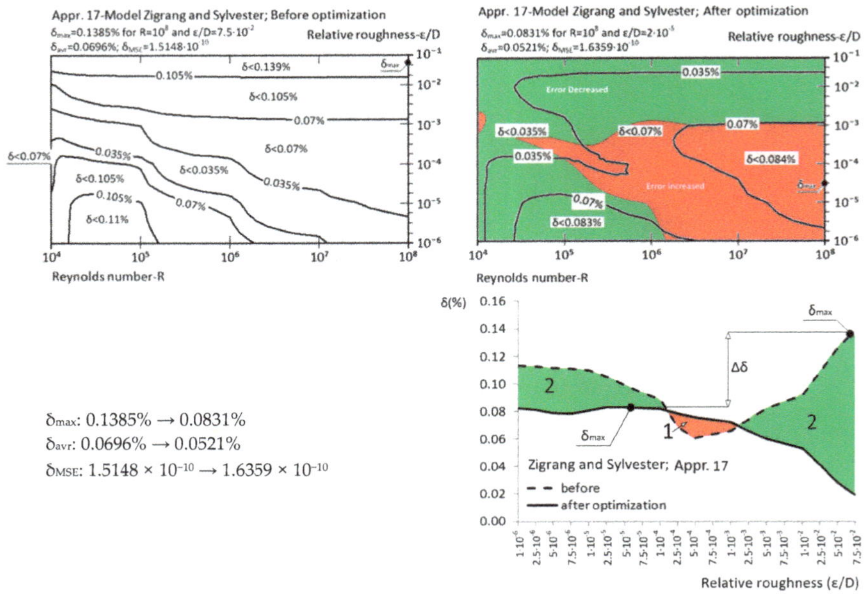

Figure 18. Relative error before and after optimization; Zigrang and Sylvester (Appr. 17; Equation (22)).

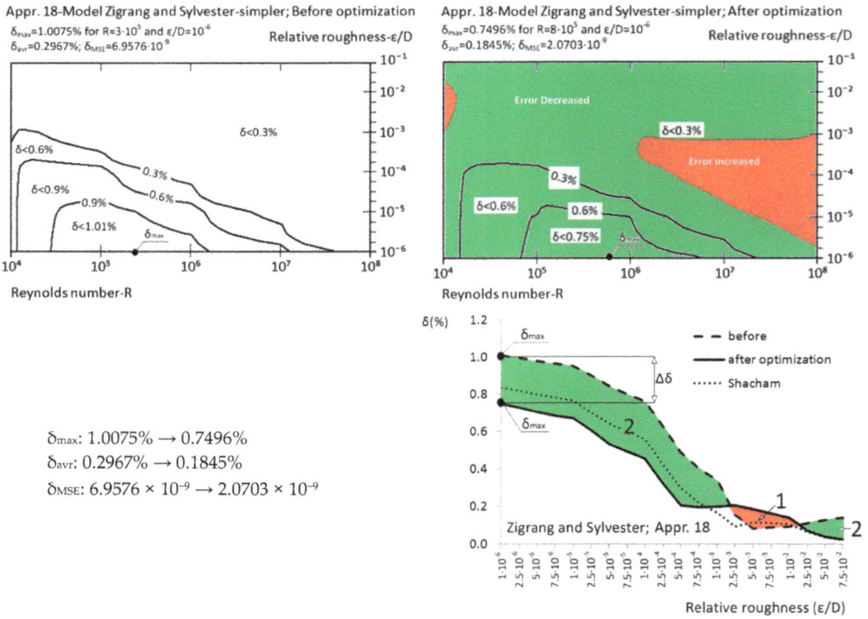

Figure 19. Relative error of simpler version before and after optimization; Zigrang and Sylvester (Appr. 18; Equation (23)).

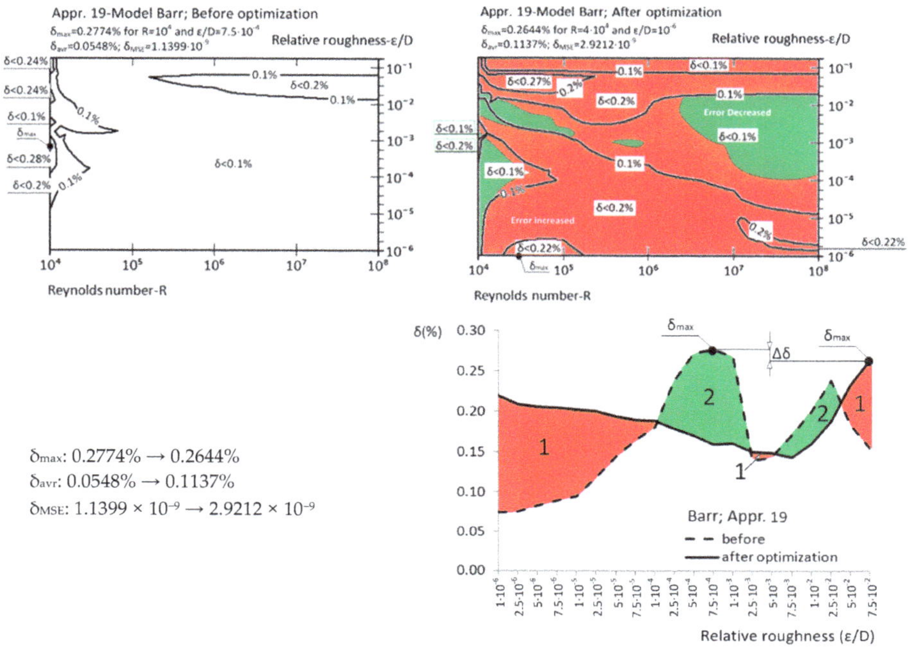

Figure 20. Relative error before and after optimization; Barr (Appr. 19; Equation (24)).

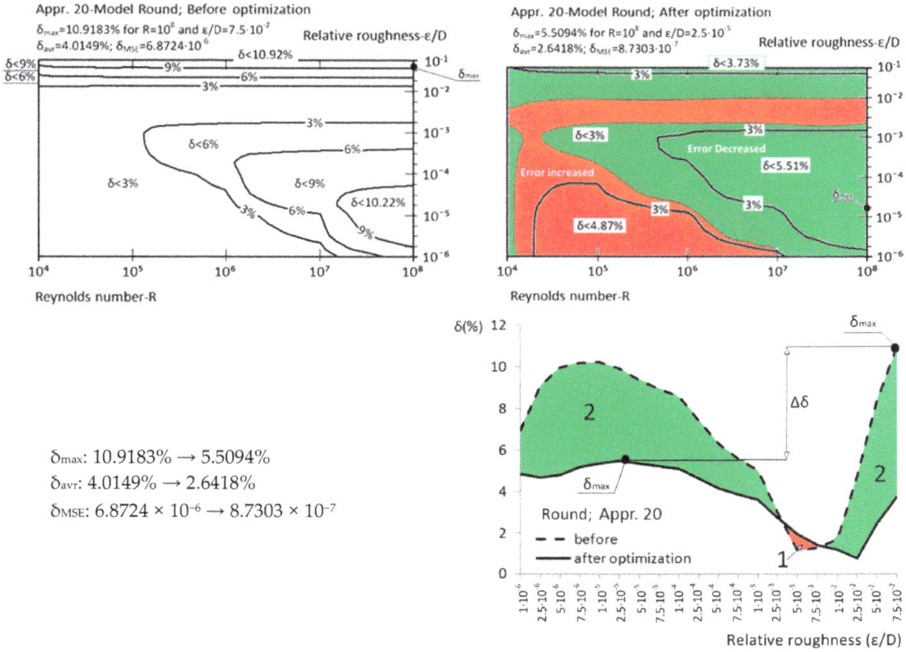

Figure 21. Relative error before and after optimization; Round (Appr. 20; Equation (25)).

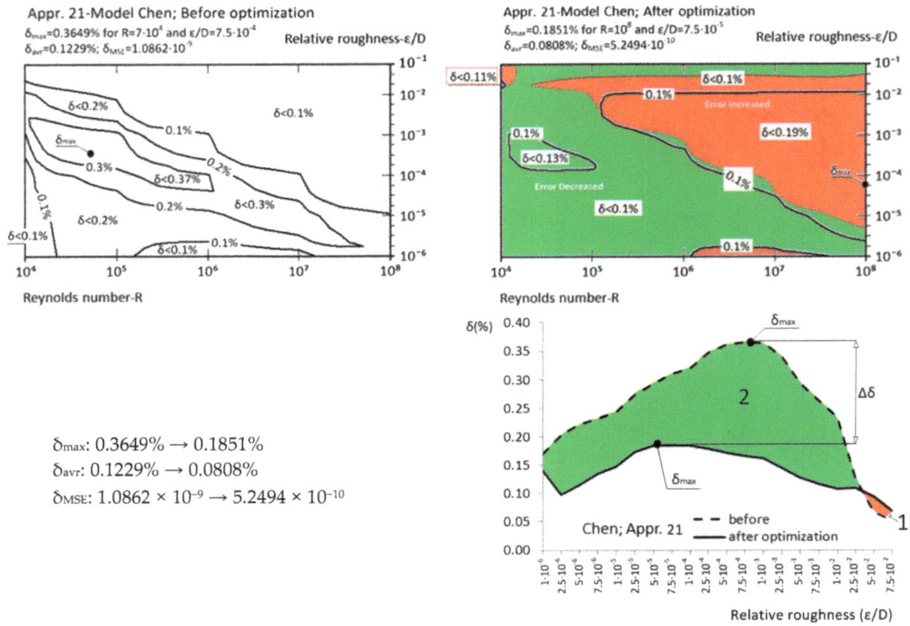

δ_{max}: 0.3649% → 0.1851%
δ_{avr}: 0.1229% → 0.0808%
δ_{MSE}: $1.0862 \times 10^{-9} \rightarrow 5.2494 \times 10^{-10}$

Figure 22. Relative error before and after optimization; Chen (Appr. 21; Equation (26)).

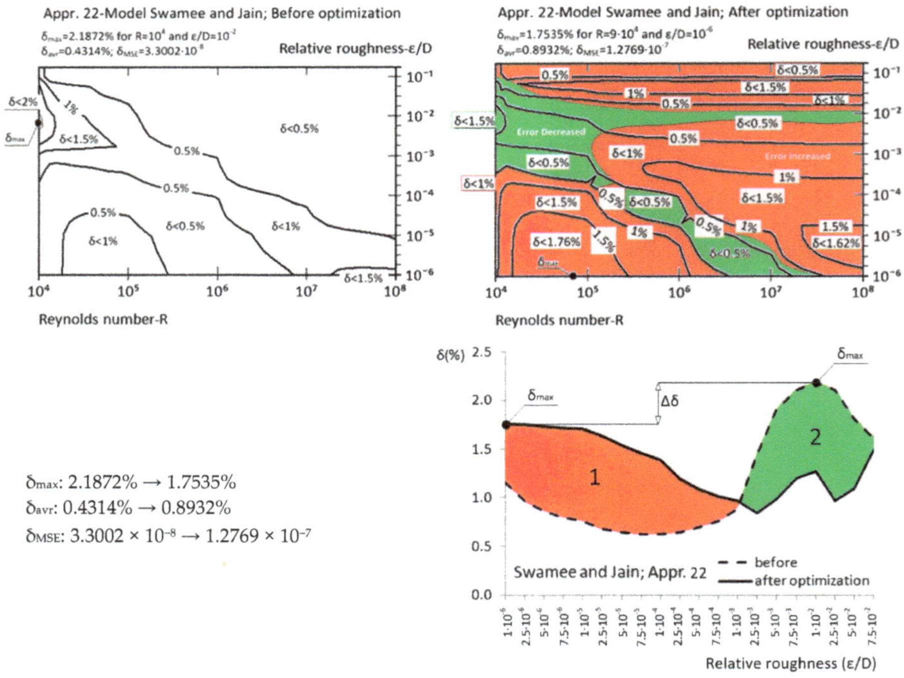

δ_{max}: 2.1872% → 1.7535%
δ_{avr}: 0.4314% → 0.8932%
δ_{MSE}: $3.3002 \times 10^{-8} \rightarrow 1.2769 \times 10^{-7}$

Figure 23. Relative error before and after optimization; Swamee and Jain (Appr. 22; Equation (27)).

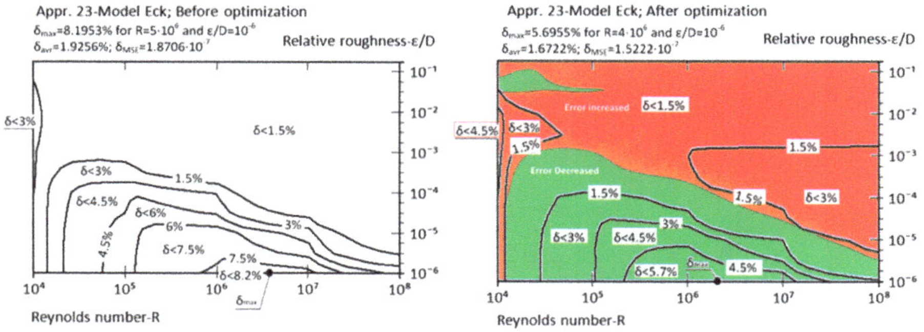

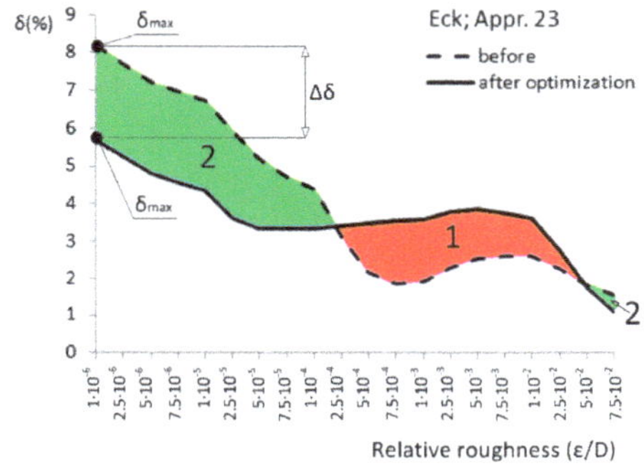

δ_{max}: 8.1953% → 5.6955%
δ_{avr}: 1.9256% → 1.6722%
δ_{MSE}: 1.18706 × 10^{-7} → 1.5222 × 10^{-7}

Figure 24. Relative error before and after optimization; Eck (Appr. 23; Equation (28)).

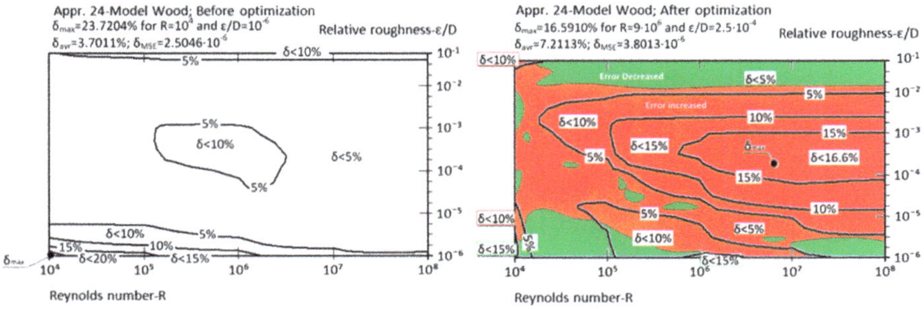

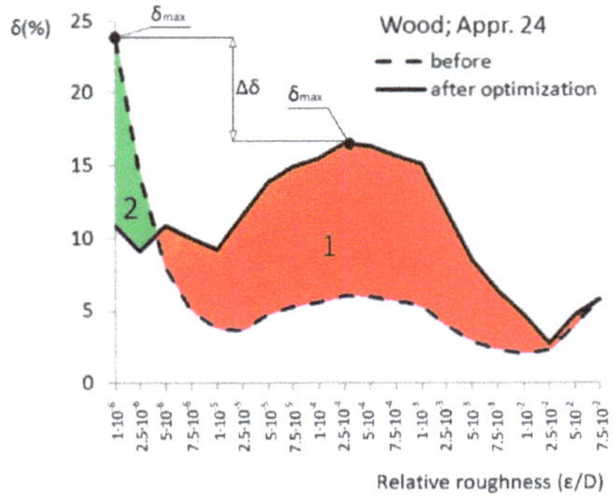

δ_{max}: 23.7204% → 16.5910%
δ_{avr}: 3.7011% → 7.2113%
δ_{MSE}: 2.5046 × 10^{-6} → 3.8013 × 10^{-6}

Figure 25. Relative error before and after optimization; Wood (Appr. 24; Equation (29)).

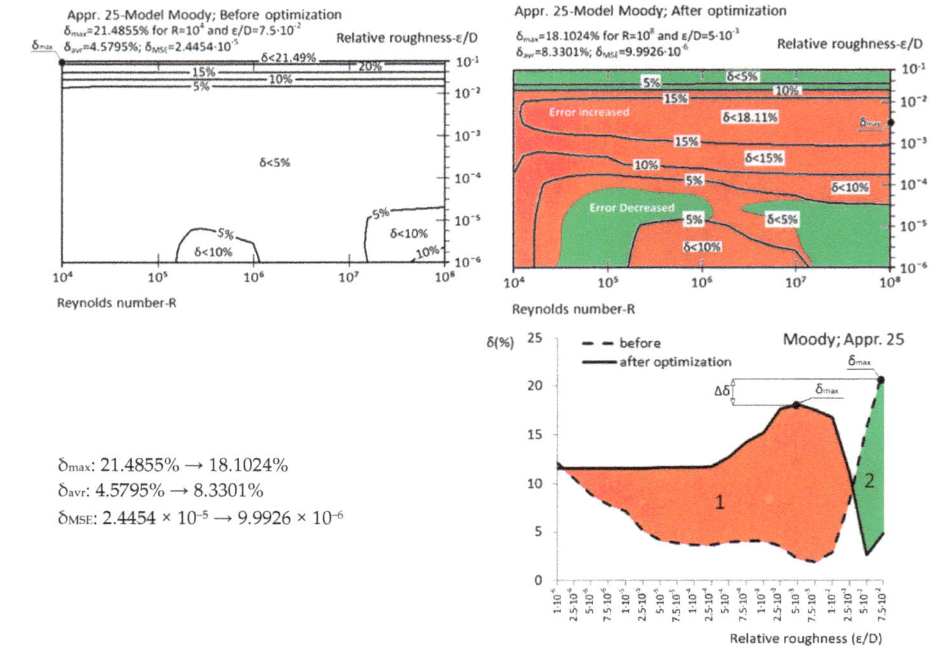

Figure 26. Relative error before and after optimization; Moody (Appr. 25; Equation (30)).

Brkić approximation (Appr. 2): Relevant parameters and errors related to the approximation by Brkić [19] after and before optimization (7) (Appr. 2) are given in Figure 3.

$$\left.\begin{array}{c}\frac{1}{\sqrt{\lambda}} \approx -2 \cdot \log_{10}\left(10^{-0.4343 \cdot a_2} + \frac{1}{3.71} \cdot \frac{\varepsilon}{D}\right) \\ a_2 \approx \ln \frac{R}{1.816 \cdot \ln\left(\frac{1.1 \cdot R}{\ln(1+1.1 \cdot R)}\right)}\end{array}\right\} \rightarrow \left.\begin{array}{c}\frac{1}{\sqrt{\lambda}} \approx -2.013 \cdot \log_{10}\left(10^{-0.43 \cdot A_2} + \frac{1}{3.71} \cdot \frac{\varepsilon}{D}\right) \\ A_2 \approx \ln \frac{R}{1.895 \cdot \ln\left(\frac{1.1 \cdot R}{\ln(1+1.1 \cdot R)}\right)}\end{array}\right\} \quad (7)$$

Brkić approximation (Appr. 3): Relevant parameters and errors related to the approximation by Brkić [20] after and before optimization (8) (Appr. 3) are given in Figure 4.

$$\left.\begin{array}{c}\frac{1}{\sqrt{\lambda}} \approx -2 \cdot \log_{10}\left(a_3 + \frac{1}{3.71} \cdot \frac{\varepsilon}{D}\right) \\ a_3 \approx \frac{150.39}{R^{0.98865}} - \frac{152.66}{R}\end{array}\right\} \rightarrow \left.\begin{array}{c}\frac{1}{\sqrt{\lambda}} \approx -2.011 \cdot \log_{10}\left(A_3 + \frac{1}{3.71} \cdot \frac{\varepsilon}{D}\right) \\ A_3 \approx \frac{147.21}{R^{0.98865}} - \frac{149.243}{R}\end{array}\right\} \quad (8)$$

Brkić approximation (Appr. 4): Relevant parameters and errors related to the approximation by Brkić [20] after and before optimization (9) (Appr. 4) are given in Figure 5.

$$\left.\begin{array}{c}\frac{1}{\sqrt{\lambda}} \approx -2 \cdot \log_{10}\left(\frac{1.25603}{R \cdot \sqrt{a_4}} + \frac{1}{3.71} \cdot \frac{\varepsilon}{D}\right) \\ a_4 \approx \frac{-0.0015702}{\ln(R)} + \frac{0.3942031}{\ln^2(R)} + \frac{2.5341533}{\ln^3(R)}\end{array}\right\} \rightarrow \left.\begin{array}{c}\frac{1}{\sqrt{\lambda}} \approx -2.013 \cdot \log_{10}\left(\frac{1.216}{R \cdot \sqrt{A_4}} + \frac{1}{3.71} \cdot \frac{\varepsilon}{D}\right) \\ A_4 \approx \frac{-0.013}{\ln(R)} + \frac{0.383}{\ln^2(R)} + \frac{2.997}{\ln^3(R)}\end{array}\right\} \quad (9)$$

Fang et al. approximation (Appr. 5): Relevant parameters and errors related to the approximation by Fang et al. [21] after and before optimization (10) (Appr. 5) are given in Figure 6.

$$\left.\begin{array}{c}\frac{1}{\sqrt{\lambda}} \approx \left(1.613 \cdot \left(\ln\left(0.234 \cdot \left(\frac{\varepsilon}{D}\right)^{1.1007} - a_5\right)\right)^{-2}\right)^{-2} \\ a_5 \approx \frac{60.525}{R^{1.1105}} + \frac{56.291}{R^{1.0712}}\end{array}\right\} \rightarrow \left.\begin{array}{c}\frac{1}{\sqrt{\lambda}} \approx \left(1.61 \cdot \left(\ln\left(0.234 \cdot \left(\frac{\varepsilon}{D}\right)^{1.1007} - A_5\right)\right)^{-2}\right)^{-2} \\ A_5 \approx \frac{61.948}{R^{1.1105}} + \frac{57.449}{R^{1.0712}}\end{array}\right\} \quad (10)$$

Ghanbari, Farshad and Rieke approximation (Appr. 6): Relevant parameters and errors related to the approximation by Ghanbari et al. [22] after and before optimization (11) (Appr. 6) are given in Figure 7.

$$\left.\begin{array}{c}\frac{1}{\sqrt{\lambda}} \approx \left(\left(-1.52 \cdot \log_{10}(a_6)\right)^{-2.169}\right)^{-2} \\ a_6 \approx \left(\frac{1}{7.21} \cdot \frac{\varepsilon}{D}\right)^{1.042} + \left(\frac{2.731}{R}\right)^{0.9152}\end{array}\right\} \rightarrow \left\{\begin{array}{c}\frac{1}{\sqrt{\lambda}} \approx \left(\left(-1.606 \cdot \log_{10}(A_6)\right)^{-2.195}\right)^{-2} \\ A_6 \approx \left(\frac{1}{7.03} \cdot \frac{\varepsilon}{D}\right)^{0.967} + \left(\frac{2.629}{R}\right)^{0.858}\end{array}\right\} \quad (11)$$

Papaevangelou, Evangelides and Tzimopoulos approximation (Appr. 7): Relevant parameters and errors related to the approximation by Papaevangelou et al. [23] after and before optimization (12) (Appr. 7) are given in Figure 8.

$$\left.\begin{array}{c}\frac{1}{\sqrt{\lambda}} \approx \left(\frac{0.2479-0.0000947\cdot(a_7)^4}{\left(\log_{10}\left(\frac{1}{3.615}\cdot\frac{\varepsilon}{D}+\frac{7.366}{R^{0.9142}}\right)\right)^2}\right)^{-2} \\ a_7 \approx 7 - \log_{10}(R)\end{array}\right\} \rightarrow \left\{\begin{array}{c}\frac{1}{\sqrt{\lambda}} \approx \left(\frac{0.249-0.0000974\cdot|A_7|^{3.769}}{\left(\log_{10}\left(\frac{1}{3.646}\cdot\frac{\varepsilon}{D}+\frac{7.484}{R^{0.919}}\right)\right)^2}\right)^{-2} \\ A_7 \approx 7.122 - \log_{10}(R)\end{array}\right\} \quad (12)$$

Avci and Karagoz approximation (Appr. 8): Relevant parameters and errors related to the approximation by Avci and Karagoz [24] after and before optimization (13) (Appr. 8) are given in Figure 9.

$$\left.\begin{array}{c}\frac{1}{\sqrt{\lambda}} \approx \left(\frac{6.4}{\left(\ln(R)-\ln(1+0.01\cdot a_8)\right)^{2.4}}\right)^{-2} \\ a_8 \approx R \cdot \frac{\varepsilon}{D} \cdot \left(1 + 10 \cdot \sqrt{\frac{\varepsilon}{D}}\right)\end{array}\right\} \rightarrow \left\{\begin{array}{c}\frac{1}{\sqrt{\lambda}} \approx \left(\frac{6.264}{\left(\ln(R)-\ln(1+0.009\cdot A_8)\right)^{2.383}}\right)^{-2} \\ A_8 \approx R \cdot \frac{\varepsilon}{D} \cdot \left(1 + 10 \cdot \sqrt{\frac{\varepsilon}{D}}\right)\end{array}\right\} \quad (13)$$

Buzzelli approximation (Appr. 9): Relevant parameters and errors related to the approximation by Buzzelli [25] after and before optimization (14) (Appr. 9) are given in Figure 10.

$$\left.\begin{array}{c}\frac{1}{\sqrt{\lambda}} \approx a_9 - \left(\frac{a_9+2\cdot\log_{10}\left(\frac{a_{10}}{R}\right)}{1+\frac{2.18}{a_{10}}}\right) \\ a_9 \approx \frac{(0.774\cdot\ln(R))-1.41}{\left(1+1.32\cdot\sqrt{\frac{\varepsilon}{D}}\right)} \\ a_{10} \approx \frac{1}{3.7} \cdot \frac{\varepsilon}{D} \cdot R + 2.51 \cdot a_9\end{array}\right\} \rightarrow \left\{\begin{array}{c}\frac{1}{\sqrt{\lambda}} \approx A_9 - \left(\frac{A_9+1.9999\cdot\log_{10}\left(\frac{A_{10}}{R}\right)}{0.9996+\frac{2.1018}{A_{10}}}\right) \\ A_9 \approx \frac{(0.7314\cdot\ln(R))-1.3163}{\left(1.0025+1.2435\cdot\sqrt{\frac{\varepsilon}{D}}\right)} \\ A_{10} \approx \frac{1}{3.7165} \cdot \frac{\varepsilon}{D} \cdot R + 2.5137 \cdot A_9\end{array}\right\} \quad (14)$$

Sonnad and Goudar approximation (Appr. 10): Relevant parameters and errors related to the approximation by Sonnad and Goudar [26] after and before optimization (15) (Appr. 10) are given in Figure 11. Vatankhah and Kouchakzadeh [43,44] by changing parameter a_{11} to A_{11}-0.31 slightly change the model of Sonnad and Goudar [26]. They used line fitting tool for optimization. We failed with further optimization using genetic algorithms.

$$\left.\begin{array}{c}\frac{1}{\sqrt{\lambda}} \approx 0.8686 \cdot \ln\left(\frac{0.4587\cdot R}{a_{11}^{a_{12}}}\right) \\ a_{11} \approx 0.124 \cdot R \cdot \frac{\varepsilon}{D} + \ln(0.4587 \cdot R) \\ a_{12} \approx \frac{a_{11}}{a_{11}+1}\end{array}\right\} \rightarrow \left\{\begin{array}{c}\frac{1}{\sqrt{\lambda}} \approx 0.8686 \cdot \ln\left(\frac{0.4587\cdot R}{(A_{11}-0.31)^{A_{12}}}\right) \\ A_{11} \approx 0.124 \cdot R \cdot \frac{\varepsilon}{D} + \ln(0.4587 \cdot R) \\ A_{12} \approx \frac{A_{11}}{A_{11}+0.9633}\end{array}\right\} \quad (15)$$

Romeo, Royo and Monzón approximation (Appr. 11): Relevant parameters and errors related to the approximation by Romeo et al. [27] after and before optimization (16) (Appr. 11) are given in Figure 12; already shown in the form of the preliminary note in Ćojbašić and Brkić [42].

$$\left.\begin{array}{c}\frac{1}{\sqrt{\lambda}} \approx -2 \cdot \log_{10}\left(\frac{1}{3.7065}\cdot\frac{\varepsilon}{D} - \frac{5.0272}{R}\cdot a_{13}\right) \\ a_{13} \approx \log_{10}\left(\frac{1}{3.827}\cdot\frac{\varepsilon}{D} - \frac{4.567}{R}\cdot a_{14}\right) \\ a_{14} \approx \log_{10}\left(\left(\frac{1}{7.7918}\cdot\frac{\varepsilon}{D}\right)^{0.9924} + \left(\frac{5.3326}{208.815+R}\right)^{0.9345}\right)\end{array}\right\} \rightarrow \left\{\begin{array}{c}\frac{1}{\sqrt{\lambda}} \approx -2 \cdot \log_{10}\left(\frac{1}{3.7106}\cdot\frac{\varepsilon}{D} - \frac{5}{R}\cdot A_{13}\right) \\ A_{13} \approx \log_{10}\left(\frac{1}{3.8597}\cdot\frac{\varepsilon}{D} - \frac{4.795}{R}\cdot A_{14}\right) \\ A_{14} \approx \log_{10}\left(\left(\frac{1}{7.646}\cdot\frac{\varepsilon}{D}\right)^{0.9685} + \left(\frac{4.9755}{206.2795+R}\right)^{0.8759}\right)\end{array}\right\} \quad (16)$$

Manadilli approximation (Appr. 12): Relevant parameters and errors related to the approximation by Manadilli [28] after and before optimization (17) (Appr. 12) are given in Figure 13.

$$\left.\begin{array}{r}\frac{1}{\sqrt{\lambda}} \approx -2 \cdot \log_{10}\left(a_{15} + \frac{1}{3.7} \cdot \frac{\varepsilon}{D}\right) \\ a_{15} \approx \frac{95}{R^{0.983}} - \frac{96.82}{R}\end{array}\right\} \rightarrow \left.\begin{array}{r}\frac{1}{\sqrt{\lambda}} \approx -1.98 \cdot \log_{10}\left(A_{15} + \frac{1}{3.949} \cdot \frac{\varepsilon}{D}\right) \\ A_{15} \approx \frac{95.974}{R^{0.986}} - \frac{96.02}{R}\end{array}\right\} \quad (17)$$

Chen approximation (Appr. 13): Relevant parameters and errors related to the approximation by Chen [29] after and before optimization (18) (Appr. 13) are given in Figure 14.

$$\left.\begin{array}{r}\frac{1}{\sqrt{\lambda}} \approx \left(0.184 \cdot \left(a_{16} + 0.7 \cdot \frac{\varepsilon}{D}\right)^{0.3}\right)^{-2} \\ a_{16} \approx \frac{1}{R^{0.67}}\end{array}\right\} \rightarrow \left.\begin{array}{r}\frac{1}{\sqrt{\lambda}} \approx \left(0.208 \cdot \left(A_{16} + 0.697 \cdot \frac{\varepsilon}{D}\right)^{0.315}\right)^{-2} \\ A_{16} \approx \frac{0.321}{R^{0.541}}\end{array}\right\} \quad (18)$$

Serghides approximation (Appr. 14): Relevant parameters and errors related to the approximation by Serghides [30] after and before optimization (19) (Appr. 14) are given in Figure 15. This optimization is already shown in the form of the preliminary note in Ćojbašić and Brkić [42].

$$\left.\begin{array}{r}\frac{1}{\sqrt{\lambda}} \approx a_{17} - \frac{(a_{18}-a_{17})^2}{a_{19}-2\cdot a_{18}+a_{17}} \\ a_{17} \approx -2 \cdot \log_{10}\left(\frac{1}{3.7} \cdot \frac{\varepsilon}{D} + \frac{12}{R}\right) \\ a_{18} \approx -2 \cdot \log_{10}\left(\frac{1}{3.7} \cdot \frac{\varepsilon}{D} + \frac{2.51\cdot a_{17}}{R}\right) \\ a_{19} \approx -2 \cdot \log_{10}\left(\frac{1}{3.7} \cdot \frac{\varepsilon}{D} + \frac{2.51\cdot a_{18}}{R}\right)\end{array}\right\} \rightarrow \left.\begin{array}{r}\frac{1}{\sqrt{\lambda}} \approx A_{17} - \frac{(A_{18}-A_{17})^2}{A_{19}-2\cdot A_{18}+A_{17}} \\ A_{17} \approx -2 \cdot \log_{10}\left(\frac{1}{3.71} \cdot \frac{\varepsilon}{D} + \frac{12.585}{R}\right) \\ A_{18} \approx -2 \cdot \log_{10}\left(\frac{1}{3.71} \cdot \frac{\varepsilon}{D} + \frac{2.51\cdot A_{17}}{R}\right) \\ A_{19} \approx -2 \cdot \log_{10}\left(\frac{1}{3.71} \cdot \frac{\varepsilon}{D} + \frac{2.51\cdot A_{18}}{R}\right)\end{array}\right\} \quad (19)$$

Serghides approximation (simpler) (Appr. 15): Relevant parameters and errors related to the approximation by Serghides (simpler) [30] after and before optimization (20) (Appr. 15) are given in Figure 16.

$$\left.\begin{array}{r}\frac{1}{\sqrt{\lambda}} \approx 4.781 - \frac{(a_{20}-4.781)^2}{a_{21}-2\cdot a_{20}+4.781} \\ a_{20} \approx -2 \cdot \log_{10}\left(\frac{1}{3.7} \cdot \frac{\varepsilon}{D} + \frac{12}{R}\right) \\ a_{21} \approx -2 \cdot \log_{10}\left(\frac{1}{3.7} \cdot \frac{\varepsilon}{D} + \frac{2.51\cdot a_{20}}{R}\right)\end{array}\right\} \rightarrow \left.\begin{array}{r}\frac{1}{\sqrt{\lambda}} \approx 4.83 - \frac{(A_{20}-4.83)^2}{A_{21}-2\cdot A_{20}+4.83} \\ A_{20} \approx -2 \cdot \log_{10}\left(\frac{1}{3.71} \cdot \frac{\varepsilon}{D} + \frac{12.585}{R}\right) \\ A_{21} \approx -2 \cdot \log_{10}\left(\frac{1}{3.71} \cdot \frac{\varepsilon}{D} + \frac{2.51\cdot A_{20}}{R}\right)\end{array}\right\} \quad (20)$$

Haaland approximation (Appr. 16): Relevant parameters and errors related to the approximation by Haaland [31] after and before optimization (21) (Appr. 16) are given in Figure 17.

$$\left.\begin{array}{r}\frac{1}{\sqrt{\lambda}} \approx -1.8 \cdot \log_{10}\left(\frac{6.9}{R} + a_{22}\right) \\ a_{22} \approx \left(\frac{1}{3.7} \cdot \frac{\varepsilon}{D}\right)^{1.11}\end{array}\right\} \rightarrow \left.\begin{array}{r}\frac{1}{\sqrt{\lambda}} \approx -1.798 \cdot \log_{10}\left(\frac{6.891}{R} + A_{22}\right) \\ A_{22} \approx \left(\frac{1}{3.755} \cdot \frac{\varepsilon}{D}\right)^{1.106}\end{array}\right\} \quad (21)$$

Zigrang and Sylvester approximation (Appr. 17): Relevant parameters and errors related to the approximation by Zigrang and Sylvester [32] after and before optimization (22) (Appr. 17) are given in Figure 18.

$$\left.\begin{array}{r}\frac{1}{\sqrt{\lambda}} \approx -2 \cdot \log_{10}\left(\frac{1}{3.7} \cdot \frac{\varepsilon}{D} - \frac{5.02}{R} \cdot a_{23}\right) \\ a_{23} \approx \log_{10}\left(\frac{1}{3.7} \cdot \frac{\varepsilon}{D} - \frac{5.02}{R} \cdot a_{24}\right) \\ a_{24} \approx \log_{10}\left(\frac{1}{3.7} \cdot \frac{\varepsilon}{D} + \frac{13}{R}\right)\end{array}\right\} \rightarrow \left.\begin{array}{r}\frac{1}{\sqrt{\lambda}} \approx -2.0012 \cdot \log_{10}\left(\frac{1}{3.7027} \cdot \frac{\varepsilon}{D} - \frac{5.0605}{R} \cdot A_{23}\right) \\ A_{23} \approx \log_{10}\left(\frac{1}{3.7027} \cdot \frac{\varepsilon}{D} - \frac{5.0605}{R} \cdot A_{24}\right) \\ A_{24} \approx \log_{10}\left(\frac{1}{3.7027} \cdot \frac{\varepsilon}{D} + \frac{12.513}{R}\right)\end{array}\right\} \quad (22)$$

Zigrang and Sylvester approximation (simpler) (Appr. 18): Relevant parameters and errors related to the approximation by Zigrang and Sylvester (simpler) [32] after and before optimization (23) (Appr. 18) are given in Figure 19.

$$\left.\begin{aligned} \tfrac{1}{\sqrt{\lambda}} &\approx -2\cdot\log_{10}\left(\tfrac{1}{3.7}\cdot\tfrac{\varepsilon}{D}-\tfrac{5.02}{R}\cdot a_{25}\right) \\ a_{25} &\approx \log_{10}\left(\tfrac{1}{3.7}\cdot\tfrac{\varepsilon}{D}+\tfrac{13}{R}\right) \end{aligned}\right\}\rightarrow\left\{\begin{aligned} \tfrac{1}{\sqrt{\lambda}} &\approx -2.0012\cdot\log_{10}\left(\tfrac{1}{3.7027}\cdot\tfrac{\varepsilon}{D}-\tfrac{5.0605}{R}\cdot A_{25}\right) \\ A_{25} &\approx \log_{10}\left(\tfrac{1}{3.7027}\cdot\tfrac{\varepsilon}{D}+\tfrac{15.202}{R}\right) \end{aligned}\right\} \quad (23)$$

Barr approximation (Appr. 19): Relevant parameters and errors related to the approximation by Barr [33] after and before optimization (24) (Appr. 19) are given in Figure 20.

$$\left.\begin{aligned} \tfrac{1}{\sqrt{\lambda}} &\approx -2\cdot\log_{10}\left(\tfrac{4.518\cdot\log_{10}\left(\frac{R}{7}\right)}{a_{26}}+\tfrac{1}{3.7}\cdot\tfrac{\varepsilon}{D}\right) \\ a_{26} &\approx R\cdot\left(1+\tfrac{R^{0.52}}{29}\cdot\left(\tfrac{\varepsilon}{D}\right)^{0.7}\right) \end{aligned}\right\}\rightarrow\left\{\begin{aligned} \tfrac{1}{\sqrt{\lambda}} &\approx -1.998\cdot\log_{10}\left(\tfrac{4.509\cdot\log_{10}\left(\frac{R}{7.049}\right)}{A_{26}}+\tfrac{1}{3.737}\cdot\tfrac{\varepsilon}{D}\right) \\ A_{26} &\approx R\cdot\left(0.999+\tfrac{R^{0.525}}{28.102}\cdot\left(\tfrac{\varepsilon}{D}\right)^{0.721}\right) \end{aligned}\right\} \quad (24)$$

Round approximation (Appr. 20): Relevant parameters and errors related to the approximation by Round [34] after and before optimization (25) (Appr. 20) are given in Figure 21.

$$\left.\begin{aligned} \tfrac{1}{\sqrt{\lambda}} &\approx 1.8\cdot\log_{10}\left(\tfrac{R}{a_{27}+6.5}\right) \\ a_{27} &\approx 0.135\cdot R\cdot\tfrac{\varepsilon}{D} \end{aligned}\right\}\rightarrow\left\{\begin{aligned} \tfrac{1}{\sqrt{\lambda}} &\approx 1.898\cdot\log_{10}\left(\tfrac{R}{A_{27}+9.779}\right) \\ A_{27} &\approx 0.202\cdot R\cdot\tfrac{\varepsilon}{D} \end{aligned}\right\} \quad (25)$$

Chen approximation (Appr. 21) Relevant parameters and errors related to the approximation by Chen [36] after and before optimization (26) (Appr. 21) are given in Figure 22.

$$\left.\begin{aligned} \tfrac{1}{\sqrt{\lambda}} &\approx -2\cdot\log_{10}\left(\tfrac{1}{3.7065}\cdot\tfrac{\varepsilon}{D}-\tfrac{5.0452}{R}\cdot a_{28}\right) \\ a_{28} &\approx \log_{10}\left(\tfrac{1}{2.8257}\cdot\left(\tfrac{\varepsilon}{D}\right)^{1.1098}+\tfrac{5.8506}{R^{0.8981}}\right) \end{aligned}\right\}\rightarrow\left\{\begin{aligned} \tfrac{1}{\sqrt{\lambda}} &\approx -2.003\cdot\log_{10}\left(\tfrac{1}{3.689}\cdot\tfrac{\varepsilon}{D}-\tfrac{4.933}{R}\cdot A_{28}\right) \\ A_{28} &\approx \log_{10}\left(\tfrac{1}{2.762}\cdot\left(\tfrac{\varepsilon}{D}\right)^{1.109}+\tfrac{5.89}{R^{0.923}}\right) \end{aligned}\right\} \quad (26)$$

Swamee and Jain approximation (Appr. 22): Relevant parameters and errors related to the approximation by Swamee and Jain [37] after and before optimization (27) (Appr. 22) are given in Figure 23.

$$\left.\begin{aligned} \tfrac{1}{\sqrt{\lambda}} &\approx -2\cdot\log_{10}\left(\tfrac{5.74}{R^{0.9}}+a_{29}\right) \\ a_{29} &\approx \tfrac{1}{3.7}\cdot\tfrac{\varepsilon}{D} \end{aligned}\right\}\rightarrow\left\{\begin{aligned} \tfrac{1}{\sqrt{\lambda}} &\approx -1.972\cdot\log_{10}\left(\tfrac{5.828}{R^{0.916}}+A_{29}\right) \\ A_{29} &\approx \tfrac{1}{4.04}\cdot\tfrac{\varepsilon}{D} \end{aligned}\right\} \quad (27)$$

Eck approximation (Appr. 23): Relevant parameters and errors related to the approximation by Eck [38] after and before optimization (28) (Appr. 23) are given in Figure 24.

$$\left.\begin{aligned} \tfrac{1}{\sqrt{\lambda}} &\approx -2\cdot\log_{10}\left(\tfrac{15}{R}+a_{30}\right) \\ a_{30} &\approx \tfrac{1}{3.715}\cdot\tfrac{\varepsilon}{D} \end{aligned}\right\}\rightarrow\left\{\begin{aligned} \tfrac{1}{\sqrt{\lambda}} &\approx -1.963\cdot\log_{10}\left(\tfrac{14.064}{R}+A_{30}\right) \\ A_{30} &\approx \tfrac{1}{4.034}\cdot\tfrac{\varepsilon}{D} \end{aligned}\right\} \quad (28)$$

Wood approximation (Appr. 24): Relevant parameters and errors related to the approximation by Wood [39] after and before optimization (29) (Appr. 24) are given in Figure 25.

$$\left.\begin{aligned} \tfrac{1}{\sqrt{\lambda}} &\approx \left(a_{31}+88\cdot\left(\tfrac{\varepsilon}{D}\right)^{0.44}\cdot R^{-a_{32}}\right)^{-2} \\ a_{31} &= 0.094\cdot\left(\tfrac{\varepsilon}{D}\right)^{0.225}+0.53\cdot\tfrac{\varepsilon}{D} \\ a_{32} &\approx 1.62\cdot\left(\tfrac{\varepsilon}{D}\right)^{0.134} \end{aligned}\right\}\rightarrow\left\{\begin{aligned} \tfrac{1}{\sqrt{\lambda}} &\approx \left(A_{31}+85.005\cdot\left(\tfrac{\varepsilon}{D}\right)^{0.33}\cdot R^{-A_{32}}\right)^{-2} \\ A_{31} &= 0.094\cdot\left(\tfrac{\varepsilon}{D}\right)^{0.209}+0.376\cdot\tfrac{\varepsilon}{D} \\ A_{32} &\approx 1.501\cdot\left(\tfrac{\varepsilon}{D}\right)^{0.101} \end{aligned}\right\} \quad (29)$$

Moody approximation (Appr. 25): Relevant parameters and errors related to the approximation by Moody [40] after and before optimization (30) (Appr. 25) are given in Figure 26.

$$\left.\begin{aligned} \tfrac{1}{\sqrt{\lambda}} &\approx \left(0.0055\cdot\left(1+a_{33}\right)\right)^{-2} \\ a_{33} &\approx \left(2\cdot 10^{4}\cdot\tfrac{\varepsilon}{D}+\tfrac{10^{6}}{R}\right)^{0.333} \end{aligned}\right\}\rightarrow\left\{\begin{aligned} \tfrac{1}{\sqrt{\lambda}} &\approx \left(0.006\cdot\left(0.775+A_{33}\right)\right)^{-2} \\ A_{33} &\approx \left(2.443\cdot 10^{4}\cdot\tfrac{\varepsilon}{D}+\tfrac{10^{6}}{R}\right)^{0.343} \end{aligned}\right\} \quad (30)$$

4. Conclusions

Today, Colebrook's equation is mostly accepted as an informal standard for the modeling of turbulent flow in hydraulically smooth and rough pipes, including the transient zone in between. Of course, approximations carry certain error compared with the iterative solution where the highest level of accuracy can be reached after enough iterations. The explicit approximations give a relatively good prediction of the friction factor (λ) and can reproduce accurately Colebrook's equation and its Moody's plot. Usually, more complex models of approximations are more accurate and vice versa. Using genetic algorithms in order to increase the accuracy of available approximations of the Colebrook equation for flow friction, the numerical values of empirical parameters in 25 existing models of approximations are changed while the computational burden remains the same. Using the value of decreased maximal relative error, $\Delta\delta$, and the change of relative error over the entire domain of the Reynolds number (R) and the relative roughness of inner pipe surface (ε/D), the success of genetic optimization is summarized in Table 1.

Table 1. Maximal relative error of the explicit approximations of the Colebrook–White equation before and after genetic optimization.

Approximation No.	With Original Parameters	After Genetic Optimization	Estimation of Improvement	Source
Appr. 11; Equation (16)	0.1345%	0.0083%	extremely successful	Romeo et al. [27,42]
Appr. 14; Equation (19)	0.1385%	0.0026%	extremely successful	Serghides [30,42]
Appr. 10; Equation (15)	0.8007%	0.1473%	successful	Sonnad and Goudar [26,43]
Appr. 2; Equation (7)	3.1560%	1.2871%	successful	Brkić [19]
Appr. 9; Equation (14)	0.1385%	0.0797%	successful	Buzzelli [25]
Appr. 15; Equation (20)	0.3543%	0.2739%	successful	Serghides [30]
Appr. 17; Equation (22)	0.1385%	0.0831%	successful	Zigrang and Sylvester [32]
Appr. 18; Equation (23)	1.0075%	0.7496%	successful	Zigrang and Sylvester [32]
Appr. 20; Equation (25)	10.9183%	5.5094%	successful	Round [34]
Appr. 21; Equation (26)	0.3649%	0.1851%	successful	Chen [36]
Appr. 1; Equation (6)	2.2065%	1.2868%	moderately successful	Brkić [19]
Appr. 3; Equation (8)	2.0715%	1.3326%	moderately successful	Brkić [20]
Appr. 4; Equation (9)	2.0111%	1.2866%	moderately successful	Brkić [20]
Appr. 5; Equation (10)	0.6167%	0.5669%	moderately successful	Fang et al. [38]
Appr. 12; Equation (17)	2.0651%	1.5018%	moderately successful	Manadilli [28]
Appr. 13; Equation (18)	27.5074%	18.4800%	moderately successful	Chen [29]
Appr. 16; Equation (21)	1.4083%	1.1098%	moderately successful	Haaland [31]
Appr. 22; Equation (27)	2.1872%	1.7535%	moderately successful	Swamee and Jain [37]
Appr. 23; Equation (28)	8.1953%	5.6955%	moderately successful	Eck [21]
Appr. 6; Equation (11)	2.8962%	2.5947%	not very successful	Ghanbari et al. [22]
Appr. 7; Equation (12)	0.8248%	0.7312%	not very successful	Papaevangelou et al. [23]
Appr. 8; Equation (13)	4.7858%	3.1259%	not very successful	Avci and Karagoz [24]
Appr. 19; Equation (24)	0.2774%	0.2644%	not very successful	Barr [33]
Appr. 24; Equation (29)	23.7204%	16.5910%	not very successful	Wood [39]
Appr. 25; Equation (30)	21.4855%	18.1024%	not very successful	Moody [40]

Since the main idea of this study was to use metaheuristic optimization to improve the accuracy of a wider set of Colebrook's turbulent flow friction approximations, as opposed to other approaches used by other authors and ourselves, Genetic Algorithms were selected as being referential among global optimization techniques. Additional accuracy for the certain approximations can be possibly reached through rearrangement of their structure (such as was done for Appr. 10 [26,43]), using the MS Excel fitting tool [49], etc. Furthermore, some further use of genetic algorithms can be encouraged. Future research directions could include also the application of other metaheuristic optimization techniques for the same task, such as particle swarm optimization (PSO), ant colony optimization (ACO), simulated annealing (SA) and others [69–71]. Further accuracy improvements might be possible, but the highest accuracy approximations have already been optimized to extremely high precision levels regarding practical application.

During this study, it is found that the criterion from Winning and Coole [16] about the accuracy of approximations using the value of mean square error should be modified as: very small error is lower than 10^{-10}; small is between 10^{-10} and 10^{-8}; medium is between 10^{-8} and 5×10^{-7}; and large is above 5×10^{-7}. The criterion of accuracy using the value of maximal relative error δ_{max} should be set as: very small error is lower than 0.2%; small is between 0.2% and 1%; medium is between 1% and 3%; and large is above 3% (extremely large above 5%). Furthermore, it is found that the error distribution, set as a criterion in Winning and Coole [16], does not depend only on the model of approximation, but changes equally with the change of the values of the parameters.

Aside from the Colebrook equation, the presented methodology can be used to fit the raw and updated measured data, all similar empirical equations that cover the same region of turbulent flow [54,55]. The friction factor curves derived from the Colebrook equation are said to be monotonic, i.e., the friction factor (λ) decreases continuously with increasing Reynolds number (R). For some tests carried out on pipes that were artificially roughened with grains of sand, the curves were inflectional in nature, i.e., the friction factor (λ) decreases to a minimum value with increasing Reynolds number (R) and then rises again to reach a constant value for complete turbulence [53,54]. The proposed optimization procedure can be used also to fit such data.

The results are relevant for all engineering fields that deal with fluid flow through pipes [72–74] and the related calculation of hydraulic flow friction.

Supplementary Materials: Excel and MATLAB codes of the approximations presented in this paper are available online at www.mdpi.com/2311-5521/2/2/15/s1.

Acknowledgments: The work of Žarko Ćojbašić has been supported by the Ministry of Education, Science and Technological Development of the Republic of Serbia under Grants TR35016 and TR35005.

Author Contributions: Dejan Brkić has scientific interest in hydraulics, and he prepared data related to flow friction, checked the accuracy of the final results, drew diagrams and wrote the manuscript. Žarko Ćojbašić has scientific interest in artificial intelligence, and he performed optimization using genetic algorithms. Both authors contributed equally to this study (the authors are listed alphabetically according to their surnames).

Conflicts of Interest: The views expressed are those of the authors and may not in any circumstances be regarded as stating an official position of the European Commission or the University of Niš.

References

1. Colebrook, C.F.; White, C.M. Experiments with fluid friction in roughened pipes. *Proc. R. Soc. Ser. A Math. Phys. Sci.* **1937**, *161*, 367–381. [CrossRef]
2. Colebrook, C.F. Turbulent flow in pipes with particular reference to the transition region between the smooth and rough pipe laws. *J. Inst. Civ. Eng. (Lond.)* **1939**, *11*, 133–156. [CrossRef]
3. Moody, L.F. Friction factors for pipe flow. *Trans. ASME* **1944**, *66*, 671–684.
4. LaViolette, M. On the history, science, and technology included in the Moody diagram. *J. Fluids Eng. ASME* **2017**, *139*, 030801. [CrossRef]
5. Mikata, Y.; Walczak, W.S. Exact analytical solutions of the Colebrook-White equation. *J. Hydraul. Eng. ASCE* **2016**, *142*. [CrossRef]
6. Brkić, D. W solutions of the CW equation for flow friction. *Appl. Math. Lett.* **2011**, *24*, 1379–1383. [CrossRef]
7. Biberg, D. Fast and accurate approximations for the Colebrook equation. *J. Fluids Eng. ASME* **2017**, *139*, 031401. [CrossRef]
8. Keady, G. Colebrook-White formula for pipe flow. *J. Hydraul. Eng. ASCE* **1998**, *124*, 96–97. [CrossRef]
9. Clamond, D. Efficient resolution of the Colebrook equation. *Ind. Eng. Chem. Res.* **2009**, *48*, 3665–3671. [CrossRef]

10. Brkić, D. Review of explicit approximations to the Colebrook relation for flow friction. *J. Petrol. Sci. Eng.* **2011**, *77*, 34–48. [CrossRef]
11. Brkić, D. Determining friction factors in turbulent pipe flow. *Chem. Eng. (N. Y.)* **2012**, *119*, 34–39.
12. Gregory, G.A.; Fogarasi, M. Alternate to standard friction factor equation. *Oil Gas J.* **1985**, *83*, 125–127.
13. Zigrang, D.J.; Sylvester, N.D. A review of explicit friction factor equations. *J. Energy Resour. Technol. ASME* **1985**, *107*, 280–283. [CrossRef]
14. Yıldırım, G. Computer-based analysis of explicit approximations to the implicit Colebrook-White equation in turbulent flow friction factor calculation. *Adv. Eng. Softw.* **2009**, *40*, 1183–1190. [CrossRef]
15. Genić, S.; Aranđelović, I.; Kolendić, P.; Jarić, M.; Budimir, N.; Genić, V. A review of explicit approximations of Colebrook's equation. *FME Trans.* **2011**, *39*, 67–71.
16. Winning, H.K.; Coole, T. Explicit friction factor accuracy and computational efficiency for turbulent flow in pipes. *Flow Turbul. Combust.* **2013**, *90*, 1–27. [CrossRef]
17. Lira, I. On the uncertainties stemming from use of the Colebrook-White equation. *Ind. Eng. Chem. Res.* **2013**, *52*, 7550–7555. [CrossRef]
18. Giustolisi, O.; Berardi, L.; Walski, T.M. Some explicit formulations of Colebrook–White friction factor considering accuracy vs. computational speed. *J. Hydroinformatics* **2011**, *13*, 401–418. [CrossRef]
19. Brkić, D. An explicit approximation of the Colebrook equation for fluid flow friction factor. *Petrol. Sci. Technol.* **2011**, *29*, 1596–1602. [CrossRef]
20. Brkić, D. New explicit correlations for turbulent flow friction factor. *Nucl. Eng. Des.* **2011**, *241*, 4055–4059. [CrossRef]
21. Fang, X.; Xu, Y.; Zhou, Z. New correlations of single-phase friction factor for turbulent pipe flow and evaluation of existing single-phase friction factor correlations. *Nucl. Eng. Des.* **2011**, *241*, 897–902. [CrossRef]
22. Ghanbari, A.; Farshad, F.F.; Rieke, H.H. Newly developed friction factor correlation for pipe flow and flow assurance. *J. Chem. Eng. Mater. Sci.* **2011**, *2*, 83–86.
23. Papaevangelou, G.; Evangelides, C.; Tzimopoulos, C. A new explicit relation for the friction factor coefficient in the Darcy–Weisbach equation. In Proceedings of the Protection and Restoration of the Environment, Corfu, Greece, 5–9 July 2010; pp. 166–172.
24. Avci, A.; Karagoz, I. A novel explicit equation for friction factor in smooth and rough pipes. *J. Fluids Eng. ASME* **2009**, *131*, 061203. [CrossRef]
25. Buzzelli, D. Calculating friction in one step. *Mach. Des.* **2008**, *80*, 54–55.
26. Sonnad, J.R.; Goudar, C.T. Turbulent flow friction factor calculation using a mathematically exact alternative to the Colebrook–White equation. *J. Hydraul. Eng. ASCE* **2006**, *132*, 863–867. [CrossRef]
27. Romeo, E.; Royo, C.; Monzón, A. Improved explicit equations for estimation of the friction factor in rough and smooth pipes. *Chem. Eng. J.* **2002**, *86*, 369–374. [CrossRef]
28. Manadilli, G. Replace implicit equations with signomial functions. *Chem. Eng. (N. Y.)* **1997**, *104*, 129–130.
29. Chen, J.J.J. A simple explicit formula for the estimation of pipe friction factor. *Proc. Inst. Civ. Eng.* **1984**, *77*, 49–55. [CrossRef]
30. Serghides, T.K. Estimate friction factor accurately. *Chem. Eng. (N. Y.)* **1984**, *91*, 63–64.
31. Haaland, S.E. Simple and explicit formulas for the friction factor in turbulent pipe flow. *J. Fluids Eng. ASME* **1983**, *105*, 89–90. [CrossRef]
32. Zigrang, D.J.; Sylvester, N.D. Explicit approximations to the solution of Colebrook's friction factor equation. *AIChE J.* **1982**, *28*, 514–515. [CrossRef]
33. Barr, D.I.H. Solutions of the Colebrook-White function for resistance to uniform turbulent flow. *Proc. Inst. Civ. Eng.* **1981**, *71*, 529–535.
34. Round, G.F. An explicit approximation for the friction factor-Reynolds number relation for rough and smooth pipes. *Can. J. Chem. Eng.* **1980**, *58*, 122–123. [CrossRef]
35. Schorle, B.J.; Churchill, S.W.; Shacham, M. Comments on: "An explicit equation for friction factor in pipe". *Ind. Eng. Chem. Fundam.* **1980**, *19*, 228–230. [CrossRef]
36. Chen, N.H. An explicit equation for friction factor in pipes. *Ind. Eng. Chem. Fundam.* **1979**, *18*, 296–297. [CrossRef]
37. Swamee, D.K.; Jain, A.K. Explicit equations for pipe flow problems. *J. Hydraul. Div. ASCE* **1976**, *102*, 657–664.
38. Eck, B. *Technische Stromungslehre*; Springer: New York, NY, USA, 1973.
39. Wood, D.J. An explicit friction factor relationship. *Civ. Eng.* **1966**, *36*, 60–61.

40. Moody, L.F. An approximate formula for pipe friction factors. *Trans. ASME* **1947**, *69*, 1005–1006.
41. Brkić, D. A note on explicit approximations to Colebrook's friction factor in rough pipes under highly turbulent cases. *Int. J. Heat Mass Tran.* **2016**, *93*, 513–515. [CrossRef]
42. Ćojbašić, Ž.; Brkić, D. Very accurate explicit approximations for calculation of the Colebrook friction factor. *Int. J. Mech. Sci.* **2013**, *67*, 10–13. [CrossRef]
43. Vatankhah, A.R.; Kouchakzadeh, S. Discussion of Turbulent flow friction factor calculation using a mathematically exact alternative to the Colebrook-White equation. *J. Hydraul. Eng. ASCE* **2008**, *134*, 1187. [CrossRef]
44. Vatankhah, A.R.; Kouchakzadeh, S. Discussion: Exact equations for pipe-flow problems. *J. Hydraul. Res. IAHR* **2009**, *47*, 537–538. [CrossRef]
45. Goldberg, D.E. *Genetic Algorithms in Search, Optimization and Machine Learning*; Addison-Wesley Inc.: Reston, VA, USA, 1989.
46. Fleming, P.J.; Purshouse, R.C. Evolutionary algorithms in control systems engineering: A survey. *Control Eng. Pract.* **2002**, *10*, 1223–1241. [CrossRef]
47. Samadianfard, S. Gene expression programming analysis of implicit Colebrook-White equation in turbulent flow friction factor calculation. *J. Petrol. Sci. Eng.* **2012**, *92–93*, 48–55. [CrossRef]
48. Brkić, D. Discussion of "Gene expression programming analysis of implicit Colebrook–White equation in turbulent flow friction factor calculation". *J. Petrol. Sci. Eng.* **2014**, *124*, 399–401. [CrossRef]
49. Vatankhah, A.R. Comment on "Gene expression programming analysis of implicit Colebrook–White equation in turbulent flow friction factor calculation". *J. Petrol. Sci. Eng.* **2014**, *124*, 402–405. [CrossRef]
50. Brkić, D.; Ćojbašić, Ž. Intelligent flow friction estimation. *Comput. Intell. Neurosci.* **2016**, *2016*, 5242596. [CrossRef] [PubMed]
51. Ćojbašić, Ž.; Nikolić, V.; Petrović, E.; Pavlović, V.; Tomić, M.; Pavlović, I.; Ćirić, I. A real time neural network based finite element analysis of shell structure. *Facta Univ. Mech. Eng.* **2014**, *12*, 149–155.
52. Dučić, N.; Ćojbašić, Ž.; Radiša, R.; Slavković, R.; Milićević, M. CAD/CAM design and genetic optimization of feeders for sand casting process. *Facta Univ. Mech. Eng.* **2016**, *14*, 147–158.
53. Brkić, D. Efficiency of Distribution and Use of Natural Gas in Households (Ефикасност дистрибуције и коришћења природног гаса у домаћинствима, In Serbian). Ph.D. Thesis, University of Belgrade, Belgrade, Serbia, 2010.
54. Allen, J.J.; Shockling, M.A.; Kunkel, G.J.; Smits, A.J. Turbulent flow in smooth and rough pipes. *Proc. R. Soc. Ser. A Math. Phys. Sci.* **2007**, *365*, 699–714. [CrossRef] [PubMed]
55. Brkić, D. A gas distribution network hydraulic problem from practice. *Petrol. Sci. Technol.* **2011**, *29*, 366–377. [CrossRef]
56. Brkić, D. Can pipes be actually really that smooth? *Int. J. Refrig.* **2012**, *35*, 209–215. [CrossRef]
57. Brkić, D. Discussion of "Jacobian matrix for solving water distribution system equations with the Darcy-Weisbach head-loss model". *J. Hydraul. Eng. ASCE* **2012**, *138*, 1000–1002. [CrossRef]
58. Brkić, D. Discussion of "Water distribution system analysis: Newton-Raphson method revisited". *J. Hydraul. Eng. ASCE* **2012**, *138*, 822–824. [CrossRef]
59. Brkić, D. Discussion of "Method to cope with zero flows in newton solvers for water distribution systems". *J. Hydraul. Eng. ASCE* **2014**, *140*, 07014003. [CrossRef]
60. Sonnad, J.R.; Goudar, C.T. Constraints for using Lambert W function-based explicit Colebrook–White equation. *J. Hydraul. Eng. ASCE* **2004**, *130*, 929–931. [CrossRef]
61. Brkić, D. Comparison of the Lambert W-function based solutions to the Colebrook equation. *Eng. Comput.* **2012**, *29*, 617–630. [CrossRef]
62. Rollmann, P.; Spindler, K. Explicit representation of the implicit Colebrook–White equation. *Case Stud. Therm. Eng.* **2015**, *5*, 41–47. [CrossRef]
63. Brkić, D. Spreadsheet-based pipe networks analysis for teaching and learning purpose. *Spreadsheets Educ. (eJSiE)* **2016**, *9*. Available online: http://epublications.bond.edu.au/ejsie/vol9/iss2/4/ (accessed on 3 April 2017).
64. Brkić, D. Iterative methods for looped network pipeline calculation. *Water Resour. Manag.* **2011**, *25*, 2951–2987. [CrossRef]
65. Simpson, A.; Elhay, S. Jacobian matrix for solving water distribution system equations with the Darcy-Weisbach head-loss model. *J. Hydraul. Eng. ASCE* **2011**, *137*, 696–700. [CrossRef]

66. Spiliotis, M.; Tsakiris, G. Water distribution system analysis: Newton-Raphson method revisited. *J. Hydraul. Eng. ASCE* **2011**, *137*, 852–855. [CrossRef]
67. Brkić, D. An improvement of Hardy Cross method applied on looped spatial natural gas distribution networks. *Appl. Energy* **2009**, *86*, 1290–1300. [CrossRef]
68. Brkić, D.; Tanasković, T. Systematic approach to natural gas usage for domestic heating in urban areas. *Energy* **2008**, *33*, 1738–1753. [CrossRef]
69. Ćojbašić, Ž.; Petković, D.; Shamshirband, S.; Tong, C.W.; Ch, S.; Janković, P.; Dučić, N.; Baralić, J. Surface roughness prediction by extreme learning machine constructed with abrasive water jet. *Precis Eng.* **2016**, *43*, 86–92. [CrossRef]
70. Shamshirband, S.; Petković, D.; Saboohi, H.; Anuar, N.B.; Inayat, I.; Akib, S.; Ćojbašić, Ž.; Nikolić, V.; Kiah, M.L.M.; Gani, A. Wind turbine power coefficient estimation by soft computing methodologies: Comparative study. *Energy Convers. Manag.* **2014**, *81*, 520–526. [CrossRef]
71. Ćojbašić, Ž.; Nikolić, V.D.; Ćirić, I.; Grigorescu, S. Advanced evolutionary optimization for intelligent modeling and control of FBC process. *Facta Univ. Ser. Mech. Eng.* **2010**, *8*, 47–56.
72. Cross, H. *Analysis of Flow in Networks of Conduits or Conductors*; University of Illinois at Urbana Champaign: Champaign, IL, USA, 1936; Volume 34, pp. 3–29.
73. Praks, P.; Kopustinskas, V.; Masera, M. Probabilistic modelling of security of supply in gas networks and evaluation of new infrastructure. *Reliab. Eng. Sys. Saf.* **2015**, *144*, 254–264. [CrossRef]
74. Pambour, K.A.; Bolado, L.R.; Dijkema, G.P.J. An integrated transient model for simulating the operation of natural gas transport systems. *J. Nat. Gas. Sci. Eng.* **2016**, *28*, 672–690. [CrossRef]

fluids

MDPI

Review

A Review of Time Relaxation Methods

Sean Breckling, Monika Neda * and Tahj Hill

Department of Mathematical Sciences, University of Nevada Las Vegas, Las Vegas, NV 89154, USA; sean.breckling@unlv.edu (S.B.); hillt12@unlv.nevada.edu (T.H.)

* Correspondence: monika.neda@unlv.edu; Tel.: +1-702-895-5170

Received: 20 June 2017; Accepted: 10 July 2017; Published: 17 July 2017

Abstract: The time relaxation model has proven to be effective in regularization of Navier–Stokes Equations. This article reviews several published works discussing the development and implementations of time relaxation and time relaxation models (TRMs), and how such techniques are used to improve the accuracy and stability of fluid flow problems with higher Reynolds numbers. Several analyses and computational settings of TRMs are surveyed, along with parameter sensitivity studies and hybrid implementations of time relaxation operators with different regularization techniques.

Keywords: time-relaxation; finite element; filtering; deconvolution

---

1. Introduction

Time relaxation models (TRM) are a novel class of regularizations of the Navier–Stokes Equations (NSE) formulated by *nudging* solutions of the NSE toward a spatially-regularized solution. This is accomplished by adding a *relaxation* term to the momentum equations of the NSE, i.e.,

$$\mathbf{v}_t - \nu\triangle\mathbf{v} + \mathbf{v}\cdot\nabla\mathbf{v} + \chi\mathbf{v}' + \nabla p = \mathbf{f}, \tag{1}$$

where $\mathbf{v}$ is fluid velocity, p is pressure, ν is kinematic viscosity, and $\mathbf{f}$ accounts for external forcing. The scaling parameter $\chi > 0$ has the units $[time]^{-1}$. For our purposes, the relaxation term $\chi\mathbf{v}'$ is constructed to exclusively dissipate (through time) scales of motion beneath a user-selected threshold.

In a general sense, the notion of nudging was first seen in [1,2], though instead of dissipating small velocity scales, the authors nudged computed velocity solutions toward recorded atmospheric data. If we momentarily let $\mathbf{v}' = (\mathbf{v} - \mathbf{v}_{data})$, where $\mathbf{v}_{data}$ is appropriately-interpolated real-world data while $\mathbf{v}$ is sufficiently smooth, the simplified ODE,

$$\mathbf{v}_t' + \chi(\mathbf{v} - \mathbf{v}_{data}) = \mathbf{0} \tag{2}$$

shows that for some constant C,

$$||\mathbf{v}'|| \leq Ce^{-\chi t}. \tag{3}$$

Over an extended time scale it becomes apparent that in this context, computed solutions to a model like (1) move toward $\mathbf{v}_{data}$. The rate and extent that this occurs is wholly determined by the problem settings, $\mathbf{v}_{data}$, and the relaxation parameter χ.

It is well understood that the principal difficulty in accurately resolving turbulent fluid flow is one of scale. In viscous, incompressible flow in three dimensions, K-41 theory suggests that, given a Reynolds number Re, a direct numerical simulation of the NSE requires spatial discretization resulting in $\mathcal{O}(Re^{9/4})$ degrees of freedom [3]. Such calculations can become challenging, and often unfeasible for practical problems at even modest Reynolds numbers. While some practitioners might be tempted to side-step the difficulty by simply using a coarse discretization, under-resolving the dissipation scale of motion can lead to unacceptable error. As a result, several techniques have been developed to relax these difficult discretization requirements.

The TRMs discussed in this paper were devised to address this difficulty by nudging a carefully under-resolved numerical approximation to the NSE toward a spatially filtered solution. This is a practical solution since the larger, dominant scales of motion contain the bulk of a flow's kinetic energy, and account for most of the momentum transport. Hence, it becomes possible to accurately resolve the largest scales of motion without the burden of resolving the smallest.

Broadly speaking, regularization terms in time relaxation models are often formulated:

$$\chi \mathbf{v}' = \chi(\mathbf{v} - G * \mathbf{v}), \tag{4}$$

where G is a smoothing kernel selected to act as a low-pass spatial filter for the fluid velocity. Hence, including such a term in (1) acts to nudge computed solutions toward the spatially-filtered velocity. The rate and scale at which this occurs is determined by χ, and the filtering process G.

The goal of this article is to effectively summarize the body of research pertaining to TRM models, and the results of their implementations in computational settings. We begin in Section 2 where we briefly introduce basic notational and function spaces, as well as the differential filtering used in TRMs and the higher-order approximate deconvolution operators. We then introduce the time relaxation techniques seen in [4–6] in Section 3. We present and briefly discuss linear and nonlinear forms of time relaxation terms, and their effects on the transfer of kinetic energy from large scales of motion to small. This discussion continues in Section 4 where a brief survey of implementations utilizing finite element methods (FEM) is included, [7–13]. We will continue surveying these results by comparing the performance of the several TRM formulations and FEM techniques on the benchmark flow past a full step problem. Section 5 presents a summary of a sensitivity analysis seen in [13]. We conclude in Section 6 with a brief summary, and a discussion of open problems.

2. Notation and Preliminaries

Throughout this article and the papers referenced, the common function spaces and norms considered are,

$$\mathbf{X} = \left(H_0^1(\Omega)\right)^d := \{\mathbf{v} \in \left(H^1(\Omega)\right)^d \,:\, \mathbf{v}|_{\partial\Omega} = 0\},$$

$$Q = L_0^2(\Omega) = \left\{q \in L^2(\Omega) \,:\, \int_\Omega q \,\mathrm{d}x = 0\right\},$$

$$\mathbf{V} := \{\mathbf{v} \in \mathbf{X} \,:\, \nabla \cdot \mathbf{v} = 0\},$$

where Q contains all pressure functions of finite energy $||p||$, and $\mathbf{X}$ contains all fluid velocities $\mathbf{v}$ with finite kinetic energy $||\mathbf{v}||$, and finite rate of energy dissipation $||\nabla \mathbf{v}||$. The space $\mathbf{V}$ is required given the added restriction on incompressible flows that $\mathbf{v}$ be divergence-free.

It is often necessary in convergence analyses to impose significant smoothness assumptions on the true, weak solutions to the incompressible NSE. Typically, we require these solutions to be bounded, and to weakly admit several spatial derivatives. For instance, the sufficient condition of Lemma 1 requires the true solution to be $L^4(0,T)$ in time, and $W^{1,4}$ in space where,

$$W^{k,p} := \left\{\mathbf{v} \in (L^p(\Omega))^d \,:\, \forall|\alpha| < k,\, \partial_{x_i}^{\alpha}\mathbf{v} \in (L^p(\Omega))^d \,\forall\, i \in [1,\ldots,d]\right\}.$$

When discussing computations, h represents a minimal mesh scale, and $\mathbf{X}_h$, Q_h, and $\mathbf{V}_h$ represent the discrete counterparts to $\mathbf{X}$, Q and $\mathbf{V}$, respectively. There have been a variety of spatial discretizations of TRMs, and in an effort to simplify the notation in this article we carry this notation between these discussions.

2.1. Differential Filtering

The notion of a differential smoothing filter was introduced by Germano [14].

Definition 1 (Continuous Differential Filter). *Given $\boldsymbol{\phi} \in \left(L^2(\Omega)\right)^d$ and a given filtering radius of $\delta > 0$, we define the filtering of $\boldsymbol{\phi}$ as $\overline{\boldsymbol{\phi}}$, where $\overline{\boldsymbol{\phi}}$ is the solution to the PDE.*

$$-\delta^2 \Delta \overline{\boldsymbol{\phi}} + \overline{\boldsymbol{\phi}} = \boldsymbol{\phi} \text{ on } \Omega, \tag{5}$$

$$\overline{\boldsymbol{\phi}} = \boldsymbol{\phi} \text{ on } \partial\Omega. \tag{6}$$

We can infer from Definition 1 that on a suitable domain Ω, given a particular quantity $\boldsymbol{\phi} \in L^2(\Omega)$, the low-pass filtered quantity $\overline{\boldsymbol{\phi}}$ is the unique $H_0^1(\Omega)$ solution to the equation $\overline{\boldsymbol{\phi}} = F\boldsymbol{\phi}$ where $F = (I - \delta^2 \triangle)^{-1}$.

The spatial filtering process behaves as local averaging. In [14], it is shown that in 3D, the Green's function for the PDE defined above is a smoothing kernel. This differs from other averaging techniques like long-time averaging, as features present beneath the spatial filter length scale will certainly be removed. In contrast, in long-time averaging persistent flow, features will remain if they are steady in time, regardless of spatial scale.

It is clear that repeated applications of the differential spatial filter will result in added truncation of sub filter-length scales. This is important in approximate deconvolution techniques, which are discussed in the following section. For notational simplicity, as it becomes necessary to filter quantities multiple times, let the number of over-bars account for this information such that if $\boldsymbol{\phi}$ is filtered two times, this is written as $\overline{\overline{\boldsymbol{\phi}}}$, and if it is filtered three times this is shown as $\overline{\overline{\overline{\boldsymbol{\phi}}}}$.

2.2. Approximate Deconvolution

Deconvolution operators are often presented as the pseudo-inverse of a particular convolution operator. Thus, as its name would suggest, approximate deconvolution operators approximate such deconvolution operators. The deconvolution problem can be stated by first letting $\boldsymbol{\phi} \in L^2(\Omega)^d$ be a function over a suitable domain Ω. Let $\overline{\boldsymbol{\phi}}$ be obtained by filtering $\boldsymbol{\phi}$ by the differential filter in Definition 1. Let F denote the filtering operator. The deconvolution problem, as discussed in [15], becomes,

$$\text{Given } \overline{\boldsymbol{\phi}}, \text{ solve } F\boldsymbol{\phi} = \overline{\boldsymbol{\phi}} \text{ for } \boldsymbol{\phi}. \tag{7}$$

This approach approximates a deconvolution operator D through fixed point iteration. Due to its ease of implementation, to date, TRM models have exclusively utilized the van Cittert approximate deconvolution algorithms, which are described below. Alternative techniques do exist, and are proposed as open problems in Section 6.

Algorithm 1 (van Cittert Approximate Deconvolution). *Given a filter F, a filtered function $\overline{\boldsymbol{\phi}}$, assign $\boldsymbol{\phi}_0 = \overline{\boldsymbol{\phi}}$, then for $n = 0, 1, 2, \ldots, N-1$ perform the fixed-point iteration,*

$$\boldsymbol{\phi}_{n+1} = \boldsymbol{\phi}_n + (\overline{\boldsymbol{\phi}} - F\boldsymbol{\phi}_n).$$

We then call $\boldsymbol{\phi}_N := D_N\overline{\boldsymbol{\phi}}$.

This algorithm's formulation can be simplified slightly, such that, given an order of deconvolution N,

$$D_N\overline{\boldsymbol{\phi}} := \sum_{n=0}^{N} (I - F)^n \overline{\boldsymbol{\phi}}. \tag{8}$$

Hence, the added precision is obtained through added extrapolation. The first few approximate-deconvolution operators can be written as,

$$\begin{aligned} D_0\overline{\boldsymbol{\phi}} &= \overline{\boldsymbol{\phi}}, \\ D_1\overline{\boldsymbol{\phi}} &= 2\overline{\boldsymbol{\phi}} - \overline{\overline{\boldsymbol{\phi}}}, \\ D_2\overline{\boldsymbol{\phi}} &= 3\overline{\boldsymbol{\phi}} - 3\overline{\overline{\boldsymbol{\phi}}} + \overline{\overline{\overline{\boldsymbol{\phi}}}}. \end{aligned}$$

3. Time Relaxation

Time relaxation techniques, in the context of turbulent fluid flow regularization, were introduced by Stolz et al. in [4,5]. Their formulation is considered a regularization of a Chapman–Enskog expansion [16,17], and was devised as a way to introduce added energy dissipation to A the approximate deconvolution models, a class of large eddy simulations. In [6], the authors instead introduced this time relaxation term to the unregularized NSE. Therein the linear TRM was first written:

$$\mathbf{v}_t - \nu\triangle\mathbf{v} + (\mathbf{v}\cdot\nabla)\mathbf{v} + \nabla p + \chi(\mathbf{v} - D_N\overline{\mathbf{v}}) = \mathbf{f}, \tag{9}$$

$$\nabla\cdot\mathbf{v} = 0. \tag{10}$$

This new class of models nudged computed solutions of the unregularized NSE toward the regularized flow $D_N\overline{\mathbf{v}}$. Existence and uniqueness were demonstrated in [6], along with regularity of strong solutions. It is also shown that the model is effective in dampening spatial fluctuations beneath $\mathcal{O}(\delta)$ without substantially altering the larger, dominant scales, particularly when high-order deconvolution is utilized. A nonlinear extension was also considered, i.e.,

$$\mathbf{v}_t - \nu\triangle\mathbf{v} + (\mathbf{v}\cdot\nabla)\mathbf{v} + \nabla p + \chi^{3/2}(I - D_N F)\left\{|\mathbf{v} - D_N\overline{\mathbf{v}}|(\mathbf{v} - D_N\overline{\mathbf{v}})\right\} = \mathbf{f}, \tag{11}$$

$$\nabla\cdot\mathbf{v} = 0. \tag{12}$$

This particular nonlinear relaxation model is a special case of the nonlinear TRM studied in [11]. It was argued that this quadratic extension presents a novel consistency with friction forcing, which is proportional to the square of the velocity.

3.1. *Convergence of Weak Solutions of the Linear TRM to the NSE*

Weak solutions to the linear TRM were shown to converge to weak solutions of the NSE in [7]. We will demonstrate that result here.

A variational formulation of the NSE can be written: *Find* $\mathbf{u} \in L^2(0,T;X)\cap L^\infty(0,T;L^2(\Omega))$, $r \in L^2(0,T;Q)$ *with* $\mathbf{u}_t \in L^2(0,T;X')$ *satisfying,*

$$(\mathbf{u}_t,\boldsymbol{\phi}) + (\mathbf{u}\cdot\nabla\mathbf{u},\boldsymbol{\phi}) - (r,\nabla\cdot\boldsymbol{\phi}) + \nu(\nabla\mathbf{u},\nabla\boldsymbol{\phi}) = (\mathbf{f},\boldsymbol{\phi})\,,\ \forall\boldsymbol{\phi}\in\mathbf{X}, \tag{13}$$

$$(q,\nabla\cdot\mathbf{u}) = 0\,,\ \forall q\in Q, \tag{14}$$

$$\mathbf{u}(0,\mathbf{x}) = \mathbf{u}_0(\mathbf{x})\,,\ \forall\mathbf{x}\in\Omega. \tag{15}$$

Next, consider a variational formulation to the linear TRM (9) and (10): *Find* $\mathbf{v} \in L^2(0,T;X)\cap L^\infty(0,T;L^2(\Omega))$, $p \in L^2(0,T;Q)$ *with* $\mathbf{v}_t \in L^2(0,T;X')$ *satisfying,*

$$(\mathbf{v}_t,\boldsymbol{\phi}) + (\mathbf{v}\cdot\nabla\mathbf{v},\boldsymbol{\phi}) - (p,\nabla\cdot\boldsymbol{\phi}) + \nu(\nabla\mathbf{v},\nabla\boldsymbol{\phi}) + \chi(\mathbf{v} - D_N\bar{\mathbf{v}},\boldsymbol{\phi}) = (\mathbf{f},\boldsymbol{\phi})\,,\ \forall\boldsymbol{\phi}\in\mathbf{X}, \tag{16}$$

$$(q,\nabla\cdot\mathbf{v}) = 0\,,\ \forall q\in Q, \tag{17}$$

$$\mathbf{v}(0,\mathbf{x}) = \mathbf{u}_0(\mathbf{x})\,,\ \forall\mathbf{x}\in\Omega. \tag{18}$$

The operator $(I - D_N F)$ is self-adjoint and positive on $\mathbf{X}$, [7]. Define B such that,

$$B^2\boldsymbol{\phi} := \delta^{-(2N+2)}(I - D_N F)\boldsymbol{\phi} = \delta^{-(2N+2)}(\boldsymbol{\phi} - D_N\bar{\boldsymbol{\phi}}). \tag{19}$$

Therefore $B = \delta^{-(N+1)}\sqrt{(I - D_N F)}$ is well-defined, positive, and bounded. Next, select $\boldsymbol{\phi}^* := \delta^{(N+1)} B\boldsymbol{\phi} \approx \boldsymbol{\phi} - \bar{\boldsymbol{\phi}}$. We see,

$$\|\boldsymbol{\phi}^*\| = (\boldsymbol{\phi} - D_N\bar{\boldsymbol{\phi}}, \boldsymbol{\phi})^{1/2} = \left(\delta^{2N+2}(B\boldsymbol{\phi}, B\boldsymbol{\phi})\right)^{1/2} = \delta^{N+1}\|B\boldsymbol{\phi}\|.$$

Define the difference $\mathbf{e}(\mathbf{x},t) := \mathbf{u}(\mathbf{x},t) - \mathbf{v}(\mathbf{x},t)$, then subtract (16) from (13) to see,

$$(\mathbf{e}_t, \boldsymbol{\phi}) + (\mathbf{e}\cdot\nabla\mathbf{u}, \boldsymbol{\phi}) + (\mathbf{v}\cdot\nabla\mathbf{e}, \boldsymbol{\phi}) + \nu(\nabla\mathbf{e}, \nabla\boldsymbol{\phi}) + \chi(\mathbf{e} - D_N\bar{\mathbf{e}}, \boldsymbol{\phi}) = \chi(\mathbf{u} - D_N\bar{\mathbf{u}}, \boldsymbol{\phi}), \forall\boldsymbol{\phi} \in \mathbf{V}. \quad (20)$$

Test this equation with $\boldsymbol{\phi} = \mathbf{e}$ to see

$$\begin{aligned} \frac{1}{2}\frac{d}{dt}\|\mathbf{e}\|^2 + (\mathbf{e}\cdot\nabla\mathbf{u},\mathbf{e}) + \nu\|\nabla\mathbf{e}\|^2 + \chi\,(\mathbf{e} - D_N\bar{\mathbf{e}},\mathbf{e}) &= \chi\,(\mathbf{u} - D_N\bar{\mathbf{u}},\mathbf{e}), \\ \frac{1}{2}\frac{d}{dt}\|\mathbf{e}\|^2 - |(\mathbf{e}\cdot\nabla\mathbf{u},\mathbf{e})| + \nu\|\nabla\mathbf{e}\|^2 + \chi\,\|\mathbf{e}^*\|^2 &\le \chi\,\delta^{2N+2}\|B\mathbf{u}\|\,\|B\mathbf{e}\|. \end{aligned} \quad (21)$$

Using Young's inequality, Equation (21) becomes,

$$\begin{aligned} |(\mathbf{e}\cdot\nabla\mathbf{u},\mathbf{e})| &\le C\sqrt{\|\mathbf{e}\|\,\|\nabla\mathbf{e}\|}\,\|\nabla\mathbf{u}\|\,\|\nabla\mathbf{e}\| \\ &\le \frac{1}{2}\nu\|\nabla\mathbf{e}\|^2 + C_1\nu^{-3}\|\nabla\mathbf{u}\|^4\,\|\mathbf{e}\|^2, \end{aligned}$$

$$\frac{d}{dt}\|\mathbf{e}\|^2 - C_1\,\nu^{-3}\|\nabla\mathbf{u}\|^4\,\|\mathbf{e}\|^2 + \nu\|\nabla\mathbf{e}\|^2 + \chi\,\|\mathbf{e}^*\|^2 \le C_2\chi\,\delta^{2N+2}\|B\mathbf{u}\|^2. \quad (22)$$

Applying Gronwall's Lemma, then multiplying through by the integrating factor, $\exp(-C_1\,\nu^{-3}\int_0^\tau \|\nabla\mathbf{u}\|^4\,ds)$, then using $\|\mathbf{e}\|(0) = 0$, we have,

$$\begin{aligned} \|\mathbf{e}\|^2 &+ \int_0^t e^{(C_1\nu^{-3}\int_\tau^t\|\nabla\mathbf{u}\|^4\,ds)}\left(\nu\|\nabla\mathbf{e}\|^2 + \chi\,\|\mathbf{e}^*\|^2\right)d\tau \\ &\le \int_0^t e^{(C_1\nu^{-3}\int_\tau^t\|\nabla\mathbf{u}\|^4\,ds)}\left(C_2\,\chi\,\delta^{2N+2}\|B\mathbf{u}\|^2\right)d\tau, \end{aligned}$$

i.e.,

$$\|\mathbf{e}\|^2 + \nu\int_0^t \|\nabla\mathbf{e}\|^2\,d\tau + \int_0^t \chi\,\|\mathbf{e}^*\|^2\,d\tau \le C_2 e^{C_1\nu^{-3}\|\mathbf{u}\|_{4,1}^4}\,\chi\,\delta^{2N+2}\int_0^t \|B\mathbf{u}\|^2\,d\tau.$$

Based on the above, the following lemma was stated.

Lemma 1. *Assuming $\mathbf{u} \in L^4(0,T;W^{1,4})$ satisfying (13)–(15) and $\mathbf{v}$ given by (16)–(18) we have that there exists constants C_1, $C_2 > 0$, such that,*

$$\begin{aligned} \|\mathbf{u} - \mathbf{v}\|^2 &+ \nu\int_0^t \|\nabla(\mathbf{u} - \mathbf{v})\|^2\,d\tau + \chi\int_0^t \|(\mathbf{u} - \mathbf{v})^*\|^2\,d\tau \\ &\le C_2 e^{C_1\nu^{-3}\|\mathbf{u}\|_{4,1}^4}\,\chi\,\delta^{2N+2}\int_0^t \|B\mathbf{u}\|^2\,d\tau. \end{aligned} \quad (23)$$

□

This result guarantees a weak sense of limit consistency, i.e., if for a particular value of χ we call the linear TRM solution $\mathbf{v}(\chi)$, then for any $\boldsymbol{\phi} \in \mathbf{X}$, $\lim_{\chi\to 0}(\mathbf{v}(\chi), \boldsymbol{\phi}) = (\mathbf{u}, \boldsymbol{\phi})$.

3.2. The Effects of Time Relaxaiton on the Energy Cascade

In the linear case (9) and (10), a similarity study was also conducted in [6] revealing that if χ is selected appropriately, one will not only observe an energy cascade, but potentially control that cascade by inducing a new effective Kolmogorov micro-scale for (9) and (10) that agrees with the filtering length δ. This was called a "perfect resolution," since in this case the extra dissipation perfectly balances the energy transfer from external power sources to the new cutoff scale while preventing a non-physical accumulation of energy there. This is accomplished by selecting,

$$\chi \simeq \frac{U}{L^{1/3}} 2^{\frac{N+1}{3}} \delta^{-2/3} \tag{24}$$

where U is the global velocity scale, and L is the characteristic length scale of the domain. This results in $\mu_{TRM} = \mathcal{O}(\delta)$ and in turn guarantees, for deconvolution order N,

$$|\chi(\mathbf{v} - D_N\overline{\mathbf{v}})| = \mathcal{O}\left(\chi\delta^{2N+2}\right) = \mathcal{O}\left(\delta^{2N+\frac{4}{3}}\right). \tag{25}$$

Forcing an accelerated dissipation to occur at a this larger length scale has the potential to substantially relax the spatial discretization requirements in numerical simulations. In 3D flows, the sufficient condition to fully resolve (9) and (10) is,

$$N_{dof} \simeq \left(\frac{L}{\delta}\right)^3,$$

which is totally independent of Re. The savings can be estimated by the ratio,

$$\left(\frac{N_{dof}^{NSE}}{N_{dof}}\right) \simeq \left(\frac{Re^{9/4}}{L^3\delta^{-3}}\right)^{4/3} = \left(\frac{\delta}{L}\right)^4 Re^3.$$

This is a significant improvement, especially for flows presenting large Reynolds' numbers.

3.3. A General Nonlinear Time Relaxation Extension

A natural extension to the nonlinear formulation seen in [6] and (11) and (12) is discussed in detail in [11]. The main motivational claim for exploring the extended nonlinear model is that it does a better job of regularizing flow at higher wave numbers. This is demonstrated with numerical results, and discussed further in Section 4.

For $2 \leq r < \infty$, the time relaxation term is scaled such that,

$$\chi\mathbf{v}' = \chi(I - D_N F^{-1})\{|\mathbf{v} - D_N\overline{\mathbf{v}}|^{r-2}(\mathbf{v} - D_N\overline{\mathbf{v}})\}. \tag{26}$$

Calculating the energy dissipation penalty for such models is a simple endeavor. If we let $\varepsilon_{model}(\mathbf{v})(t)$ represent the energy dissipation penalty functional, in inviscid L-periodic flow on the cube $\Omega = [0, L]^3$, we find that for the linear case,

$$\varepsilon_{TRM}(\mathbf{v})(t) = \frac{1}{L^3}\int_\Omega \chi(\mathbf{v} - D_N\overline{\mathbf{v}}) \cdot \mathbf{v}\, d\mathbf{x}. \tag{27}$$

While this is fairly simple, if we note that the approximate deconvolution operators D_N are self-adjoint, in the same settings, the energy dissipated by the generalized nonlinear models is then given by,

$$\varepsilon_{NLTRM}(\mathbf{v})(t) = \frac{1}{L^3}\int_\Omega \chi(I - D_N F^{-1})\{|\mathbf{v} - D_N\overline{\mathbf{v}}|^{r-2}(\mathbf{v} - D_N\overline{\mathbf{v}})\} \cdot \mathbf{v}\, d\mathbf{x} \tag{28}$$

$$= \frac{1}{L^3}\int_\Omega \chi|\mathbf{v} - D_N\overline{\mathbf{v}}|^r d\mathbf{x}. \tag{29}$$

Hence, it can be surmised from (27) and (29) that if $\mathbf{v}$ differs *significantly* from a spatially-filtered $D_N\overline{\mathbf{v}}$, a greater penalty will be contributed to the energy dissipation rate; i.e., more energy will be dissipated. The nonlinear formulation stands to easily exaggerate that process for larger values of r.

4. Finite Element Implementations of the Navier Stokes Equations with Time Relaxation

The first finite element analysis and implementation of a TRM model of the form (9) and (10) was seen in [7]. In that study, an algorithm was presented for the linear TRM model using the (P_n, P_{n-1}) Taylor–Hood finite elements in space, with a Crank–Nicolson time-stepping scheme. A complete velocity stability and convergence analysis is included.

A similar analysis is performed in [8] for the linear TRM utilizing the $(P_n^{disc}, P_{n-1}^{disc})$ finite elements in space, and Crank–Nicolson in time. Such a finite element formulation provides a number of distinct advantages. In particular, mesh refinement and derefinement are straightforward, the divergence-free condition can be enforced locally, and complicated geometries can be accommodated easily with unstructured meshes.

These analyses were expanded in [10], where the authors performed an Aubin–Nitsche lift technique, proving the optimal error convergence for velocity solutions seen in their computations. This is a noteworthy result, as it is often the case in finite element analysis of CFD problems that authors will neglect to prove the highest optimal convergence rates if they are better than the interpolation order. In addition to a thorough demonstration of these rates, the authors presented the results of the van Karman vortex street experiment with $0 \leq Re \leq 200$. Good results were reported when verified against the benchmark values seen in [18]. This experiment is discussed further in Section 4.3.

A thorough investigation of how introducing linear time relaxation operators to finite element approximations of convective fluid-flow problems affects the over-all error asymptotics was performed in [19]. Specifically, in the case of the advection equation, they considered whether or not the regularization provided an increase in accuracy, stability, or both. The analysis demonstrated that a slight, $\mathcal{O}(h^{1/2})$, increase in the error convergence rate is possible for well-selected values of χ and δ. Their numerical experiments in both 1D and 2D demonstrated a substantial increase in stability.

Further progress was made in [9], where the standard (P_n, P_{n-1}) finite element discretization implemented in [7] was augmented in two key ways. The first technique adjusts the differential filter discussed in Definition (1), requiring it to be locally divergence-free,

$$\begin{aligned} -\delta^2 \triangle \overline{\boldsymbol{\phi}} + \overline{\boldsymbol{\phi}} + \nabla \lambda &= \boldsymbol{\phi} \\ \nabla \cdot \overline{\boldsymbol{\phi}} &= 0. \end{aligned}$$

This condition, when included in a finite element scheme, guarantees that the filtered velocities conserve mass discretely. Enforcing the constraint requires a Lagrange multiplier $\nabla \lambda$, sometimes referred to as a "filtering pressure." If the filtered velocities conserve mass, the linear time relaxation term will conserve mass, which will help guarantee computed solutions to the TRM enjoy a higher degree of physical accuracy. The second consideration includes a second-order accurate linear extrapolation of the convection velocity. Previous finite element implementations address the nonlinearity by computing expensive Picard iterations. This modification removes the necessity of fixed-point iterations, at the cost of storing the velocity solution at a previous time-step in memory.

A complete velocity stability and convergence analysis of the finite element scheme is included.

The first finite element analysis of a nonlinear TRM formulation can be seen in [11]. While the results of numerical experiments could be seen in [7], the algorithm itself was not addressed. In this case, analysis of the stability and convergence of velocity solutions is included, along with a calculation of those convergence rates. The authors use the 3D Etheir–Steinman problem to demonstrate that the quadratic nonlinear TRM, (26) with $r = 3$, can be significantly more accurate than a linear TRM. This experiment is discussed in greater detail in Section 4.2.

The time relaxed, discrete approximation on the time interval $(0, T]$, is given by the following algorithm.

Algorithm 2. *For $n = 1, 2, \ldots, M-1$, where $M := \frac{T}{\Delta t}$, $\mathbf{v}_h^0 := P_{V_h}(\mathbf{v}_0))$, find $\mathbf{v}_h^{n+1} \in X_h$, $p_h^{n+1} \in Q_h$, such that,*

$$\begin{aligned}
\frac{1}{\Delta t}\left(\mathbf{v}_h^{n+1} - \mathbf{v}_h^n, \mathbf{v}_h\right) + b^*\left(\frac{3}{2}\mathbf{v}_h^n - \frac{1}{2}\mathbf{v}_h^{n-1}, \mathbf{v}_h^{n+\frac{1}{2}}, \mathbf{v}_h\right) \quad &+ \quad \nu\left(\nabla \mathbf{v}_h^{n+\frac{1}{2}}, \nabla \mathbf{v}_h\right) - \left(p_h^{n+1}, \nabla \cdot \mathbf{v}_h\right) \\
+\chi\left(\mathbf{v}_h^{n+\frac{1}{2}} - D_N^h \overline{\mathbf{v}_h^{n+\frac{1}{2}}}^h, \mathbf{v}_h\right) + \gamma h\left(\nabla\left(\mathbf{v}_h^{n+1} - \mathbf{v}_h^n\right), \nabla \mathbf{v}_h\right) \quad &= \quad \left(\mathbf{f}(t^{n+\frac{1}{2}}), \mathbf{v}_h\right), \ \forall \mathbf{v}_h \in X_h, \\
\left(q_h, \nabla \cdot \mathbf{v}_h^{n+1}\right) \quad &= \quad 0, \ \forall q_h \in Q_h.
\end{aligned}$$

where $b^*(\cdot,\cdot,\cdot) : \mathbf{X} \times \mathbf{X} \times \mathbf{X} \to \mathbb{R}$ is the *skew-symmetric trilinear form,*

$$[H] b^*(\mathbf{u}, \mathbf{v}, \mathbf{w}) := \frac{1}{2}(\mathbf{u} \cdot \nabla \mathbf{v}, \mathbf{w}) - \frac{1}{2}(\mathbf{u} \cdot \nabla \mathbf{w}, \mathbf{v}).$$

The added energy dissipation penalty of a time relaxation operator can be helpful when paired with other regularization techniques. One such example can be found in [12] where the time relaxation operator is paired with a model to calculate flow ensembles. This hybrid model is implemented using the Taylor–Hood (P_n, P_{n-1}) finite elements with a backward-Euler time-stepping scheme, and contains a complete stability and convergence analysis. At best, Taylor–Hood finite elements can only enforce the divergence-free condition weakly, since in this case $\nabla \cdot \mathbf{X}_h \not\subseteq Q_h$. This shortcoming of the Taylor–Hood elements directly contributes to non-physical errors. The authors address this successfully through the use of grad-div stabilization.

4.1. Two-Dimensional Full-Step Benchmark with Taylor–Hood Finite Elements

The implementations from [7,9,11] each perform the 2D flow about a full-step obstruction benchmark. This is an excellent experiment to test regularization techniques like TRMs, as it develops a highly-rotational flow, and on a coarse mesh a direct numerical simulation of the NSE produces an unacceptable amount of error. In this section we have collected and discussed the results from [7,9,11], summarizing a selection of their results.

The calculations presented below, as well as most of the calculations performed in the articles we have cited in this section, were completed using FreeFem++ [20]. This particular experiment considers a 10×40 2D rectangular domain, with a 1×1 rectangular obstruction along the bottom. The problem settings are established with no-slip boundary conditions along the top and bottom, and about the obstruction; i.e., $\mathbf{v}_h = \mathbf{0}$. Given $\mathbf{v}_h(\mathbf{x}, t) = (v_1, v_2)^T$, parabolic inflow is prescribed on the left boundary such that,

$$v_1(0, y) = (1/25)y(10 - y), \tag{30}$$

$$v_2(0, y) = 0. \tag{31}$$

The right boundary is assigned a "do-nothing", or zero-traction boundary. The initial velocity is set to zero, i.e., $\mathbf{v}_h(\mathbf{x}, t = 0) = \mathbf{0}$. The kinematic viscosity ν is fixed at $1/600$, therefore the Reynolds number varies $0 \leq Re \leq 600$, by assuming a characteristic length of the flow $L = 1$, the height of the obstruction.

The mesh discretization levels are enumerated in Table 1. An example discretization of the domain can be seen in Figure 1. These meshes were used consistently in [7] and [11]. The "coarse" mesh in [9] is essentially the same as the level-2 mesh, whereas the "fine" mesh has a tighter discretization than the level-3 mesh. Levels 0 and 1 substantially under-resolve the flow in the unregularized NSE in these settings. It is the goal of any regularization model to improve on the performance of these results.

Table 1. These four meshes are utilized consistently in [7,11]. The N_{dof} are calculated for the (P_3, P_2) Taylor–Hood finite elements.

Mesh Scale	N_{dof}
Level 0	2072
Level 1	5433
Level 2	13,899
Level 3	41,502

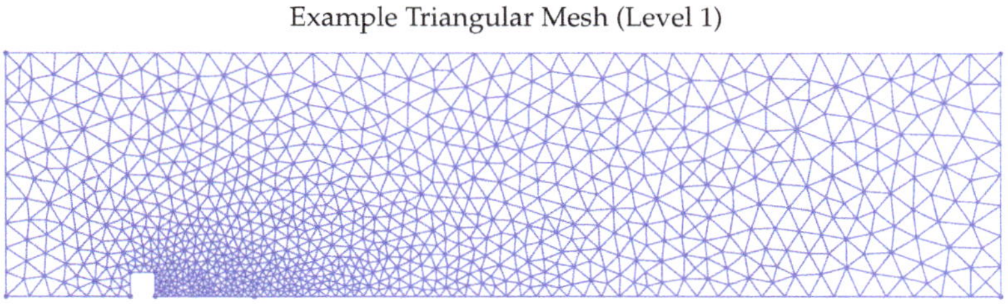

Figure 1. Typically, a tighter discretization is used behind the block to resolve flow rotations.

The true behavior is characterized by a vortex growing behind the obstruction, eventually splitting and shedding through time. Typically, the experiment is conducted over the time interval $0 \leq t \leq 40$. An example "correct" solution can be seen at t = 40 s in Figure 2. Contrast this with the under-resolved flow seen Figure 3, where while we still see growing and separating spatial eddies, strong erroneous oscillating features begin to destroy the calculation as time advances. These figures are reported in [11].

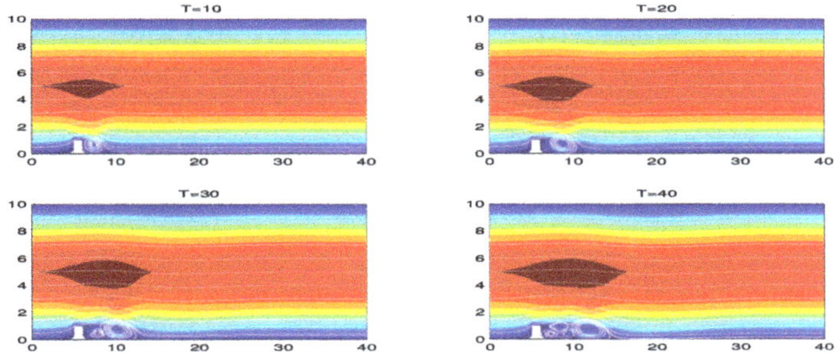

Figure 2. Four snapshots (t = 10, 20, 30, and 40) of a full-resolution direct numerical simulation of the Navier–Stokes Equations (NSE). Shown are streamlines overlaid above speed contours $|\mathbf{v}_h(\mathbf{x}, t)|$. This figure was originally published in [11].

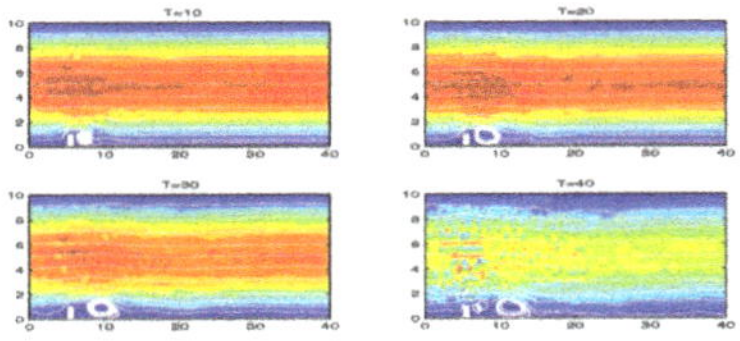

Figure 3. Four snapshots ($t = 10, 20, 30,$ and 40) of an under-resolved direct numerical simulation of the NSE, plagued with spurious oscillations. Shown are streamlines overlaid above speed contours $|\mathbf{v}_h(\mathbf{x}, t)|$. This figure was originally published in [11].

On a level-1 mesh, the linear time relaxation model (9) and (10), with deconvolution order $N = 0$, relaxation term $\chi = 0.01$, and differential filtering length $\delta = 1.5$ shows a considerable improvement over the unregularized NSE. No erroneous speed oscillations are observed, and vortices are seen to be growing and separating. This is seen in Figure 4. Note that Figure 4 comes from [7], where they used a different color-map for the speed contours. It is also worth mentioning that this result comes from the simplest formulation of the considered time relaxation models, paired with well-selected parameter values. Increasing the order of deconvolution to $N = 1$ produces a nearly indistinguishable result (which also can be seen in [7]).

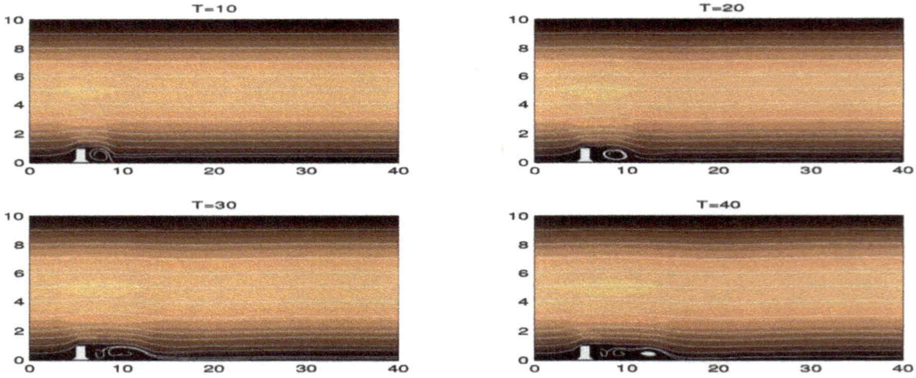

Figure 4. Four snapshots ($t = 10, 20, 30,$ and 40) of the linear time relaxation model (TRM) with deconvolution order $N = 0$, filter length $\delta = 1.5$, and relaxation term $\chi = 0.01$. Shown are streamlines overlaid above speed contours $|\mathbf{v}_h(\mathbf{x}, t)|$. Lighter shades represent higher speeds. This figure was originally published in [7].

As mentioned briefly at the end of Section 3.3, if the relaxation term is increased (be it by a high degree of disagreement measured in the $\left(\mathbf{v}_h - D_N^h \overline{\mathbf{v}}_h\right)$ term, or by substantially increasing χ) the energy dissipation penalty increases proportionally. If the dissipation penalty is too high, the flow can over-regulate. This is observed in [11], where they utilize a nonlinear formulation of the form (26) with $r = 3$ (quadratic). On a level-1 mesh, with deconvolution order $N = 0$, filtering length $\delta = 0.5$, and $\chi = 3.0$, we see in Figure 5 that the smaller-scale fluctuations have been over-smoothed such that the vortices will not separate. Increasing to a first-order deconvolution ($N = 1$) effectively decreases the dissipation rate, allowing for some of the smaller rotations to be resolved. This growth and splitting can be observed in Figure 6. In both Figures 5 and 6 we see erroneous oscillations growing in time within the speed contours. To address this, one would could increase the filter length δ.

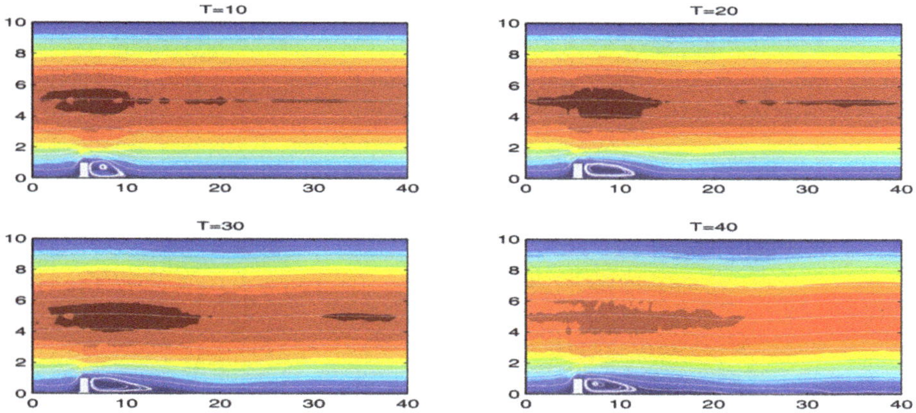

Figure 5. Four snapshots ($t = 10, 20, 30,$ and 40) of the nonlinear TRM with deconvolution order $N = 0$, filter length $\delta = 0.5$, and relaxation term $\chi = 3.0$. Shown are streamlines overlaid above speed contours $|\mathbf{v}_h(\mathbf{x}, t)|$. This figure was originally published in [11].

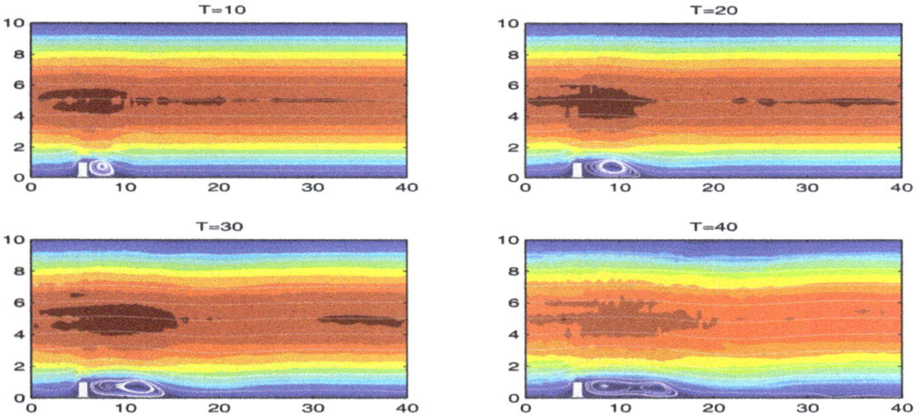

Figure 6. Four snapshots ($t = 10, 20, 30,$ and 40) of the nonlinear TRM with deconvolution order $N = 1$, filter length $\delta = 0.5$, and relaxation term $\chi = 3.0$. Shown are streamlines overlaid above speed contours $|\mathbf{v}_h(\mathbf{x}, t)|$. This figure was originally published in [11].

4.2. 3D Ethier-Steinman Problem for the Nonlinear Time Relaxation

This numerical computation was performed in [11]. The velocity and pressure of the Eithier–Steinman problem satisfy Navier–Stokes equations exactly. The domain is $[-1,1]^3$ and for given parameters a, d and viscosity ν, the solution is given by,

$$u_1 = -a\left(e^{ax}\sin(ay+dz)+e^{az}\cos(ax+dy)\right)e^{-\nu d^2 t} \quad (32)$$

$$u_2 = -a\left(e^{ay}\sin(az+dx)+e^{ax}\cos(ay+dz)\right)e^{-\nu d^2 t} \quad (33)$$

$$u_3 = -a\left(e^{az}\sin(ax+dy)+e^{ay}\cos(az+dx)\right)e^{-\nu d^2 t} \quad (34)$$

$$\begin{aligned} p = & -\frac{a^2}{2}(e^{2ax}+e^{2ay}+e^{2az}+2\sin(ax+dy)\cos(az+dx)e^{a(y+z)} \\ & +2\sin(ay+dz)\cos(ax+dy)e^{a(z+x)} \\ & +2\sin(az+dx)\cos(ay+dz)e^{a(x+y)})e^{-\nu d^2 t} \end{aligned} \quad (35)$$

The parameter values were selected such that $a = 1.25$, $d = 1$, kinematic viscosity $\nu = 0.0001$, time-step $\Delta t = 0.005$, final time $T = 1$, time relaxation parameter $\chi = 1$, and order of deconvolution $N = 0$ (for both linear and nonlinear cases). The initial velocity is given by the above true solution, i.e., $\mathbf{u}_0 = (u_1(0), u_2(0), u_3(0))^T$. Dirichlet boundary conditions were similarly enforced, given (32)–(34). The mesh considered contained 3072 tetrahedral elements. The finite element discretization used the (P_2, P_1) Taylor–Hood finite elements.

Figure 7 presents graphs of the velocity errors for the linear (full line) and nonlinear (dotted line) time relaxation model in time. The graphs show that the nonlinear relaxation method develops less error as time progresses in time.

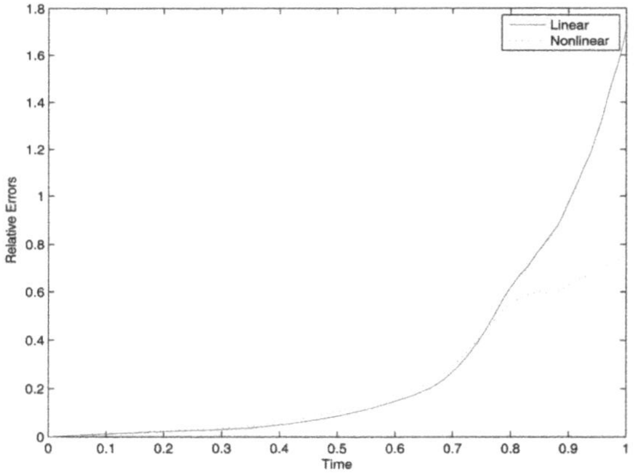

Figure 7. Comparison of the relative L^2 error.

4.3. Two-Dimensional Flow about a Cylinder Problem

In this study from [10] the creation of the vortex street and calculation of maximum drag and lift, and the change in pressure for a body of rotational form immersed in a fluid was studied. These calculations require accurate estimation of pressure and derivatives of velocity on the flow boundary, which is more difficult than accurately predicting fluid flow velocity. This numerical experiment has been studied by Schafer and Turek in [21] and John in [18]. The domain for the flow is given in Figure 8. The inflow is specified as $0.41^{-2} \cdot \sin(\pi t/8) \cdot 6y(0.41-y)$ m/s, "do-nothing" at the outflow boundary and no slip boundary conditions at the top and bottom of the domain and around the cylinder. Since the mean inflow velocity $U(t) = \sin(\pi t/8)$ ranges from 0 to 1, the Reynolds number $Re = \frac{LU}{\nu} \in [0, \frac{L}{\nu}] = [0, 100]$ as the diameter $L = 0.1$ m. The meshes were generated using Delaunay triangulation and three mesh levels are used in simulation: level 1 with 7500 dof, level 2 with 28,806 dof, and level 3 with 114,042 dof. Mesh level 1 is shown in Figure 9. The Taylor–Hood elements are used for space discretization. More details can be found in [10].

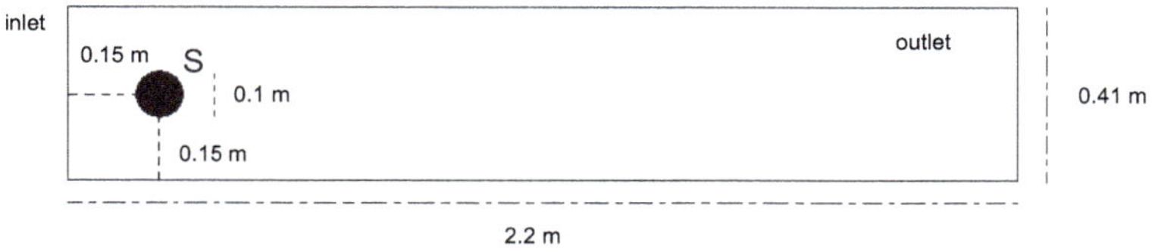

Figure 8. Domain.

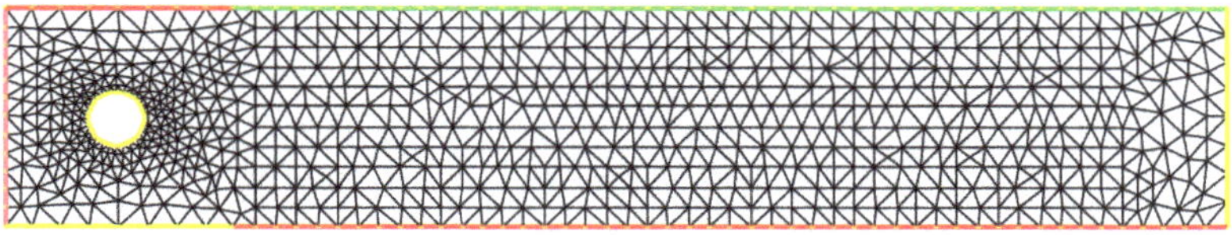

Figure 9. Mesh Level 1.

The indication of a successful simulation is the formation of a vortex street, as seen in Figure 10, and the correct profiles in time for drag $c_d(t)$, lift $c_l(t)$, and pressure drop $\Delta p(t)$ may be found in Figure 11.

The following reference values are given by John in [18]:

$$t(c^{ref}_{d,max}) = 3.93625, \quad c^{ref}_{d,max} = 2.950921575$$
$$t(c^{ref}_{l,max}) = 5.693125, \quad c^{ref}_{l,max} = 0.47795$$
$$\Delta p^{ref}(8s) = -0.1116$$

Furthermore, Schäfer and Turek have provided the following reference intervals of:

$$c^{ref}_{d,max} \in [2.93, 2.97], \; c^{ref}_{l,max} \in [0.47, 0.49], \text{ and } \triangle p^{ref}(8s) \in [-0.115, -0.105]$$

for the maximum drag, lift, and difference in pressure, respectively [21]. The results of this experiment fall within these reference values on mesh levels 2 and 3 for all the tested time steps, (see Table 2 for order of deconvolution $N = 0$).

The usefulness of the time relaxation model at higher Reynolds numbers was tested. With $\nu = 10^{-5}$, a value of $\chi = 20$ produced a solution with visible vortex street (see Figure 12), while the Navier–Stokes simulations were unsuccessful (i.e., the fixed point iteration of the nonlinearity did not converge).

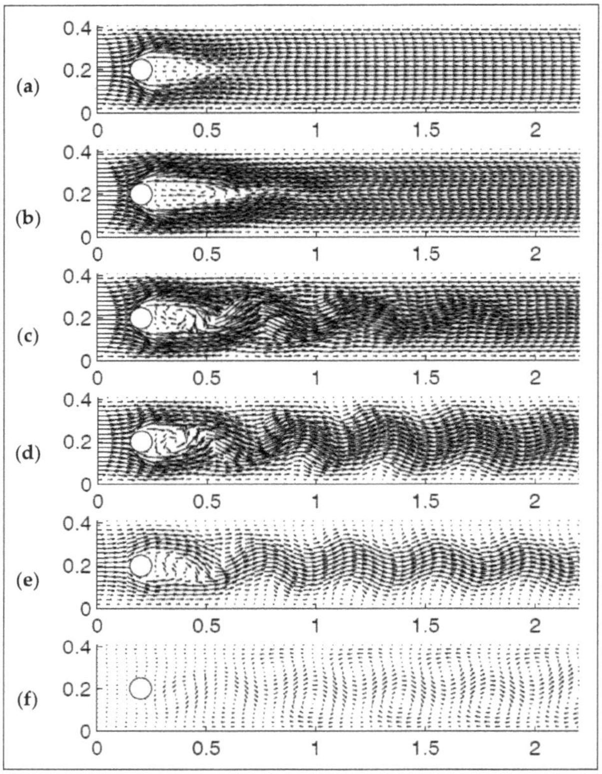

Figure 10. Velocity snapshots presented from (**a–f**) at t = 2, 4, 5, 6, 7, and 8 for level-3 mesh with $\triangle t = 0.00125$, viscosity $\nu = 10^{-3}$, $\chi = 0.01$ and $N = 0$.

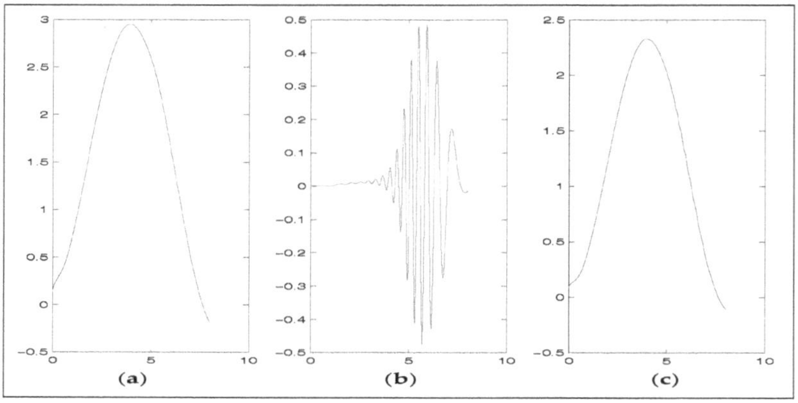

Figure 11. The development of $c_d(t)$, $c_l(t)$, and $\Delta p(t)$ (from to (**a–c**), respectively) for level-3 mesh with $\triangle t = 0.00125$, viscosity $\nu = 10^{-3}$, $\chi = 0.01$ and $N = 0$.

Table 2. Maximal drag, maximal lift, and change in pressure, order of deconvolution $N = 0$.

Level	$\triangle t$	t ($c_{d,max}$)	$c_{d,max}$	t ($c_{l,max}$)	$c_{l,max}$	αp (8 s)
1	0.04	3.96	2.85876	6.36	0.242883	−0.100637
1	0.02	3.94	2.86052	6.18	0.293106	−0.109724
1	0.01	3.93	2.86136	6.13	0.305785	−0.109681
1	0.005	3.93	2.86177	6.12	0.308338	−0.108987
1	0.0025	3.9275	2.86197	6.115	0.308694	−0.108891
1	0.00125	3.925	2.86208	6.11375	0.308617	−0.108924
2	0.04	3.96	2.94203	6.12	0.393238	−0.10326
2	0.02	3.96	2.94399	5.98	0.464875	−0.105554
2	0.01	3.94	2.94497	5.94	0.477624	−0.110691
2	0.005	3.94	2.94544	5.925	0.48004	−0.111333
2	0.0025	3.9375	2.94567	5.92	0.479669	−0.111552
2	0.00125	3.9375	2.94578	5.91875	0.479152	−0.111665
3	0.04	3.96	2.94543	6.12	0.40593	−0.102126
3	0.02	3.96	2.94741	5.98	0.468239	−0.106822
3	0.01	3.94	2.9484	5.93	0.485103	−0.110926
3	0.005	3.94	2.94887	5.915	0.486785	−0.111227
3	0.0025	3.9375	2.94909	5.9125	0.486471	−0.111363
3	0.00125	3.9375	2.9492	5.9125	0.485944	−0.111455

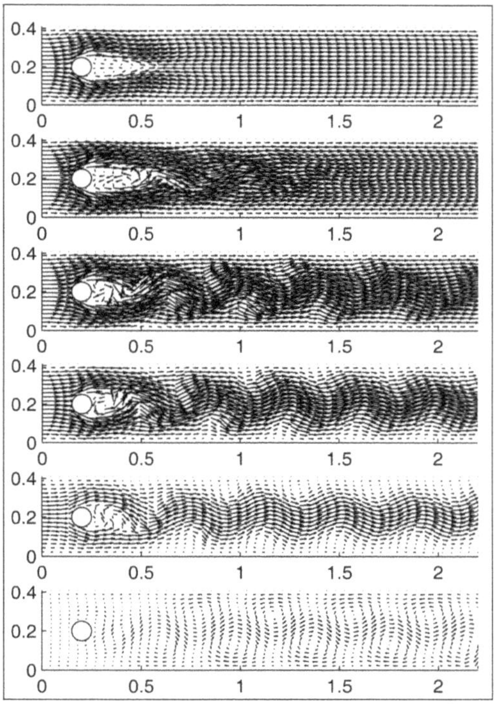

Figure 12. Velocity snapshots presented from (**a–f**) at t = 2, 4, 5, 6, 7, and 8 for level-2 mesh with $\triangle t = 0.00125$, viscosity $\nu = 10^{-5}$, $\chi = 20$, and $N = 0$.

5. Parameter-Sensitivity Study of the Time Relaxation Model

The "optimal" parameter choices for non-physical quantities like the relaxation coefficient χ, nonlinear time relaxation order r, and the filtering length δ are discussed [6,11,19]. That said, the notion of "optimal" scales for such parameters are inherently problem- and goal-specific. This is further complicated by the fact computed solutions are inherently sensitive to these parameters.

The notion of $\chi-$sensitivity for the linear time relaxation model was addressed in [22]. Since it is understood that a velocity solution $\mathbf{v}$ to the linear model is sensitive to perturbations of χ, we denote velocity solutions to the linear TRM for a given χ as $\mathbf{v}(\chi)$. The $\chi = 0$ solution results in a direct numerical simulation of the NSE. We can estimate $\chi-$sensitivity by the forward finite difference (FFD)

$$\mathbf{s}_{FFD}(\chi) = \frac{\mathbf{v}(\chi) - \mathbf{v}(0)}{\chi}. \tag{36}$$

Further, if we assume $\mathbf{v}(\chi)$ varies smoothly with χ, we can define the sensitivity as the implicit derivative,

$$\mathbf{s}_{SEM}(\chi) = \frac{\partial}{\partial\chi}\mathbf{v}(\chi). \tag{37}$$

We similarly define the sensitivity of pressure as $\psi(\chi)$, and filtered velocity $\mathbf{w}(\chi)$.

The $\chi-$sensitivity is evaluated in two ways. Firstly, the FFD method (36) can be used and secondly, the sensitivity equation method (SEM) can also be used. The SEM is developed by taking an implicit derivative of (9) and (10), as well a the filter Equation (5). For simplicity, we let $\mathbf{s}_{SEM} = \mathbf{s}$ then by selecting $N = 0$, the $\chi-$sensitivity equations are written:

$$\mathbf{s}_t + \mathbf{v}\cdot\nabla\mathbf{s} + \mathbf{s}\cdot\nabla\mathbf{v} + \nabla\psi - \nu\triangle\mathbf{s} + (\mathbf{v} - \overline{\mathbf{v}}) + \chi(\mathbf{s} - \mathbf{w}) = \mathbf{0}, \tag{38}$$

$$\nabla\cdot\mathbf{s} = 0 \tag{39}$$

$$-\delta^2\triangle\mathbf{w} + \mathbf{w} = \mathbf{s}. \tag{40}$$

Initial and boundary conditions for (9) and (10) are inherently insensitive to χ, hence for all implementations,

$$\mathbf{s}(\mathbf{x}, 0) = \mathbf{0},$$
$$\mathbf{s}(\partial\Omega, t) = \mathbf{0}.$$

The new PDE (38)–(40) is discretized and computed in conjunction with the model's discretization. In [22] the SEM equations are discretized using the same scheme, Taylor–Hood (P_n, P_{n-1}) finite elements, and Crank–Nicolson time stepping. For notational clarity, in the discussion of the Crank–Nicolson temporal discretization, we let $\mathbf{v}(t^{n+1/2}) = \mathbf{v}((t^{n+1} + t^n)/2)$ for the continuous variable and $\mathbf{v}^{n+1/2} = (\mathbf{v}^{n+1} + \mathbf{v}^n)/2$ for both, continuous and discrete variables. Thus, we obtain the following discretized finite element variational formulations.

Given $(\mathbf{X}_h, Q_h)$, end-time $T > 0$, the time step is chosen $\triangle t < T = M\triangle t$, find the approximated TRM solution $(\mathbf{v}_h^{n+1}, p_h^{n+1}) \in (\mathbf{X}_h, Q_h)$, for $n = 0, 1, 2....M-1$ satisfying,

$$\frac{1}{\triangle t}(\mathbf{v}_h^{n+1} - \mathbf{v}_h^n, \boldsymbol{\phi}_h) + \nu a(\mathbf{v}_h^{n+1/2}, \boldsymbol{\phi}_h) + b^*(\mathbf{v}_h^{n+1/2}, \mathbf{v}_h^{n+1/2}, \boldsymbol{\phi}_h) - (p_h^{n+1}, \nabla\cdot\boldsymbol{\phi}_h)$$
$$+\chi(\mathbf{v}_h^{n+1/2} - \overline{\mathbf{v}}_h^{n+1/2}, \boldsymbol{\phi}_h) = (\mathbf{f}^{n+1/2}, \boldsymbol{\phi}_h), \quad \forall\boldsymbol{\phi}_h \in \mathbf{X}_h \tag{41}$$

$$(\nabla\cdot\mathbf{v}_h^{n+1}, q_h) = 0, \quad \forall q_h \in Q_h \tag{42}$$

$$\delta^2(\nabla\overline{\mathbf{v}}_h^{n+1}, \nabla\boldsymbol{\phi}_h) + (\overline{\mathbf{v}}_h^{n+1}, \boldsymbol{\phi}_h) = (\mathbf{v}_h^{n+1}, \boldsymbol{\phi}_h), \quad \forall\boldsymbol{\phi}_h \in \mathbf{X}_h \tag{43}$$

and the sensitivity solution $(\mathbf{s}_h^{n+1}, v_h^{n+1}) \in (\mathbf{X}_h, Q_h)$, for $n = 0, 1, 2....M-1$ satisfying,

$$\frac{1}{\triangle t}(\mathbf{s}_h^{n+1} - \mathbf{s}_h^n, \boldsymbol{\phi}_h) + \nu a(\mathbf{s}_h^{n+1/2}, \boldsymbol{\phi}_h) + b^*(\mathbf{s}_h^{n+1/2}, \mathbf{v}_h^{n+1/2}, \boldsymbol{\phi}_h) + b^*(\mathbf{v}_h^{n+1/2}, \mathbf{s}_h^{n+1/2}, \boldsymbol{\phi}_h)$$
$$-(\psi_h^{n+1}, \nabla \cdot \boldsymbol{\phi}_h) + (\mathbf{v}_h^{n+1/2} - \overline{\mathbf{v}}_h^{h\,n+1/2}, \boldsymbol{\phi}_h) + \chi(\mathbf{s}_h^{n+1/2} - \mathbf{u}_h^{n+1/2}, \boldsymbol{\phi}_h) = 0, \quad \forall \boldsymbol{\phi}_h \in \mathbf{X}_h \tag{44}$$
$$(\nabla \cdot \mathbf{s}_h^{n+1}, q_h) = 0, \quad \forall q_h \in Q_h \tag{45}$$
$$\delta^2(\nabla \mathbf{u}_h^{n+1}, \nabla z_h) + (\mathbf{u}_h^{n+1}, \boldsymbol{\phi}_h) = (\mathbf{s}_h^{n+1}, \boldsymbol{\phi}_h). \quad \forall \boldsymbol{\phi}_h \in \mathbf{X}_h \tag{46}$$

Equations (41)–(43) and (44)–(46) can be rewritten equivalently in the space V^h as given below. Find $\mathbf{v}_h^{n+1} \in V^h$, for $n = 0, 1, 2....M-1$ satisfying,

$$\frac{1}{\triangle t}(\mathbf{v}_h^{n+1} - \mathbf{v}_h^n, \boldsymbol{\phi}_h) + \nu a(\mathbf{v}_h^{n+1/2}, \boldsymbol{\phi}_h) + b^*(\mathbf{v}_h^{n+1/2}, \mathbf{v}_h^{n+1/2}, \boldsymbol{\phi}_h)$$
$$+\chi(\mathbf{v}_h^{n+1/2} - \overline{\mathbf{v}}_h^{h\,n+1/2}, \boldsymbol{\phi}_h) = (\mathbf{f}^{n+1/2}, \boldsymbol{\phi}_h), \quad \forall \boldsymbol{\phi}_h \in V^h \tag{47}$$
$$\delta^2(\nabla \overline{\mathbf{v}}_h^{h\,n+1}, \nabla \boldsymbol{\phi}_h) + (\overline{\mathbf{v}}_h^{h\,n+1}, \boldsymbol{\phi}_h) = (\mathbf{v}_h^{n+1}, \boldsymbol{\phi}_h), \quad \forall \boldsymbol{\phi}_h \in V^h \tag{48}$$

and for the sensitivity solution, find $\mathbf{s}_h^{n+1} \in V^h$, for $n = 0, 1, 2....M-1$ satisfying,

$$\frac{1}{\triangle t}(\mathbf{s}_h^{n+1} - \mathbf{s}_h^n, \boldsymbol{\phi}_h) + \nu a(\mathbf{s}_h^{n+1/2}, \boldsymbol{\phi}_h) + b^*(\mathbf{s}_h^{n+1/2}, \mathbf{v}_h^{n+1/2}, \boldsymbol{\phi}_h) + b^*(\mathbf{v}_h^{n+1/2}, \mathbf{s}_h^{n+1/2}, \boldsymbol{\phi}_h)$$
$$+(\mathbf{v}_h^{n+1/2} - \overline{\mathbf{v}}_h^{h\,n+1/2}, \boldsymbol{\phi}_h) + \chi(\mathbf{s}_h^{n+1/2} - \mathbf{u}_h^{n+1/2}, \boldsymbol{\phi}_h) = 0, \quad \forall \boldsymbol{\phi}_h \in V^h \tag{49}$$
$$\delta^2(\nabla \mathbf{u}_h^{n+1}, \nabla \boldsymbol{\phi}_h) + (\mathbf{u}_h^{n+1}, \boldsymbol{\phi}_h) = (\mathbf{s}_h^{n+1}, \boldsymbol{\phi}_h), \quad \forall \boldsymbol{\phi}_h \in V^h \tag{50}$$

The $\chi-$sensitivity is evaluated on several benchmark problems that consider a variety of mesh scales and Reynolds numbers. Each sensitivity calculation is computed by both the FFD and SEM techniques. One such benchmark is the 2D lid-driven cavity problem where $\Omega = [0,1]^2$, and $0 \le t \le 1.0$. The initial and boundary conditions for the linear TRM velocity are given by,

$$\begin{aligned}
\mathbf{v}(\mathbf{x}, t) &= (16x^2(1-x)^2, 0)^T, \ \mathbf{x} \in \partial\Omega \cap \{y = 1\}, \\
\mathbf{v}(\mathbf{x}, t) &= \mathbf{0}, \ \mathbf{x} \in \partial\Omega \cap \{y < 1\}, \\
\mathbf{v}(\mathbf{x}, 0) &= (3y^2 - 2y, 0)^T.
\end{aligned}$$

The $\chi-$sensitivity estimates are enumerated in Tables 3 and 4. It is with these values that one can determine the reliability of a the linear time relaxation model when $\chi \in [0.1, 10.0]$.

Table 3. Computations of $\chi \, ||\mathbf{s}_{SEM}||_{L^2([0,1];L^2(\Omega))}$ for the lid-driven cavity problem.

χ	$Re = 1000$	$Re = 5000$	$Re = 10{,}000$
0.01	1.068×10^{-4}	1.913×10^{-4}	2.248×10^{-4}
0.1	1.036×10^{-3}	1.840×10^{-3}	2.155×10^{-3}
1.0	8.033×10^{-3}	1.314×10^{-2}	1.493×10^{-2}
10.0	5.038×10^{-2}	6.006×10^{-2}	6.215×10^{-2}

Table 4. Computations of $\chi\,||\mathbf{s}_{FFD}||_{L^2([0,1];L^2(\Omega))}$ for the lid-driven cavity problem.

χ	$Re = 1000$	$Re = 5000$	$Re = 10{,}000$
0.01	1.075×10^{-4}	1.996×10^{-4}	2.402×10^{-4}
0.1	1.056×10^{-3}	1.935×10^{-3}	2.317×10^{-3}
1.0	8.994×10^{-3}	1.466×10^{-2}	1.675×10^{-2}
10.0	3.173×10^{-2}	3.650×10^{-2}	3.734×10^{-2}

5.1. Comparing the SEM and FFD Methods

In [22], the authors used both the SEM and FFD techniques, under the assumption they would provide similar results in the tested settings. This appears to have been done as a form of sanity check. Neither technique demonstrates an overall advantage. For instance, both methods require effectively doubling all computational efforts by solving a second system of equations. Both methods are also easily implemented in working code bases.

The only situation that would necessitate the use of FFD over the SEM is if there is good reason to doubt the existence of the implicit derivatives in (37). On the other hand, a FFD computation is effectively meaningless if the $\mathbf{v}(\chi = 0)$ case is unstable or otherwise unreliable, which is not uncommon in under-resolved flows at higher Reynolds numbers. The choice of implementing one technique over the other is largely problem-specific, and a question best left to the good judgment of future authors.

5.2. Determining Reliability of Time-Relaxation Models χ through Sensitivity Study

As mentioned at the beginning of this section, determining an optimal choice of χ is a difficult endeavor. Sensitivity studies can be used with great effect to help quantify the reliability of such calculations. While [22] did not explore particular methods to directly quantify model reliability, examples can be found in [23,24]. One such technique is to quantify reliability by using the linear estimate,

$$||\mathbf{v}_h(0) - \mathbf{v}_h(\chi)|| \leq \chi||\mathbf{s}_h|| + \mathcal{O}(\chi^2). \tag{51}$$

The above estimate holds for any norm.

If, for example, the total kinetic energy of a particular flow is known or can be accurately estimated, enforcing a %−accuracy as an upper bound can determine permissible intervals of model-reliability. Likewise, if for a given norm, the goal is for the computed results of the TRM model to be accurate to within 5% of $||\mathbf{u}_{true}||$, we simply attempt to find an interval $0 < a \leq \chi \leq b$ such that,

$$\chi||\mathbf{s}_h|| \leq 0.05||\mathbf{u}_{true}||. \tag{52}$$

To date, this type of study has not been performed on time relaxation models, and would make for interesting future work.

6. Conclusions and Open Problems

Herein we have presented a summary of a selection of published results about and relevant to time relaxation models. We briefly introduced the technique and the motivations for their development, as well as the results from several computational implementations. While numerous studies have been performed in literature, a number of open questions remain.

Much of the numerical results surveyed herein use the Crank–Nicolson time-stepping scheme. The behavior of linear and nonlinear TRMs with alternate time-stepping schemes, like BDF2 or schemes that permit a variable time-step, remains an open problem. Higher-order time-stepping schemes have also not been considered.

It would be interesting to see the extent to which the energy dissipation penalty for the nonlinear TRM parameter, as determined by χ and r, affects the energy cascade. Furthermore, numerical experiments have not been seen for nonlinear orders greater than the quadratic case, $r > 3$.

In fluid flow problems, approximate deconvolution has been largely accomplished by the van Cittert approximate deconvolution technique. Other techniques exist, i.e., Guerts approximate inverse operators [25], and others. Time relaxation models would be an excellent test bed for such studies, given their simplicity.

So far, the only parameter sensitivity investigation for TRMs was performed in [22], wherein analysis and experiments to estimate $\chi-$sensitivity were done. It would be interesting to see a full analysis of the δ, r, and χ sensitivities for both the linear and nonlinear TRMs. It would also be interesting to see how these parameters affect the outcomes of important functionals, i.e., forces acting on submerged bodies like lift and drag, or physical quantities like energy, enstrophy, helicity, and their dissipations. Finally, a full investigation into the reliability of time relaxation models in challenging computational settings would be of particular interest. This is briefly discussed in Section 5.2.

Conflicts of Interest: The authors declare no conflict of interest.

References

1. Kistler, R. A Study of Data Assimilation Techniques in an Autobarotropic Primitive Equation Channel Model. Master's Thesis, Penn State University, State College, PA, USA, 1974.
2. Hoke, J.; Anthes, R. The initialization of numerical models by a dynamic-initialization technique. *Mon. Weather Rev.* **1976**, *104*, 1551–1556.
3. Kolmogorov, A. The local structure of turbulence in incompressible viscous fluids for very large Reynolds numbers. *Dokl. Akad. Nauk SSR* **1941**, *30*, 9–13.
4. Stolz, S.; Adams, N.; Kleiser, L. The approximate deconvolution model for large-eddy simulations of compressible flows and its application to shock-turbulent-boundary-layer interaction. *Phys. Fluids* **2001**, *13*, 2985–3001.
5. Stolz, S.; Adams, N.; Kleiser, L. An approximate deconvolution model for large-eddy simulation with application to incompressible wall-bounded flows. *Phys. Fluids* **2001**, *13*, 997–1015.
6. Layton, W.; Neda, M. Truncation of scales by time relaxation. *J. Math. Anal. Appl.* **2007**, *325*, 788–807.
7. Ervin, V.; Layton, W.; Neda, M. Numerical analysis of a higher order time relaxation model of fluids. *Int. J. Numer. Anal. Model.* **2007**, *4*, 648–670.
8. Neda, M. Discontinuous time relaxation method for the time-dependent Navier-Stokes equations. *Adv. Numer. Anal.* **2010**, *2010*, 21.
9. Neda, M.; Sun, X.; Yu, L. Increasing accuracy and efficiency for regularized Navier-Stokes equations. *Acta Appl. Math.* **2012**, *118*, 57–79.
10. De, S.; Hannasch, D.; Neda, M.; Nikonova, E. Numerical analysis and computations of a high accuracy time relaxation fluid flow model. *Int. J. Comput. Math.* **2012**, *89*, 2353–2373.
11. Dunca, A.; Neda, M. Numerical analysis of a nonlinear time relaxation model of fluids. *J. Math. Anal. Appl.* **2014**, *420*, 1095–1115.
12. Takhirov, A.; Neda, M.; Waters, J. Time relaxation algorithm for flow ensembles. *Numer. Methods Partial Differ. Equ.* **2016**, *32*, 757–777.
13. Neda, M.; Waters, J. Finite element computations of time relaxation algorithm for flow ensembles. *Appl. Eng. Lett.* **2016**, *1*, 51–56.
14. Germano, M. Differential filters for the large eddy numerical simulation of turbulent flows. *Phys. Fluids* **1986**, *29*, 1755–1757.
15. Bertero, M.; Boccacci, P. *Introduction to Inverse Problems in Imaging*; CRC Press: Boca Raton, FL, USA, 1998.
16. Rosenau, P. Extending hydrodynamics via the regularization of the Chapman–Enskog expansion. *Phys. Rev. A* **1989**, *40*, 7193–7196.
17. Schochet, S.; Tadmor, E. The regularized Chapman–Enskog expansion for scalar conservation laws. *Arch. Ration. Mech. Anal.* **1992**, *119*, 95–107.

18. Volker, J. Reference values for drag and lift of a two-dimensional time-dependent flow around a cylinder. *Int. J. Numer. Methods Fluids* **2004**, *44*, 777–788.
19. Connors, J.; Layton, W. On the Accuracy of the Finite Element Method Plus Time Relaxation. *Math. Comput.* **2010**, *79*, 619–648.
20. Hecht, F. New development in FreeFem++. *J. Numer. Math.* **2012**, *20*, 251–265.
21. Schäfer, M.; Turek, S. The benchmark problem 'flow around a cylinder'. In *Flow Simulation with High-Performance Computers II, Notes on Numerical Fluid Mechanics*; Vieweg + Teubner Verlag: Wiesbaden, Germany, 1996; Volume 52, pp. 547–566.
22. Neda, M.; Pahlevani, F.; Waters, J. Sensitivity Analysis of the Time Relaxation Model. *Appl. Math. Mech.* **2015**, *7*, 89–115.
23. Pahlevani, F.; Davis, L. Parameter Sensitivity of an Eddy Viscosity Model: Analysis, Computation, and It's Application to Quantifying Model Reliability. *Int. J. Uncertain. Quantif.* **2013**, *3*, 397–419.
24. Breckling, S.; Neda, M.; Pahlevani, F. Sensitivity Analyses of the Navier-Stokes-α Model. **2017**, submitted.
25. Geurts, B.J. Inverse modeling for large-eddy simulation. *Phys. Fluids* **1997**, *9*, 3585–3587.

fluids

MDPI

Article

Improving Accuracy in α-Models of Turbulence through Approximate Deconvolution

Argus A. Dunca

Department of Mathematics and Informatics, University Politehnica of Bucharest, București 060042, Romania; argus_dunca@yahoo.com or adunca@mathem.pub.ro

Received: 29 June 2017; Accepted: 11 October 2017; Published: 27 October 2017

Abstract: In this report, we present several results in the theory of α-models of turbulence with improved accuracy that have been developed in recent years. The α-models considered herein are the Leray-α model, the zeroth Approximate Deconvolution Model (ADM) turbulence model, the modified Leray-α and the Navier–Stokes-α model. For all of the models from above, the accuracy is limited to α^2 in smooth flow regions. Better accuracy requires decreasing the filter radius α, which, in turn, requires a smaller mesh width that will lead in the end to a higher computational cost. Instead, one can use approximate deconvolution (without decreasing the mesh size) to attain better accuracy. Such deconvolution methods have been considered recently in many studies that show the efficiency of this approach. For smooth flows, periodic boundary conditions and van Cittert deconvolution operator of order N, the expected accuracy is α^{2N+2}. In a bounded domain, such results are valid only in case special conditions are satisfied. In more general conditions, the author has recently proved that, in the case of the ADM, the expected accuracy of the finite element method with Taylor–Hood elements and Crank–Nicolson time stepping method is $\Delta t^2 + h^2 + K^N \alpha^2$, where the constant $K < 1$ depends on the ratio α/h, which is assumed constant. In this study, we present the extension of the result to the rest of the models.

Keywords: large eddy simulation; turbulence; approximate deconvolution; discrete Stokes filter; finite element method

1. Introduction

At a high Reynolds number, turbulence is not efficient to simulate using the Navier–Stokes equations because they require a very fine mesh and a high computational cost [1]. Therefore, in practical applications, reduced order turbulence models, such as the α-models of turbulence discussed herein, have to be used.

The α-models of turbulence are regularizations of the Navier–Stokes equations (NSE) whose goal is to allow stable computations on coarser grids than NSE. They are given by

$$\begin{aligned} \mathbf{w}_t + N(\mathbf{w}) - \nu\Delta\mathbf{w} + \nabla p &= \mathbf{f} && \text{in } \Omega, \\ \nabla\cdot\mathbf{w} &= 0 && \text{in } \Omega, \\ \mathbf{w}(\mathbf{x},0) &= \mathbf{w}_0(\mathbf{x}) && \text{in } \Omega. \end{aligned} \tag{1}$$

Here, the nonlinearity N is $N(\mathbf{w}) = \overline{\mathbf{w}}\cdot\nabla\mathbf{w}$ in the Leray-α model [2], $N(\mathbf{w}) = \mathbf{w}\cdot\nabla\overline{\mathbf{w}}$ in the case of the modified Leray-α model [3], $N(\mathbf{w}) = \overline{\mathbf{w}}\cdot\nabla\overline{\mathbf{w}}$ in the case of the zeroth Approximate Deconvolution Model (ADM) model [4] (here, the ADM is formulated as in [5]), and $N(\mathbf{w}) = (\nabla\times\mathbf{w})\times\overline{\mathbf{w}}$ in the case of the Navier–Stokes-α model.

In the above formula, the average $\overline{\mathbf{w}}$ is the solution of the BVP [6]

$$\begin{aligned} \overline{\mathbf{w}} - \alpha^2\Delta\overline{\mathbf{w}} + \nabla p &= \mathbf{w} \text{ in } \Omega, \\ \overline{\mathbf{w}} &= \mathbf{w} \text{ on } \partial\Omega. \end{aligned} \tag{2}$$

The study of α-model of turbulence started with the work of Leray who introduced the Leray-α model in [7,8] (with the filter defined as the convolution with the Gaussian) and later studied in [2,9,10]. The other three models have been introduced and studied in papers such as [3,4,11–20].

One property that all four models have in common is that their accuracy $||\mathbf{u}_{NSE} - \mathbf{w}_{\alpha-model}||$ is limited to $\mathcal{O}(\alpha^2)$ for smooth flows. In practice, this may lead to over-diffusivity and a lack of accuracy that requires smaller filter radius α and, as a consequence, a smaller mesh size that in turn causes higher computational costs. One way to correct this deficiency is to use an approximate deconvolution operator D having the approximation property $\mathbf{u} \approx D\overline{\mathbf{u}}$. One example of such deconvolution operator is the van Cittert deconvolution operator that approximates $\mathbf{u}$ by $D_N\overline{\mathbf{u}}$ defined as

$$D_N := \sum_{n=0}^{N} (I - G)^n , \quad N = 0, 1, 2, \dots . \tag{3}$$

For $N = 0, 1, 2$, the velocity field $\mathbf{u}$ will be approximated by

$$\begin{array}{lcll} \mathbf{u} & \approx & \overline{\mathbf{u}} & N = 0 \\ \mathbf{u} & \approx & 2\overline{\mathbf{u}} - \overline{\overline{\mathbf{u}}} & N = 1 \\ \mathbf{u} & \approx & 3\overline{\mathbf{u}} - 3\overline{\overline{\mathbf{u}}} + \overline{\overline{\overline{\mathbf{u}}}} & N = 2 \end{array} \tag{4}$$

The finite element analysis of the van Cittert deconvolution method has been studied in [21]. The α-models enhanced using approximate deconvolution become

$$\begin{array}{rll} \mathbf{w}_t + DN(\mathbf{w}) - \nu\Delta\mathbf{w} + \nabla p & = \mathbf{f} & \text{in } \Omega, \\ \nabla \cdot \mathbf{w} & = 0 & \text{in } \Omega, \\ \mathbf{w}(\mathbf{x}, 0) & = \mathbf{w}_0(\mathbf{x}) & \text{in } \Omega, \end{array} \tag{5}$$

where the nonlinearity DN is $DN(\mathbf{w}) = D\overline{\mathbf{w}} \cdot \nabla\mathbf{w}$ in the Leray-deconvolution model, $DN(\mathbf{w}) = \mathbf{w} \cdot \nabla D\overline{\mathbf{w}}$ in the case of the modified Leray-deconvolution model, $DN(\mathbf{w}) = D\overline{\mathbf{w}} \cdot \nabla D\overline{\mathbf{w}}$ in the case of the ADM model (formulated as in [21]) and $DN(\mathbf{w}) = (\nabla \times \mathbf{w}) \times D\overline{\mathbf{w}}$ in the case of the enhanced Navier–Stokes-α model.

The Leray-deconvolution model has been introduced in [11] and later studied numerically in [22]. The ADM model (together with its zeroth version) has been introduced and studied in [13,14,23] and later studied in [20,24–29] (see also the monograph [30]). The deconvolution-enhanced Navier–Stokes-α model has been introduced and studied in [19] and also investigated numerically in [31–33]. To the author's knowledge, the modified Leray-deconvolution model has not been studied so far.

Most of the analysis for the deconvolution-enhanced α-models has been carried out in the context of van Cittert deconvolution. For periodic boundary conditions and assuming a smooth NSE solution, the modeling error is

$$||\mathbf{u}_{NSE} - \mathbf{w}||_{L^\infty(L^2)} + ||\mathbf{u}_{NSE} - \mathbf{w}||_{L^2(H^1)} = \mathcal{O}(\alpha^{2N+2}) \tag{6}$$

where $\mathbf{w}$ is the solution of the deconvolution enhanced Leray-α or the ADM or the deconvolution enhanced Navier–Stokes-α model [5,11,19,24,34]. Here, N is the order of the van Cittert deconvolution operator D_N. In a bounded domain, such estimates are valid only if special boundary conditions are satisfied by the exact NSE solution $\mathbf{u}$ [35].

The finite element analysis has been carried out for the Leray-deconvolution and the deconvolution enhanced Navier–Stokes-α in [19,22] with Crank–Nicolson time discretization and $P2/P1$ elements and the main error estimate is

$$||\mathbf{u}_{NSE} - \mathbf{w}_h||_{L^\infty(L^2)} + ||\mathbf{u}_{NSE} - \mathbf{w}_h||_{L^2(H^1)} = \mathcal{O}(\Delta t^2 + h^2 + \alpha^{2N+2}) \tag{7}$$

in the case of periodic boundary conditions. Here, the filter radius α satisfies $\alpha = \mathcal{O}(h)$. In the above formula, the term α^{2N+2} is induced by the deconvolution modeling. The case $N = 0$ corresponds to the classical α turbulence models. In the case of a bounded domain, the formula above is no longer valid [35], and one can only prove that, for all N, the error satisfies

$$||\mathbf{u}_{NSE} - \mathbf{w}_h||_{L^\infty(L^2)} + ||\mathbf{u}_{NSE} - \mathbf{w}_h||_{L^2(H^1)} = \mathcal{O}(\Delta t^2 + h^2 + \alpha^2) \tag{8}$$

Recently, in [29], the following estimate has been proved for the ADM:

$$||\mathbf{u}_{NSE} - \mathbf{w}_h||_{L^\infty(L^2)} + ||\mathbf{u}_{NSE} - \mathbf{w}_h||_{L^2(H^1)} = \mathcal{O}(\Delta t^2 + h^2 + K^N\alpha^2) \tag{9}$$

In the above estimate, N is the order of the van Cittert deconvolution operator and $K < 1$ is a constant that only depends on the ratio α/h (which is assumed constant) and the sequence of quasiuniform meshes used in the computation. The term $K^N\alpha^2$ is due to the deconvolution and it can be made small by increasing N and decreasing the ratio α/h [21]. This estimate supports the claim that high order deconvolution operators indeed improve accuracy, a behavior that has been observed in the numerical tests in [22,29].

We believe that this is the general behavior of deconvolution enhanced α-models of turbulence in case the van Cittert deconvolution procedure is used. Using similar techniques such as those in [29], similar estimates can be obtained for the Leray-deconvolution model, the Navier–Stokes α deconvolution model and the modified Leray-α deconvolution model.

2. Mathematical Context

The mathematical notations and concepts are similar to the ones used in [19,22,29]. $\Omega \subset \mathbb{R}^d$, $d = 2,3$, is regular, bounded, polyhedral domain and $\|\cdot\|$ and $(\cdot,\cdot)$ is the $L^2(\Omega)$ norm and inner product. Lebesgue spaces and their norms are denoted by $L^p(\Omega)$, $\|\cdot\|_{L^p}$. We use standard notations for the Lebesque and Sobolev spaces and their norms.

In the variational problem, we use the functional spaces [36]:

$$\begin{aligned} \text{Velocity Space} - X &:= [H_0^1(\Omega)]^d \\ \text{Pressure Space} - Q &:= L_0^2(\Omega) = \{q \in L^2(\Omega) : \textstyle\int_\Omega q\,dx = 0\} \\ \text{Divergence} - \text{free Space} - V &:= \{v \in X : \textstyle\int_\Omega q\nabla\cdot\mathbf{v}\,dx = 0,\ \forall\, q \in Q\} \end{aligned} \tag{10}$$

The finite element analysis is carried out on a family of triangulations $(\mathcal{T}_h)_h$ on Ω that are quasiuniform [37]. We also consider an inf-sup stable pair of finite elements (X_h, Q_h) where $X_h \subset X$, $Q_h \subset Q$, which are assumed to satisfy [37]:

$$\begin{aligned} \|\mathbf{u} - I\mathbf{u}_h\| &\le Ch^{k+1}\|\mathbf{u}\|_{k+1}, \ \ \mathbf{u} \in H^{k+1}(\Omega)^d \\ \|\mathbf{u} - I\mathbf{u}_h\|_1 &\le Ch^k\|\mathbf{u}\|_{k+1}, \ \ \mathbf{u} \in H^{k+1}(\Omega)^d \\ \|p - Ip_h\| &\le Ch^k\|p\|_k, \ \ p \in H^k(\Omega) \end{aligned} \tag{11}$$

where $I\mathbf{u}_h \in X_h$ is an interpolant of $\mathbf{u}$ and $Ip_h \in Q_h$ is an interpolant of p.

The discretely divergence-free space V_h is

$$V_h := \{\mathbf{v} \in X_h : (q, \nabla\cdot\mathbf{v}) = 0,\ \forall q \in Q_h\} \tag{12}$$

We assume that V_h satisfies the first two approximation properties in the inequalities (11) above if $\mathbf{u} \in V$.

We let $t_n = n\Delta t$, $t_{n+1/2} = (n+1/2)\Delta t$, $T := N_T\Delta t$. Here, Δt is the chosen time step, N_T is the number of time steps, and T is the final time. We will also use the notations $\mathbf{v}^n = \mathbf{v}(n\Delta t)$ and $\mathbf{v}^{n+1/2} = (\mathbf{v}^n + \mathbf{v}^{n+1})/2$ for a function $\mathbf{v}$.

We will assume that the L^2 projection operator onto V_h denoted by $P_{V_h} : L^2(\Omega) \to V_h \subset L^2(\Omega)$ is H^1 stable. The analysis also use the inverse inequality [37]:

Lemma 1.
$$||\nabla \mathbf{v}_h|| \le \frac{C_{\mathcal{T}}}{h}||\mathbf{v}_h||, \ \ \forall \mathbf{v}_h \in X^h \tag{13}$$

The constant $C_{\mathcal{T}}$ does not depend on h.

Definition 1. *[30] The discrete Stokes operator $A_h : V_h \to V_h$ satisfies $(A_h v, v_h) = (\nabla v, \nabla v_h)$ for any $v_h \in V_h$.*

One may easily show that A_h is a bijective, self-adjoint, positive operator with respect to the inner products $(\cdot,\cdot)$ and $(\nabla\cdot,\nabla\cdot)$ on V_h.

The eigenvalues of A_h are denoted by $\lambda_1 < \lambda_2 < \ldots < \lambda_M$.

One may show using the inverse inequality (13) that

$$\lambda_M \le \frac{C_{\mathcal{T}}^2}{h^2}, \ \ ||A_h\mathbf{v}|| \le \frac{C_{\mathcal{T}}}{h}||\nabla\mathbf{v}||, \ \ ||A_h\mathbf{v}|| \le \frac{C_{\mathcal{T}}^2}{h^2}||\mathbf{v}|| \tag{14}$$

for $\mathbf{v} \in V_h$.

Definition 2. *(Discrete Stokes average [30]) For given $\alpha > 0$, we let $G : V_h \to V_h$,*

$$G\mathbf{v} = \overline{\mathbf{v}}^h = (I + \alpha^2 A_h)^{-1}\mathbf{v} \tag{15}$$

When $\mathbf{v} \in X$, we set $\overline{\mathbf{v}}^h := GP_{V_h}\mathbf{v}$.

Theorem 1. *[29] Due to the spectral mapping theorem, the eigenvalues of the discrete Stokes filter G are $\frac{1}{1+\alpha^2\lambda_i}$, $i = 1,2,...,M$, and they satisfy*

$$\frac{1}{1 + C_{\mathcal{T}}^2\frac{\alpha^2}{h^2}} \le \frac{1}{1+\alpha^2\lambda_M} < \frac{1}{1+\alpha^2\lambda_2} < \ldots < \frac{1}{1+\alpha^2\lambda_1} < 1 \tag{16}$$

Therefore, $||\overline{\mathbf{v}}^h|| \le ||\mathbf{v}||$, $||\nabla\overline{\mathbf{v}}^h|| \le ||\nabla\mathbf{v}||$ for all $\mathbf{v} \in V_h$.

One may also notice that the largest eigenvalue of $I - G$ is bounded by

$$K = \frac{C_{\mathcal{T}}^2\frac{\alpha^2}{h^2}}{1 + C_{\mathcal{T}}^2\frac{\alpha^2}{h^2}} < 1 \tag{17}$$

Therefore,

$$||I - G|| \le K \tag{18}$$

Definition 3. *[30] The N-th order discrete van Cittert deconvolution operator $D_N : V_h \to V_h$ is given by:*

$$D_N := \sum_{n=0}^{N}(I - G)^n, \ \ N = 0,1,2,\ldots. \tag{19}$$

Several properties of D_N that can be found in [30] are listed below. D_N is self-adjoint, positive with respect to the inner products $(\cdot,\cdot)$ and $(\nabla\cdot,\nabla\cdot)$. $D_N G = I - (I-G)^{N+1}$; therefore, the eigenvalues of $D_N G$ are $1 - \left(\frac{\alpha^2\lambda_i}{1+\alpha^2\lambda_i}\right)^{N+1}$, $i = 1\ldots M$, and can be bounded as [29]:

$$1-\left(\frac{C_{\mathcal{T}}^2\frac{\alpha^2}{h^2}}{1+C_{\mathcal{T}}^2\frac{\alpha^2}{h^2}}\right)^{N+1}<1-\left(\frac{\alpha^2\lambda_M}{1+\alpha^2\lambda_M}\right)^{N+1}<\ldots<1-\left(\frac{\alpha^2\lambda_1}{1+\alpha^2\lambda_1}\right)^{N+1}<1 \quad (20)$$

Following [19], we define two norms on V_h

$$||\mathbf{v}||_E := (\mathbf{v}, D_N G\mathbf{v})^{1/2}, \ ||\mathbf{v}||_\epsilon := (\nabla\mathbf{v}, \nabla D_N G\mathbf{v})^{1/2} \quad (21)$$

Using inequalities (20), it follows that, see also for more details [29],

$$\begin{array}{lclcl} \sqrt{1-K^{N+1}}||\mathbf{v}|| & \leq & ||\mathbf{v}||_E & \leq & ||\mathbf{v}|| \\ \sqrt{1-K^{N+1}}||\nabla\mathbf{v}|| & \leq & ||\mathbf{v}||_\epsilon & \leq & ||\nabla\mathbf{v}|| \end{array} \quad (22)$$

and

$$\begin{array}{l} ||D_N G\mathbf{v}|| \leq ||\mathbf{v}||_E, \ ||\nabla D_N G\mathbf{v}|| \leq ||\mathbf{v}||_\epsilon, \ \forall \mathbf{v} \in V_h \\ ||D_N G P_{V_h}\mathbf{v}|| \leq ||\mathbf{v}||, \ ||\nabla D_N G P_{V_h}\mathbf{v}|| \leq C||\nabla\mathbf{v}||, \ \forall \mathbf{v} \in X \end{array} \quad (23)$$

We will also use the next estimate of the discrete deconvolution error $\mathbf{v} - D_N\overline{\mathbf{v}}^h$ that has been obtained in [21].

Theorem 2. *[21] If $\mathbf{v} \in H^{k+1}(\Omega) \cap V$, $k \geq 1$ and $\alpha = \mathcal{O}(h)$, then*

$$\begin{array}{lcl} ||\mathbf{v} - D_N\overline{\mathbf{v}}^h|| & \leq & C(h^2K^N||\Delta\mathbf{v}|| + h^{k+1}|\mathbf{v}|_{k+1}) \\ ||\nabla\mathbf{v} - \nabla D_N\overline{\mathbf{v}}^h|| & \leq & C(hK^N||\Delta\mathbf{v}|| + h^k|\mathbf{v}|_{k+1}) \end{array} \quad (24)$$

Here, N is the order of the discrete deconvolution operator and $K < 1$ has been previously defined in (17).

Proof. We recall the main ideas of the proof presented in [21]. First, we split the error $||\mathbf{v} - D_N\overline{\mathbf{v}}^h||$ using the triangle inequality

$$||\mathbf{v} - D_N\overline{\mathbf{v}}^h|| \leq ||\mathbf{v} - P_{V_h}\mathbf{v}|| + ||P_{V_h}\mathbf{v} - D_N\overline{\mathbf{v}}^h|| \quad (25)$$

The term $||\mathbf{v} - P_{V_h}\mathbf{v}||$ is bounded by $Ch^{k+1}|\mathbf{v}|_{k+1}$ due to the approximation properties of V_h. It remains then to estimate $||P_{V_h}\mathbf{v} - D_N\overline{\mathbf{v}}^h||$. Since $D_N G = I - (I-G)^{N+1}$, this term is written as

$$||P_{V_h}\mathbf{v} - D_N\overline{\mathbf{v}}^h|| = ||(I-G)^{N+1}P_{V_h}\mathbf{v}|| \quad (26)$$

It follows that

$$[||P_{V_h}\mathbf{v} - D_N\overline{\mathbf{v}}^h|| \leq ||I-G||^N||(I-G)P_{V_h}\mathbf{v}|| \leq K^N\alpha^2||A_h\overline{\mathbf{v}}^h|| \quad (27)$$

The discrete Laplacian $||A_h\overline{\mathbf{v}}^h||$ can be estimated as

$$||A_h\overline{\mathbf{v}}^h|| = ||\overline{A_h P_{V_h}\mathbf{v}}^h|| \leq C||A_h P_{V_h}\mathbf{v}|| \leq C||\Delta\mathbf{v}|| \quad (28)$$

The last inequality from above can be found in [38]. It follows that

$$||P_{V_h}\mathbf{v} - D_N\overline{\mathbf{v}}^h|| \leq CK^N\alpha^2||\Delta\mathbf{v}|| \quad (29)$$

This proves the first inequality in inequalities (24).

The second inequality in inequalities (24) is immediatelly proved using the same arguments combined with inverse inequalities.

□

3. Numerical Scheme for the α-Models of Turbulence

In our analysis $b^*(\cdot,\cdot,\cdot) : X \times X \times X \to \mathbb{R}$ will denote the standard skew-symmetrized trilinear form [36]:

$$b^*(\mathbf{u},\mathbf{v},\mathbf{w}) := \frac{1}{2}(\mathbf{u}\cdot\nabla\mathbf{v},\mathbf{w}) - \frac{1}{2}(\mathbf{u}\cdot\nabla\mathbf{w},\mathbf{v}) \tag{30}$$

One can show that the trilinear form satisfies [39]:

$$|b^*(\mathbf{u},\mathbf{v},\mathbf{w})| + |b^*(\mathbf{v},\mathbf{u},\mathbf{w})| \le C||\mathbf{u}||\,(||\mathbf{v}||_{L^\infty} + ||\nabla\mathbf{v}||_{L^3})\,||\nabla\mathbf{w}|| \tag{31}$$

It follows then using inequalities (35) and (36) in [38] that

$$|b^*(\mathbf{u},\mathbf{v},\mathbf{w})| + |b^*(\mathbf{v},\mathbf{u},\mathbf{w})| \le C||\mathbf{u}||\,||A_h\mathbf{v}||\,||\nabla\mathbf{w}|| \tag{32}$$

for $\mathbf{u},\mathbf{w} \in X$, $\mathbf{v} \in V_h$.

In the case of the Navier–Stokes in rotational form, the trilinear form is

$$\tilde{b}(\mathbf{u},\mathbf{v},\mathbf{w}) = ((\nabla\times\mathbf{v})\times\mathbf{u},\mathbf{w}) \tag{33}$$

For body force $\mathbf{f} \in L^2((0,T],X)$ and initial velocity $\mathbf{u}_0 \in X$, the discrete solution $\mathbf{w}_h^{n+1} \in V_h$ of the accuracy enhanced α model of turbulence at step $n+1$ for $n = 0,1,2,...,N_T-1$ satisfies:

$$\begin{aligned}
\left(\frac{\mathbf{w}_h^{n+1}-\mathbf{w}_h^n}{\Delta t},\mathbf{v}\right) &+ B(\mathbf{w}_h^{n+1/2},\mathbf{v}) - (p_h^{n+1/2},\nabla\cdot\mathbf{v})\\
+\nu(\nabla\mathbf{w}_h^{n+1/2},\nabla\mathbf{v}) &= (\mathbf{f}((n+1/2)\Delta t),\mathbf{v}),\ \forall\mathbf{v}\in X_h\\
(q,\nabla\cdot\mathbf{w}_h^{n+1}) &= 0\,,\ \forall q\in P_h\\
(\mathbf{w}_h^0,\mathbf{v}) &= (\mathbf{w}_0,\mathbf{v})\,,\forall\mathbf{v}\in X_h
\end{aligned} \tag{34}$$

where

$$\begin{aligned}
B(\mathbf{w}_h^{n+1/2},\mathbf{v}) &= b^*(D_N\overline{\mathbf{w}_h^{n+1/2}}^h,\mathbf{w}_h^{n+1/2},\mathbf{v}),\ \text{Leray-deconvolution model [22]},\\
B(\mathbf{w}_h^{n+1/2},\mathbf{v}) &= b^*(\mathbf{w}_h^{n+1/2},D_N\overline{\mathbf{w}_h^{n+1/2}}^h,\mathbf{v}),\ \text{modified Leray-deconvolution model},\\
B(\mathbf{w}_h^{n+1/2},\mathbf{v}) &= b^*(D_N\overline{\mathbf{w}_h^{n+1/2}}^h,D_N\overline{\mathbf{w}_h^{n+1/2}}^h,\mathbf{v}),\ \text{ADM [5,29]},\\
B(\mathbf{w}_h^{n+1/2},\mathbf{v}) &= \tilde{b}(D_N\overline{\mathbf{w}_h^{n+1/2}}^h,\mathbf{w}_h^{n+1/2},\mathbf{v}),\ \text{NS-}\alpha\text{ deconvolution model [19]}.
\end{aligned} \tag{35}$$

Thorough the analysis, we assume that α/h is constant.

Lemma 2. *(Stability) A finite element solution $\mathbf{w}_h^{n+1}$ of the model (34) exists at each time step $n = 0,\dots N_T-1$ and satisfies the stability estimate*

$$\begin{aligned}
\sup_{0\le n<N_T} ||\mathbf{w}_h^{n+1}||^2 + \Delta t\,\nu \textstyle\sum_{n=0}^{N_T-1} ||\nabla\mathbf{w}_h^{n+1/2}||^2&\\
\le C\left(\frac{\Delta t}{\nu}\sum_{n=0}^{N_T-1}||f(t^{n+1/2})||_{-1}^2 + ||\mathbf{w}_0||^2\right)&
\end{aligned} \tag{36}$$

where C does not depend on h,α (but it depends on α/h).

Proof. The Leray-deconvolution model case has been studied (for a slightly different filter operator, but the argument is still valid) in [22]. The NS-α-deconvolution case in [19]. The modified Leray-deconvolution case can be proved using arguments similar to the NS α-deconvolution (since it requires a similar multiplier to cancel the nonlinearity).

The ADM case has been studied in [29] and shares great similarity with the stability proofs for the Leray-deconvolution and the NS-α-deconvolution models mentioned previously. We outline the ideas from [29] here.

To prove the existence of a discrete solution $\mathbf{u}_h^{n+1} \in V_h$, we consider the operator $L : V_h \to V_h$, $L\varphi = \psi$ such that φ, ψ satisfy the equality

$$\begin{aligned} 2(\psi, \mathbf{v})_E + \Delta t \nu (\psi, \mathbf{v})_\epsilon = 2(\mathbf{u}_h^n, \mathbf{v})_E - \Delta t b^*(D_N\overline{\varphi}^h, D_N\overline{\varphi}^h, D_N G\mathbf{v}) \\ + \Delta t(\mathbf{f}(t^{n+1/2}), D_N G\mathbf{v}) \quad \forall \mathbf{v} \in V_h \end{aligned} \tag{37}$$

We notice that, if φ is a fixed point of L, then $\mathbf{u}_h^{n+1} = 2\varphi - \mathbf{u}_h^n$ is a solution of the ADM model (34).

In order to find a fixed point of φ, we apply the Leray–Schauder fixed point theorem. To show that L is a compact operator, we notice that it is the composition of a continuous, linear operator $T : V_h \to V_h$, $T\zeta = \psi$, where

$$2(\psi, \mathbf{v})_E + \Delta t \nu (\psi, \mathbf{v})_\epsilon = (\zeta, \mathbf{v}) \quad \forall \mathbf{v} \in V_h \tag{38}$$

and a nonlinear, continuous and bounded operator $N : V_h \to V_h$, $N\varphi = \zeta$, where

$$(\zeta, \mathbf{v}) = 2(\mathbf{u}_h^n, \mathbf{v})_E - \Delta t b^*(D_N\overline{\varphi}^h, D_N\overline{\varphi}^h, D_N G\mathbf{v}) + \Delta t(\mathbf{f}(t^{n+1/2}), D_N G\mathbf{v}) \quad \forall \mathbf{v} \in V_h \tag{39}$$

To show that N is bounded and continuous, we use the equivalence of all norms on the finite element space V_h, similar to the proof in [19].

It remains only to show that the set

$$\{\varphi \in V_h | \varphi = \lambda L\varphi \text{ for some } \lambda \in [0,1)\} \tag{40}$$

is bounded in the L^2 norm independent of λ.

Assume $\varphi = \lambda L\varphi \in V_h$, i.e.,

$$\begin{aligned} 2(\varphi, \mathbf{v})_E + \Delta t \nu (\varphi, \mathbf{v})_\epsilon = \lambda(2(\mathbf{u}_h^n, \mathbf{v})_E - \Delta t b^*(D_N\overline{\varphi}^h, D_N\overline{\varphi}^h, D_N G\mathbf{v}) \\ + \Delta t(\mathbf{f}(t^{n+1/2}), D_N G\mathbf{v})) \quad \forall \mathbf{v} \in V_h \end{aligned} \tag{41}$$

and set $\mathbf{v} = \varphi$, cancel the nonlinearity and get

$$2||\varphi||_E^2 + \Delta t \nu ||\varphi||_\epsilon^2 = \lambda(2(\mathbf{u}_h^n, \varphi)_E + \Delta t(\mathbf{f}(t^{n+1/2}), D_N G\varphi)) \quad \forall \mathbf{v} \in V_h \tag{42}$$

Next, using the standard Cauchy–Schwartz and Yound inequalities on the right terms, we get that

$$||\varphi||_E^2 + \Delta t \nu ||\varphi||_\epsilon^2 \leq C(||\Delta t \nu^{-1} \mathbf{f}(t^{n+1/2})||_*^2 + ||\mathbf{u}_h^n||_E^2) \tag{43}$$

Therefore, by the Leray–Schauder Theorem, it follows that the operator L has a fixed point $\varphi = \mathbf{u}_h^{n+1/2}$ and the discrete solution $\mathbf{u}_h^{n+1}$ of the ADM system exists.

To obtain the stability estimate from above, we set $D_N\overline{\mathbf{u}_h^{n+1/2}}^h$ as a test function in the ADM model (34) and the nonlinear term will vanish:

$$\frac{1}{2\Delta t}||(\mathbf{u}_h^{n+1}||_E^2 - ||\mathbf{u}_h^n||_E^2) + \nu||\mathbf{u}_h^{n+1/2}||_\epsilon^2 = (\mathbf{f}(t^{n+1/2}), D_N G\mathbf{u}_h^{n+1/2})) \tag{44}$$

Using the standard Cauchy–Schwartz and Young inequalities on the right term, we get

$$\frac{1}{\Delta t}||(\mathbf{u}_h^{n+1}||_E^2 - ||\mathbf{u}_h^n||_E^2) + \nu||\mathbf{u}_h^{n+1/2}||_\epsilon^2 \leq C\nu^{-1}||\mathbf{f}(t^{n+1/2})||_*^2 \tag{45}$$

Summing up from $n = 0$ to $n = N_T - 1$, we get the required inequality

$$\sup_{0 \le n < N_T} ||\mathbf{w}_h^{n+1}||_E^2 + \Delta t\, \nu \sum_{n=0}^{N_T-1} ||\nabla \mathbf{w}_h^{n+1/2}||_\epsilon^2 \le C \left(\frac{\Delta t}{\nu} \sum_{n=0}^{N_T-1} ||f(t^{n+1/2})||_{-1}^2 + ||\mathbf{w}_0||_E^2 \right) \tag{46}$$

Here, $C = C(\Omega)$.

The norms $|| \cdot ||_E, || \cdot ||_\epsilon$ can be replaced by the standard L^2, respectively H^1 norm via the inequality (22), and this will add a dependence of C on K, (see inequality 17), i.e., on the ratio α/h.

□

The convergence of the discrete solution $\mathbf{w}^h$ of the enhanced α-model (34) to the NSE exact solution $\mathbf{u}$ is stated in the next theorem.

Theorem 3. *We let $(\mathbf{u},p)$ be a smooth strong solution of the NSE and $\mathbf{w}_h^n$, $n = 1...N_T$ be the discrete solution of (34). We assume α/h is constant and $\mathbf{w}_0 = \mathbf{u}_0$. Then, if Δt is picked small enough, there holds*

$$\sup_{0 \le n < N_T} ||\mathbf{u}^{n+1} - \mathbf{w}_h^{n+1}||^2 + \Delta t\, \nu \sum_{n=0}^{N_T-1} ||\nabla \mathbf{u}^{n+1/2} - \nabla \mathbf{w}_h^{n+1/2}||^2 \le C \left(h^{2k} + \Delta t^4 + K^{2N} \alpha^4 \right) \tag{47}$$

where C is a constant that does not depend on h, α (though it depends on α/h) and K is given in inequality (17).

Remark 1. *The proof uses the discrete Gronwall's inequality [38], which requires that Δt satisfies $\Delta t = \mathcal{O}((\nu^{-3}|||\nabla \mathbf{u}|||_{\infty,0} + 1)^{-1})$.*

Proof. The convergence proof for the Leray-deconvolution model is done in [22] and for the NS-α deconvolution model in [19], but the two proofs do not use the improved deconvolution error estimate (24) and the error is proved to be in the bounded domain case on the order of

$$\mathcal{O}\left(h^{2k} + \Delta t^4 + \alpha^4 \right) \tag{48}$$

In the two proofs in [19,22], one bounds

$$\sup_{0 \le n < N_T} ||\mathbf{u}^{n+1} - \mathbf{w}_h^{n+1}||^2 + \Delta t\, \nu \sum_{n=0}^{N_T-1} ||\nabla \mathbf{u}^{n+1/2} - \nabla \mathbf{w}_h^{n+1/2}||^2 \tag{49}$$

by several terms which can be left unchanged herein and also by the modelling error (the Intp term in [22], the Interp term in [19]), which itself contains several terms that can be left unchanged but also a term that is bounded by the deconvolution error $||\mathbf{u}^{n+1/2} - D_N \mathbf{u}^{n+1/2}||$. Applying the improved deconvolution estimate (24) provides the improved $K^{2N}\alpha^4$ instead of α^4.

The proof of the above result for the ADM is done in [29]. We outline here the main ideas of the proof presented in [29], which are very similar to the ones in [22] except that the nonlinearity is handled differently.

Similar to [19,22], the NSE is written in the form

$$\begin{aligned} &\left(\frac{\mathbf{u}^{n+1} - \mathbf{u}^n}{\Delta t}, \mathbf{v}\right) + b^*(D_N \overline{\mathbf{u}^{n+1/2}}^h, D_N \overline{\mathbf{u}^{n+1/2}}^h, \mathbf{v}) + \nu(\nabla \mathbf{u}^{n+1/2}, \nabla \mathbf{v}) \\ &- (p^{n+1/2}, \nabla \cdot \mathbf{v}) = (\mathbf{f}((n+1/2)\Delta t), \mathbf{v}) - Interp(\mathbf{u}, p, n, \mathbf{v}) \ \ \forall \mathbf{v} \in X_h \end{aligned} \tag{50}$$

where

$$\begin{aligned} Interp(\mathbf{u},p,n,\mathbf{v}) &= (\mathbf{u}_t(t^{n+1/2}) - \frac{\mathbf{u}^{n+1}-\mathbf{u}^n}{\Delta t},\mathbf{v}) + b^*(\mathbf{u},\mathbf{u},\mathbf{v}) \\ &-b^*(D_N\overline{\mathbf{u}^{n+1/2}}^h, D_N\overline{\mathbf{u}^{n+1/2}}^h,\mathbf{v}) + \nu(\nabla\mathbf{u}(t^{n+1/2}) - \nabla\mathbf{u}^{n+1/2},\nabla\mathbf{v}) \\ &+(p(t^{n+1/2}) - p^{n+1/2},\nabla\cdot\mathbf{v}) \end{aligned} \tag{51}$$

Next, one derives the error equation for $\mathbf{e}^{n+1} = \mathbf{u}^{n+1} - \mathbf{u}_h^{n+1}$ by subtracting (34) from (50). We let $\mathbf{U}^{n+1} \in V_h$ be the L^2 projection of $\mathbf{u}^{n+1}$ onto V_h, split the error as $\mathbf{e}^{n+1} = (\mathbf{u}^{n+1} - \mathbf{U}^{n+1}) - (\mathbf{u}_h^{n+1} - \mathbf{U}^{n+1}) := \eta^{n+1} - \varphi_h^{n+1}$ and set $\mathbf{v} = D_N\overline{\varphi_h^{n+1/2}}^h$ in the error equation. Upon rearranging terms, we obtain

$$\begin{aligned} &\frac{1}{2}(||\varphi_h^{n+1}||_E^2 - ||\varphi_h^n||_E^2 + \nu\Delta t||\varphi_h^{n+1/2}||_\epsilon^2 = (\eta^{n+1} - \eta^n, D_N\overline{\varphi_h^{n+1/2}}^h) + \\ &\nu\Delta t(\nabla\eta^{n+1/2},\nabla D_N\overline{\varphi_h^{n+1/2}}^h) - \Delta t b^*(D_N\overline{\eta^{n+1/2}}^h, D_N\overline{\mathbf{u}^{n+1/2}}^h, D_N\overline{\varphi_h^{n+1/2}}^h) \\ &+\Delta t b^*(D_N\overline{\varphi_h^{n+1/2}}^h, D_N\overline{\mathbf{u}^{n+1/2}}^h, D_N\overline{\varphi_h^{n+1/2}}^h) \\ &\quad -\Delta t b^*(D_N\overline{\mathbf{u}_h^{n+1/2}}^h, D_N\overline{\eta^{n+1/2}}^h, D_N\overline{\varphi_h^{n+1/2}}^h) \\ &\qquad +\Delta t(p^{n+1/2},\nabla\cdot D_N\overline{\varphi_h^{n+1/2}}^h) + \Delta t Interp(\mathbf{u},p,n,D_N\overline{\varphi_h^{n+1/2}}^h) \end{aligned} \tag{52}$$

On the right side, all above terms excepting the two trilinear terms appearing in the interpolating term (51) will be estimated exactly as in [22]. In all of these terms, wherever needed (to get rid of $D_N G$ terms and switch to the norms $||\cdot||_E, ||\cdot||_\epsilon$) to keep the arguments similar to the ones in [22], inequalities (23) will be used.

The two trilinear terms in *Interp* that are treated differently compared to [22] take the form

$$\begin{aligned} &b^*(\mathbf{u}^{n+1/2},\mathbf{u}^{n+1/2}, D_N\overline{\varphi_h^{n+1/2}}^h) - b^*(D_N\overline{\mathbf{u}^{n+1/2}}^h, D_N\overline{\mathbf{u}^{n+1/2}}^h, D_N\overline{\varphi_h^{n+1/2}}^h) = \\ &\quad b^*(\mathbf{u}^{n+1/2} - D_N\overline{\mathbf{u}^{n+1/2}}^h,\mathbf{u}^{n+1/2}, D_N\overline{\varphi_h^{n+1/2}}^h) \\ &\qquad +b^*(D_N\overline{\mathbf{u}^{n+1/2}}^h,\mathbf{u}^{n+1/2} - D_N\overline{\mathbf{u}^{n+1/2}}^h, D_N\overline{\varphi_h^{n+1/2}}^h) \end{aligned} \tag{53}$$

The first term on the right side above is estimated similar to inequality 3.26 in [22]:

$$\begin{aligned} &|b^*(\mathbf{u}^{n+1/2} - D_N\overline{\mathbf{u}^{n+1/2}}^h,\mathbf{u}^{n+1/2}, D_N\overline{\varphi_h^{n+1/2}}^h)| \\ &\qquad \le \frac{\nu}{24}||\varphi_h^{n+1/2}||_\epsilon^2 + C\nu^{-1}||\mathbf{u}^{n+1/2} - D_N\overline{\mathbf{u}^{n+1/2}}^h||^2 \end{aligned} \tag{54}$$

For the second term, we first use the inequality (32) to get

$$\begin{aligned} &|b^*(D_N\overline{\mathbf{u}^{n+1/2}}^h,\mathbf{u}^{n+1/2} - D_N\overline{\mathbf{u}^{n+1/2}}^h, D_N\overline{\varphi_h^{n+1/2}}^h)| \\ &\quad \le C||\mathbf{u}^{n+1/2} - D_N\overline{\mathbf{u}^{n+1/2}}^h||||A_h D_N\overline{\mathbf{u}^{n+1/2}}^h||||\nabla D_N\overline{\varphi_h^{n+1/2}}^h|| \end{aligned} \tag{55}$$

Using again the argument in inequality (28) to estimate the middle term and Young's inequality, we get further

$$\begin{aligned} &C||\mathbf{u}^{n+1/2} - D_N\overline{\mathbf{u}^{n+1/2}}^h||||A_h D_N\overline{\mathbf{u}^{n+1/2}}^h||||\nabla D_N\overline{\varphi_h^{n+1/2}}^h|| \\ &\quad \le C||\mathbf{u}^{n+1/2} - D_N\overline{\mathbf{u}^{n+1/2}}^h||||\Delta\mathbf{u}^{n+1/2}||||\varphi_h^{n+1/2}||_\epsilon \\ &\quad \le C\nu^{-1}||\mathbf{u}^{n+1/2} - D_N\overline{\mathbf{u}^{n+1/2}}^h||^2 + \frac{\nu}{24}||\varphi_h^{n+1/2}||_\epsilon^2 \end{aligned} \tag{56}$$

Therefore, both trilinear terms have been bounded by the deconvolution error and some terms that will eventually be hidden on the left side of Equation (52). The proof continues as in [22] by applying the discrete Gronwall's inequality. The application of the Gronwall's inequality conditions the time-step to satisfy $\Delta t \leq C(\nu^{-3}|||\nabla \mathbf{u}|||_{\infty,0} + 1)^{-1}$.

For the modified Leray-deconvolution model, the proof follows the steps outlined above down to equality (53) which, in the case of the ML-α, will take the form

$$b^*(\mathbf{u}^{n+1/2}, \mathbf{u}^{n+1/2}, D_N \overline{\varphi_h^{n+1/2}}^h) - b^*(\mathbf{u}^{n+1/2}, D_N \overline{\mathbf{u}^{n+1/2}}^h, D_N \overline{\varphi_h^{n+1/2}}^h) = b^*(\mathbf{u}^{n+1/2}, \mathbf{u}^{n+1/2} - D_N \overline{\mathbf{u}^{n+1/2}}^h, D_N \overline{\varphi_h^{n+1/2}}^h) \tag{57}$$

which is further estimated as in inequality (54) from above. From here on, the proof proceeds as in the ADM case.

□

4. A Numerical Experiment

In this section, we verify the theoretical convergence rates on a 2d NSE problem from [29] with exact solution

$$\begin{aligned} u_1(x,y,t) &= sin(2\pi y)e^{-4\nu\pi^2 t} \\ u_2(x,y,t) &= sin(\pi x)e^{-\nu\pi^2 t} \\ p(x,y,t) &= 0 \end{aligned} \tag{58}$$

therefore, $\mathbf{f} = (f_1, f_2)$ is equal to

$$\begin{aligned} f_1(x,y,t) &= 2\pi cos(2\pi y)sin(\pi x)e^{-5\nu\pi^2 t} \\ f_2(x,y,t) &= \pi cos(\pi x)sin(2\pi y)e^{-5\nu\pi^2 t} \end{aligned} \tag{59}$$

Herein, we check the Leray-deconvolution model with $N = 0, 1$ (thus suplementing the numerical results presented in [22]) and the modified Leray-deconvolution model with $N = 0, 1$. The ADM with $N = 0, 1$ has been checked in this problem in [29] on a slightly different mesh and is therefore omitted here. The NS-α deconvolution model has been checked numerically in [19].

The scope of the test is to verify the convergence rates and also to check numerically that, as N increases, the corresponding model enters the predicted convergent regime faster.

The test is carried out with the FreeFEM++ package [40]. The computational domain is the unit square $\Omega = (0,1) \times (0,1)$. The kinematic viscosity is set equal to $\nu = 1.0$ and the final time T is set equal to $T = 0.1$. An initial mesh M_0 is generated with five even nodes on the edges $x = 0, x = 1$, 4 even nodes on $y = 0$ and 6 even nodes on $y = 1$. Then, the edges are succesivelly halved to generate the computational meshes $M_1, M_2, \ldots, M_5$, see Figure 1. For the space discretization, we use $P2/P1$ Taylor–Hood finite elements. In our computation, we set the filter radius α equal to the mesh size h, i.e., $\alpha = h$.

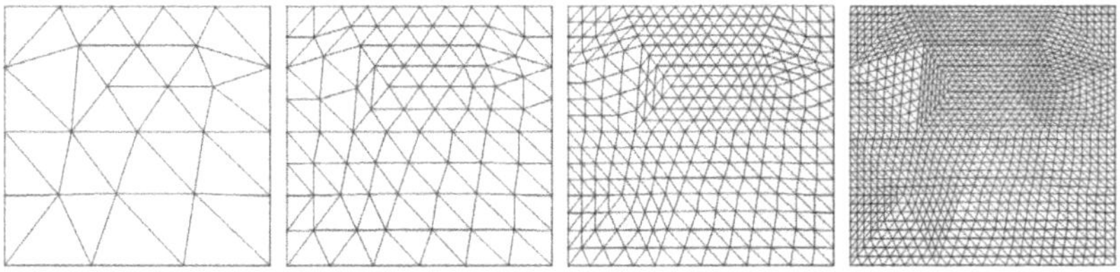

(a) M_0 **(b)** M_1 **(c)** M_2 **(d)** M_3

Figure 1. The initial mesh **(a)** M_0 and and the first three level meshes **(b)** M_1, **(c)**M_2, **(d)**$M3$.

First, we solve the Leray-deconvolution and modified Leray-deconvolution with $N = 0, 1$ on M_1 with $\Delta t = 1/80$ and eight iterations. When solving the next mesh M_2, the time step Δt is halved and we double the number of time iterations. We keep applying this procedure up to the last mesh M_5.

To stabilize pressure, the model and the filter equation are augmented with an L^2 pressure stabilization term with parameter 10^{-9}. The resulting algebraic system corresponding to the discrete model is solved using a fixed point iteration until the L^∞ norm of two successive iterates is less than 10^{-12}. The rate presented in the Tables 1–4 is computed as the log_2 of the quotient of two successive errors. The theoretical rate in the $L^\infty(L^2)$ and $L^2(H^1)$ norms for the Leray-deconvolution and modified Leray-deconvolution models is 2 and is confirmed by this numerical test. Moreover, the predicted convergent regime is reached faster for $N = 1$.

Table 1. $L^\infty(L^2)$ and $L^2(H^1)$ errors and rates for the the Leray-α model. The final time is $T = 0.1$. The predicted rate order is 2.

Level	Nr Iter	Δt	h	$L^\infty(L^2)$ Error		$L^2(H^1)$ Error	
M_1	8	1/80	0.198493	0.00349919	rate	0.00635556	rate
M_2	16	1/160	0.0992465	0.000966758	1.85	0.0017491	1.86
M_3	32	1/320	0.0496232	0.000267232	1.85	0.000472424	1.88
M_4	64	1/640	0.0248116	6.78601×10^{-5}	1.97	0.000118784	1.99
M_5	128	1/1280	0.0124058	1.68665×10^{-5}	2.00	2.9308×10^{-5}	2.01

Table 2. $L^\infty(L^2)$ and $L^2(H^1)$ errors and rates for the Leray-deconvolution model with $N = 1$. The final time is $T = 0.1$. The predicted rate order is 2.

Level	Nr Iter	Δt	h	$L^\infty(L^2)$ Error		$L^2(H^1)$ Error	
M_1	8	1/80	0.198493	0.00321879	rate	0.00599466	rate
M_2	16	1/160	0.0992465	0.000735394	2.12	0.00131779	2.18
M_3	32	1/320	0.0496232	0.000174567	2.07	0.000291998	2.17
M_4	64	1/640	0.0248116	4.29922×10^{-5}	2.02	6.89069×10^{-5}	2.08
M_5	128	1/1280	0.0124058	1.07004×10^{-5}	2.00	1.67512×10^{-5}	2.04

Table 3. $L^\infty(L^2)$ and $L^2(H^1)$ errors and rates for the modified Leray-α model. The final time is $T = 0.1$. The predicted rate order is 2.

Level	Nr Iter	Δt	h	$L^\infty(L^2)$ Error		$L^2(H^1)$ Error	
M_1	8	1/80	0.198493	0.0059773	rate	0.00995501	rate
M_2	16	1/160	0.0992465	0.00259845	1.20	0.00434837	1.19
M_3	32	1/320	0.0496232	0.000900568	1.52	0.00149832	1.53
M_4	64	1/640	0.0248116	0.000257406	1.80	0.000427468	1.80
M_5	128	1/1280	0.0124058	6.76934×10^{-5}	1.92	0.000112259	1.92

Table 4. $L^\infty(L^2)$ and $L^2(H^1)$ errors and rates for the modified Leray-deconvolution model with $N = 1$. The final time is $T = 0.1$. The predicted rate order is 2.

Level	Nr Iter	Δt	h	$L^\infty(L^2)$ Error		$L^2(H^1)$ Error	
M_1	8	1/80	0.198493	0.00453629	rate	0.00788281	rate
M_2	16	1/160	0.0992465	0.00111931	2.01	0.00198618	1.98
M_3	32	1/320	0.0496232	0.000209968	2.41	0.000373953	2.40
M_4	64	1/640	0.0248116	4.41691×10^{-5}	2.24	7.65631×10^{-5}	2.28
M_5	128	1/1280	0.0124058	1.06889×10^{-5}	2.04	1.80013×10^{-5}	2.08

5. Conclusions

This report presents some results on the theory of α models of turbulence that have been obtained in recent years. They show that the accuracy of these models can be improved using higher order approximate deconvolution operators. Some numerical tests are also presented to support the theoretical results. The technique to estimate the deconvolution error can be used in the larger context of models using approximate deconvolution such as the time-relaxation model investigated in [41] or for other models, such as the MHD, that have been recently investigated using approximate deconvolution methods.

Acknowledgments: This work has been funded by University Politehnica of Bucharest, through the "Excellence Research Grants" Program, UPB-GEX. No. 26/2016, internal No. MA 52.16.01.

Conflicts of Interest: The author declares no conflict of interest.

References

1. Kolmogorov, A.V. The Local Structure of Turbulence in Incompressible Viscous Fluids for Very Large Reynolds Number. *Dokl. Akad. Nauk SSSR* **1941**, *30*, 209–303.
2. Cheskidov, A.; Holm, D.; Olson, E.; Titi, E. On a Leray-alpha model of turbulence. *Proc. Ser. A Math. Phys. Eng. Sci.* **2005**, *461*, 629–649.
3. Ilyin, A.; Lunasin, E.; Titi, E. A modified-Leray-α subgrid scale model of turbulence. *Nonlinearity* **2006**, *19*, 879–897.
4. Layton, W.; Lewandowski, R. A simple and stable scale similarity model for large scale eddy simulation: Energy balance and existence of weak solutions. *Appl. Math. Lett.* **2003**, *16*, 1205–1209.
5. Dunca, A. On an energy inequality for the approximate deconvolution models. *Nonlinear Anal. Real World Appl.* **2016**, *32*, 294–300.
6. Germano, M. Differential filters for the large eddy numerical simulation of turbulent flows. *Phys. Fluids* **1986**, *29*, 1755–1757.
7. Leray, J. Essay sur les mouvements plans d'une liquide visqueux que limitent des parois. *J. Math. Pures Appl. Paris Ser. IX* **1934**, *13*, 331–418.
8. Leray, J. Sur les mouvements d'une liquide visqueux emplissant l'espace. *ACTA Math.* **1934**, *63*, 193–248.
9. Chepyzhov, V.V.; Titi, E.S.; Vishik, M.I. On the convergence of solutions of the Leray-α model to the trajectory attractor of the 3D Navier-Stokes system. *J. Discrete Contin. Dyn. Syst.-Ser. A* **2007**, *17*, 33–52.
10. Vishik, M.I.; Titi, E.S.; Chepyzhov, V.V. Trajectory attractor approximations of the 3D Navier-Stokes system by the Leray-α model. *Russ. Math. Dokl.* **2005**, *71*, 91–95.
11. Layton, W.; Lewandowski, R. A high accuracy Leray-deconvolution model of turbulence and its limiting behavior. *Anal. Appl. (Singap.)* **2008**, *6*, 23–49.
12. Foias, C.; Holm, D.; Titi, E. The three dimensional viscous Camassa-Holm equations, and their relation to the Navier-Stokes equations and turbulence theory. *J. Dyn. Differ. Equ.* **2002**, *14*, 1–35.
13. Stolz, S.; Adams, N. An approximate deconvolution procedure for large eddy-simulation. *Phys. Fluids* **1999**, *11*, 1699–1701.
14. Adams, N.; Stolz, S. Deconvolution methods for subgrid-scale approximation in large-eddy simulation. In *Modern Simulation Strategies for Turbulent Flow*; Geurts, B., Ed.; R.T. Edwards: Philadelphia, PA, USA, 2001; pp. 21–41.
15. Foias, C.; Holm, D.; Titi, E. The Navier-Stokes-alpha model of fluid turbulence. *Phys. D* **2001**, *152–153*, 505–519.
16. Layton, W.; Lewandowski, R. On a well-posed turbulence model. *Discrete Contin. Dyn. Syst. Ser. B* **2006**, *6*, 111–128.
17. Cao, Y.; Lunasin, E.M.; Titi, E.S. Global well-posedness of the three-dimensional viscous and inviscid simplified Bardina turbulence models. *Commun. Math. Sci.* **2006**, *4*, 823–848.
18. Hernandez, M.M.; Rebholz, L.; Tone, C.; Tone, F. On the Stability of the Crank-Nicolson-Adams-Bashforth Scheme for the 2D Leray-alpha model. *Numer. Methods Partial Differ. Equ.* **2016**, *32*, 1155–1183.

19. Miles, W.; Rebholz, L. An enhanced physics based scheme for the NS-alpha turbulence model. *Numer. Methods Partial Differ. Equ.* **2010**, *26*, 1530–1555.
20. Kaya, S.; Manica, C.C. Convergence Analysis of the Finite Element Method for a Fundamental Model in Turbulence. *Math. Models Methods Appl. Sci.* **2012**, *22*, doi:10.1142/S0218202512500339.
21. Dunca, A. Estimates of the discrete van Cittert deconvolution error in approximate deconvolution models of turbulence. **2017**, submitted.
22. Layton, W.; Manica, C.; Neda, M.; Rebholz, L. Numerical Analysis and Computational Testing of a high-order Leray-deconvolution turbulence model. *Numer. Methods Partial Differ. Equ.* **2008**, *24*, 555–582.
23. Stolz, S.; Adams, N.A.; Kleiser, L. An approximate deconvolution model for large-eddy simulation with application to incompressible wall-bounded flows. *Phys. Fluids* **2001**, *13*, 997–1015.
24. Dunca, A.; Epshteyn, Y. On the Stolz-Adams deconvolution model for the large-eddy simulation of turbulent flows. *SIAM J. Math. Anal.* **2006**, *37*, 1890–1902.
25. Stanculescu, I. Existence theory of abstract approximate deconvolution models of turbulence. *Ann. Dell'Univ. Ferrara Sez. VII Sci. Mat.* **2008**, *54*, 145–168.
26. Kaya, S.; Manica, C.; Rebholz, L. On Crank-Nicolson Adams-Bashforth timestepping for approximate deconvolution models in two dimensions. *Appl. Math. Comput.* **2014**, *246*, 23–38.
27. Galvin, K.; Rebholz, L.; Trenchea, C. Efficient, unconditionally stable, and optimally accurate FE algorithms for approximate deconvolution models. *SIAM J. Numer. Anal.* **2014**, *52*, 678–707.
28. Rebholz, L. Well-posedness of a reduced order approximate deconvolution turbulence model. *J. Math. Anal. Appl.* **2013**, *405*, 738–741.
29. Dunca, A. Numerical analysis and testing of a stable and convergent finite element scheme for approximate deconvolution turbulence models. *Comput. Math. Appl.* **2017**, doi:10.1016/j.camwa.2017.09.035.
30. Layton, W.; Rebholz, L. *Approximate Deconvolution Models of Turbulence: Analysis, Phenomenology and Numerical Analysis*; Springer: Berlin/Heidelberg, Germany, 2012.
31. Cuff, V.; Dunca, A.; Manica, C.; Rebholz, L. The reduced order NS-α model for incompressible flow: Theory, numerical analysis and benchmark testing. *ESAIM: Math. Model. Numer. Anal. (M2AN)* **2015**, *49*, 641–662.
32. Rebholz, L.; Kim, T.Y.; Byon, Y. On an accurate α model for coarse mesh turbulent channel flow simulation. *Appl. Math. Model.* **2017**, *43*, 139–154.
33. Rebholz, L.; Zerfas, C.; Zhao, K. Global in time analysis and sensitivity analysis for the reduced NS-α model of incompressible flow. *J. Math. Fluid Mech.* **2017**, *19*, 445–467.
34. Dunca, A. Estimates of the modeling error of the α- models of turbulence in two and three space dimensions. **2017**, submitted.
35. Layton, W. The interior error of van Cittert deconvolution of differential filters is optimal. *Appl. Math. E-Notes* **2012**, *12*, 88–93.
36. Layton, W. *Introduction to the Numerical Analysis of Incompressible Viscous Flows, Viscous Flows;* SIAM Publications: Philadelphia, PA, USA, 2008; 213p, ISBN: 978-0-898716-57-3.
37. Brenner, S.; Scott, L. *The Mathematical Theory of Finite Element Methods,* 3rd ed.; Springer: New York, NY, USA, 2008.
38. Heywood, J.; Rannacher, R. Finite element approximation of the nonstationary Navier-Stokes problem. Part IV. Error analysis for the second order time discretization. *SIAM J. Numer. Anal.* **1990**, *2*, 353–384.
39. Guermond, J.L.; Quartapelle, L. On the approximation of the unsteady Navier-Stokes equations by finite element projection methods. *Numer. Math.* **1998**, *80*, 207–238.
40. Hecht, F. New development in FreeFem++. *J. Numer. Math.* **2012**, *20*, 251–265.
41. Dunca, A.; Neda, M. Numerical Analysis of a Nonlinear Time Relaxation Model of Fluids. *J. Math. Anal. Appl.* **2014**, *420*, 1095–1115.

fluids

MDPI

Article

Database of Near-Wall Turbulent Flow Properties of a Jet Impinging on a Solid Surface under Different Inclination Angles

Florian Ries [1,*], Yongxiang Li [1], Martin Rißmann [2], Dario Klingenberg [1], Kaushal Nishad [1], Benjamin Böhm [2], Andreas Dreizler [2], Johannes Janicka [1] and Amsini Sadiki [1]

1. Institute of Energy and Power Plant Technology, Technische Universität Darmstadt, 64287 Darmstadt, Germany; yongxiang.li@ekt.tu-darmstadt.de (Y.L.); dario.klingenberg@gmail.com (D.K.); nishad@ekt.tu-darmstadt.de (K.N.); janicka@ekt.tu-darmstadt.de (J.J.); sadiki@ekt.tu-darmstadt.de (A.S.)
2. Institute of Reactive Flows and Diagnostics, Technische Universität Darmstadt, 64287 Darmstadt, Germany; martin.rissmann@vibratec.fr (M.R.); bboehm@ekt.tu-darmstadt.de (B.B.); dreizler@rms.tu-darmstadt.de (A.D.)

* Correspondence: ries@ekt.tu-darmstadt.de; Tel.: +49-6151-16-28756

Received: 27 November 2017 ; Accepted: 25 December 2017; Published: 2 January 2018

Abstract: In the present paper, direct numerical simulation (DNS) and particle image velocimetry (PIV) have been applied complementarily in order to generate a database of near-wall turbulence properties of a highly turbulent jet impinging on a solid surface under different inclination angles. Thereby, the main focus is placed on an impingement angle of 45°, since it represents a good generic benchmark test case for a wide range of technical fluid flow applications. This specific configuration features very complex flow properties including the presence of a stagnation point, development of the shear boundary layer and strong streamline curvature. In particular, this database includes near-wall turbulence statistics along with mean and rms velocities, budget terms in the turbulent kinetic energy equation, anisotropy invariant maps, turbulent length/time scales and near-wall shear stresses. These properties are useful for the validation of near-wall modeling approaches in the context of Reynolds-averaged Navier–Stokes (RANS) and large-eddy simulations (LES). From this study, in which further impingement angles (0°, 90°) have been considered in the experiments only, it turns out that (1) the production of turbulent kinetic energy appears negative at the stagnation point for an impingement angle other than 0° and is balanced predominantly by pressure-related diffusion, (2) quasi-coherent thin streaks with large characteristic time scales appear at the stagnation region, while the organization of the flow is predominantly toroidal further downstream, and (3) near-wall shear stresses are low at the stagnation region and intense in regions where the direction of the flow changes suddenly.

Keywords: database; impinging jet; direct numerical simulation; particle image velocimetry

1. Introduction

Impinging jets are used in a variety of engineering applications as they enable localized heat and mass transfer, e.g., cooling of electronic components, quenching of metals and glass, cooling of turbine-blades or drying of paper and other materials. Given their practical relevance, several jet geometries and flow conditions were examined, like nozzle shapes, Reynolds number effects, the influence of jet-to-plate spacing, pulsed jets, flame impingement, ribbed walls, jet impingement angle and many more. From the gained insights, various empirical correlations for the practical use of impinging flows were derived. Reviews of experimental studies, numerical modeling, general uses and performance of impinging jets can be found in [1–4].

Characterized by a strong wall/flow interaction process, impinging jets feature very complex flow properties including the presence of a stagnation point, shear boundary layer development in the free jet region and strong streamline curvature. In this respect, several experimental and direct numerical simulations (DNS) studies have been carried out in the past (e.g., [5–17]) in order to provide a deeper understanding about the underlying physical effects in such flows. The majority of these studies were focused on a single jet impinging normally on a heated solid surface using well-defined, fully-developed turbulent inlet conditions. Today, this specific jet configuration serves as a model geometry for a wide range of engineering application and is often used to validate turbulence models in the context of Reynolds-averaged Navier–Stokes (RANS) [18–20] and large-eddy simulations (LES) [21–23].

Regarding experimental studies focusing on the general flow characteristics of fully-developed turbulent jets impinging normally on a solid surface, Copper et al. [7] used hot-wire anemometry (HWA) and laser doppler velocimetry (LDV) techniques to determine mean velocities and Reynolds stresses at different wall-normal traverses. Fairweather and Hargrave [10] applied particle image velocimetry (PIV) to analyze the recirculation zone within the flow that carries material from the periphery of the wall-jet back to its initial regions. Regarding turbulent flow dynamics at the stagnation region, Tummers et al. [5] reported on detailed near-wall measurements of mean velocities and Reynolds stresses using two-component LDV and PIV. Furthermore, features of the budget of turbulent kinetic energy and turbulent stress anisotropy in the stagnation region were examined by Nishino et al. [8] by means of particle tracking velocimetry (PTV). This study revealed that the turbulence is almost in an axisymmetric state at the stagnation point and that negative production of turbulent kinetic energy takes place in the vicinity of the wall, which is compensated by the pressure diffusion. Besides this, the dynamics of coherent structures in a single impinging jet were examined by Hall and Ewing [24], among others, using pressure transducer and microphone measurements of the instantaneous pressure field at the wall. The measurements indicate that large-scale ring structures are present at the stagnation and wall-jet regions, which act to promote the heat transfer. Other aspects like Reynolds number effects, heat transport phenomena or the influence of jet-to-plate spacing have been also addressed in various experimental studies, e.g., [6,25,26].

With respect to numerical investigations, existing DNS studies of fully-developed turbulent jets impinging perpendicularly on a solid surface are mostly limited to moderate Reynolds numbers. Satake and Kunugi [13] analyzed the flow mechanism by which eddies are generated at the edge of the round nozzle and transported into the impingement region for a Reynolds number of $Re = 5300$. Flow characteristics and heat transport phenomena in a plane turbulent impinging jet ($Re = 9120$) were addressed by Hattori and Nagano [15]. The authors provided turbulence statistics of the velocity and temperature, turbulent heat fluxes, local Nusselt numbers, budget terms of turbulent kinetic energy among other turbulent quantities. Recently, DNS of impinging jet flows at $Re = 10{,}000$ and $Re = 8000$ were performed with high spatial resolution and high order numerical methods by Dairay et al. [27] and by Wilke and Sesterhenn [28], respectively. The first study focused on the role of unsteady processes to explain the spatial distribution of the heat transfer coefficient at the wall, while the latter analyzed the influence of Mach number, Reynolds number and ambient temperature on the mean velocity and temperature fields. Several Reynolds analogies were also assessed. Besides a better understanding about the underlying physical effects in fully-developed turbulent impinging jets, comprehensive datasets for validation purposes were made available by means of DNS studies, which are difficult to obtain experimentally, especially budget terms of turbulent kinetic energy and other quantities.

While fully-developed jets impinging perpendicularly on a solid surface have drawn the interest of many researchers, little attention has been paid to impinging flows that are not fully developed and impinge at a particular angle, even though such flow conditions can be found in several technical applications. This is for example the case of hollow jets impinging on the cylinder wall in internal combustion (IC) engines [29] or fluid flow inside valves [30].

The present paper therefore intends to examine impinging flows more closely related to such technical applications at moderate Reynolds number. For this purpose, DNS and PIV techniques are applied complementarily with two objectives: first to examine and deepen the understanding of the turbulent flow features in oblique jets impinging on a solid surface; secondly to generate and provide a comprehensive database for model development and validation. Thereby, the jets' inflow is highly turbulent (turbulent intensity of $\sim$10%) and not fully developed. Three different impingement angles of $0°, 45°$ and $90°$ are experimentally investigated, while DNS is only focused on the $45°$-inclination configuration, which includes the most flow features of both extremal cases of $0°$ and $90°$ inclination. In this context, a comprehensive database including fluid flow statistics, budget terms of the turbulent kinetic energy, wall shear stresses and turbulence scales are made available.

This paper is organized as follows. In Section 2, the applied measurement techniques and numerical approach are introduced. Next, the PIV and DNS setups of the oblique impinging jets are described (Section 3). Then, experimental results are reported and discussed (Section 4), for the three impingement angles of $0°, 45°$ and $90°$. Subsequently, DNS results of the $45°$-inclination configuration are presented and compared with the experiment in Section 5. Finally, some concluding remarks are provided in the last section. For the sake of completeness, a detailed code verification study is provided in the Appendix A.

2. Methods

In this work, an isothermal fluid flow jet impinging on a solid flat wall is investigated using both experiment and DNS. In this section, the applied methods are introduced.

2.1. Direct Numerical Simulation

For a viscous Newtonian fluid flow with constant physical properties, the applied governing equations in the present DNS study are the continuity:

$$\frac{\partial U_i}{\partial x_i} = 0 \tag{1}$$

and momentum equations:

$$\frac{\partial U_i}{\partial t} = -\frac{\partial}{\partial x_j}\left(U_i\, U_j\right) - \frac{\partial p}{\partial x_i} + \frac{\partial}{\partial x_j}\left(\nu\left(\frac{\partial U_i}{\partial x_j} + \frac{\partial U_j}{\partial x_i}\right)\right), \tag{2}$$

where U_i is the velocity field, p the kinematic pressure and ν the kinematic viscosity.

The governing Equations (1) and (2) are solved numerically using a low-dissipative projection method proposed by [31], which was added to the open source C++ library OpenFOAM v1612+ (OpenCFD Ltd., Reading, UK). In contrast to other pressure correction methods, (e.g., pressure implicit with splitting of operator (PISO) [32] or semi-implicit method for pressure-linked equations (SIMPLE) [33]), no corrector loop is required, which significantly speed up the calculation. Moreover, it was shown by [34] that it is further less dissipative than the standard methods in OpenFOAM and well suited for time-resolved numerical simulations. The method is applied with a three-stage explicit Runge–Kutta scheme of second order accuracy for time integration [35].

The velocity-pressure coupling method can be summarized in three steps. First an intermediate velocity $U_i^{*,k}$ is computed explicitly for each Runge–Kutta step k using the momentum equation (Equation (2)), whereby the pressure term is omitted:

$$U_i^{*,k} = U_i^n + \Delta t \cdot \alpha_k \cdot \Re\{U_i^{*,k-1}\}. \tag{3}$$

Here, Δt is the time increment, U_i^n the velocity at the n-th time step, α_k the step size of the Runge–Kutta stages and $\Re\{.\}$ the right-hand term of the momentum equation excluding the pressure gradient.

Subsequently, the pressure is calculated in such a way that the velocity field satisfies the divergence free condition. It follows for the pressure Poisson equation that:

$$\frac{\partial}{\partial x_i}\left(\frac{\partial p}{\partial x_i}\right) = \frac{1}{\Delta t \cdot \alpha_k}\frac{\partial U_i^{*,k}}{\partial x_i}. \tag{4}$$

In the last step, the intermediate velocity field is corrected to obtain the final value of the velocity as:

$$U_i^k = U_i^{*,k} - \Delta t \cdot \alpha_k \frac{\partial p}{\partial x_i}. \tag{5}$$

In the present three-stage Runge–Kutta method, $\alpha_1 = 1/3$, $\alpha_1 = 1/2$ and $\alpha_3 = 1$ are chosen, leading to second order accuracy in time. It stands that $U_i^0 = U_i^n$ and $U_i^3 = U_i^{n+1}$.

A second order central differencing scheme is applied for the convection term of the momentum equation, and a second order, conservative scheme is used for the Laplacian and gradient terms. Since the numerical resolution of the pressure Poisson equation is the crucial step in the present approach [36], convergence optimization and acceleration techniques are incorporated. In particular, the geometric agglomerated algebraic multigrid solver is applied for the resolution of the pressure equation including a diagonal-based incomplete Cholesky preconditioner. To reduce the mesh dependency, a smoother based on the Gauss–Seidel method is utilized.

2.2. Experimental Methods

Measurements of the flow field using planar two-component particle image velocimetry have been carried out. For this purpose, a frequency doubled-pulsed neodymium yttrium vanadate laser ($Nd : YVO_4$, $\lambda = 532$ nm, 4.0 mJ/pulse, pulse separation 100 µs, Edgewave IS 4II-DE (EdgeWave GmbH, Würselen, Germany)) is applied to illuminate the aluminum oxide (Al_2O_3) seeding particles ($d \sim 1$ µm). Mie-scattering from the particles is recorded with a digital complementary metal-oxide-semiconductor (CMOS) camera (Phantom v711, Vision Research, Wayne, NJ, USA) and a 180-mm f/3.5 macro camera lens. A repetition rate of 20 Hz is used to ensure statistically independent samples. Three thousand PIV images were recorded for each region of interest to obtain reliable statistical results. All PIV images are processed with LaVision DaVis 8.2.1 software (LaVision GmbH, Göttingen, Germany). Interrogation windows of 24 pixels × 24 pixels and 75% overlap are used, resulting to a vector spacing of ~200 µm.

3. Test Case

The experimental and numerical setups of the inclined impinging jet configurations are outlined in this section. Regarding the experiment, three different impingement angles of 0°, 45° and 90° are investigated, while only the 45°-configuration is computed in the DNS.

3.1. Experimental Setup

A representation of the measurement setup of the inclined impinging jet configuration is sketched in Figure 1. In the test section, dry air ($T = 298.15$ K, $p = 1$ atm) seeded with Al_2O_3 particles ($d \sim 1$ µm) enters a settling chamber and streams through honeycombs followed by two screens in order to homogenize the flow. Then, the air stream is accelerated by means of a contraction nozzle and, before exiting, encounters a turbulence generating grid. Finally, the generated turbulent air stream leaves the square nozzle and flows along or impinges on a solid surface, respectively, according to the angle of inclination of the plate.

The contraction nozzle has a length of 150 mm, an entrance area of 120 mm × 120 mm and an exit area of 40 mm × 40 mm. It results in a contraction ratio, $N_{contr.}$, between the entrance and exit section areas of $N_{contr.} = 9$. At the contraction inlet section, the flow is inherently laminar with $Re = 1650$, while it reaches a Reynolds number of $Re = 5000$ at the exit section of the nozzle. Thereby, due to the shape of the nozzle geometry, the flow field resembles a plug flow at the contraction exit with a very low turbulence level. A detailed design description of the nozzle geometry and estimations of the pressure drop and flow turbulence can be found in [37].

The turbulence generating grid is installed directly after the contraction. A 2.5 mm-thick perforated plate with hole diameters of ∅4 mm is utilized. Note that the turbulence generating grid is not symmetric. The orientation of the grid with respect to the impingement wall, laser sheet and digital CMOS camera is shown in Figure 1c.

As indicated in Figure 1a,b, the inclination of the solid wall can be adjusted according to the desired impingement angle. Three different angles of $\gamma = 0^\circ, 45^\circ$ and 90° are investigated as depicted in Figure 2. Thereby, the jet-to-plate spacing at the centerline of the jet (45° and 90° inclination) equals one hydraulic diameter of the nozzle exit. Figure 2 shows the regions of interest (highlighted in red) where PIV measurements are performed.

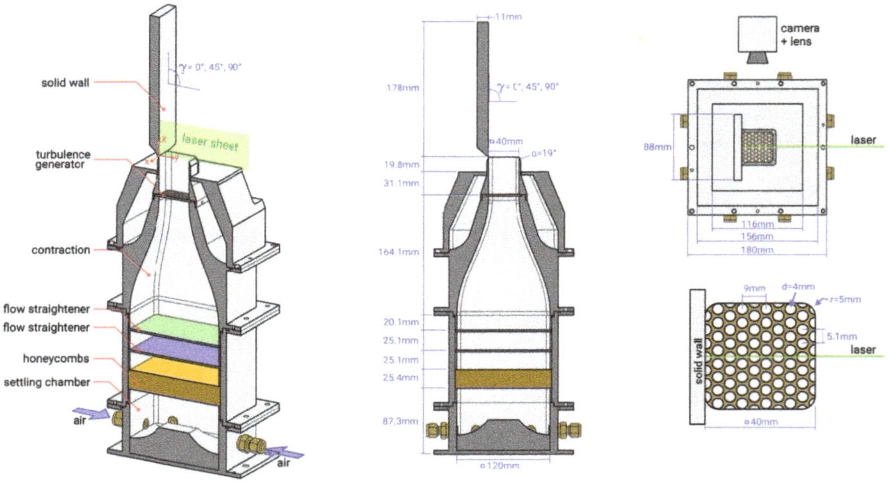

(**a**) (**b**) (**c**)

Figure 1. Experimental setup of the inclined impinging jet configuration. (**a**) Trimetric view, (**b**) side view and (**c**) top view of the 0°-configuration and turbulence generating grid.

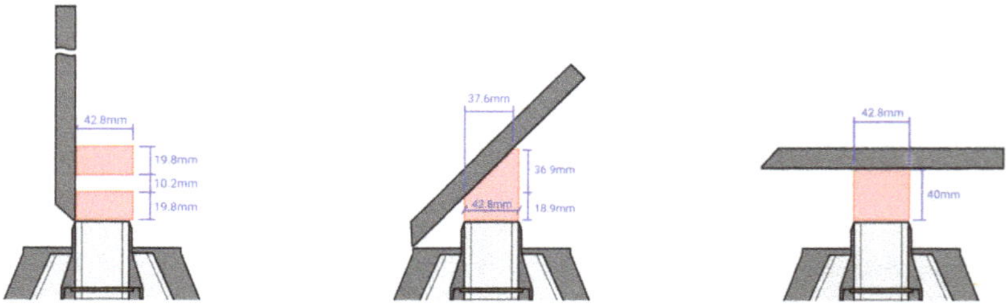

(**a**) (**b**) (**c**)

Figure 2. Regions of interest for measurements. Red shaded areas: particle image velocimetry (PIV) measurements at (**a**) 0°-, (**b**) 45°- and (**c**) 90°-inclination, respectively.

3.2. Numerical Setup

As pointed out above, only the 45°-inclination angle case is numerically investigated. Figure 3 shows: (a) the portion of the experimental section that is numerically investigated, (b) the corresponding computational domain and (c) a map of the ratio Δ/η_K between the local mesh size $\Delta = (\Delta_x \Delta_y \Delta_z)^{1/3}$ and the Kolmogorov length scale $\eta_K = (\nu^3/\epsilon_{tke})^{1/4}$. Thereby, η_K is computed for each location using the predicted local turbulent kinetic energy dissipation rate $\epsilon_{tke} = \nu < (\partial u_i'/\partial x_j)(\partial u_i'/\partial x_j) >$ and the kinematic viscosity ν.

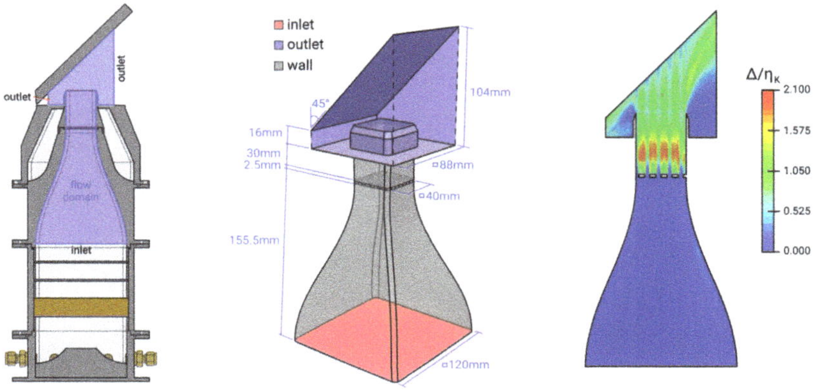

(a) (b) (c)

Figure 3. Numerical setup of the 45°-inclined impinging jet configuration. (**a**) Portion of the experimental section that is numerically investigated, (**b**) computational domain and boundary conditions and (**c**) map of the ratio Δ/η_K between the local mesh size $\Delta = (\Delta_x \Delta_y \Delta_z)^{1/3}$ and the Kolmogorov length scale η_K.

In line with the experimental setup, the computational domain (Figure 3a,b) consists of a contraction section, turbulence generating grid and impinging region after the nozzle. The settling chamber, honeycombs and flow straighteners are excluded in the numerical simulation. At the contraction inlet, a uniform velocity field is imposed, and no-slip conditions are utilized at the walls. Regarding the outflows, a velocity inlet/outlet boundary condition is used to allow entrainment of air from the surroundings. Thereby, the incoming fluid velocity is obtained by the internal cell value, while Neumann conditions are applied in the case of outflow. In contrast to the experimental setup, the small enclosed corner at the lower-left-hand side of the plate (see Figure 3a) is omitted in the numerical study. Here, an outflow condition is applied in order to allow an unrestricted fluid flow directed outward along the impinged wall in all directions. As will be shown later, this boundary condition has no influence on the region of interest around the stagnation point.

A block-structured, three-dimensional grid is employed in the present study. It consists of approximately 109 million control volumes and is refined around the perforated plate and towards the walls. Considering the commonly-used DNS spatial resolution criterion [38], the ratio of local mesh size and Kolmogorov length scale is below $\Delta/\eta_K < 2.1$ in almost the entire domain, as shown in Figure 3c, which ensures sufficient spatial resolution.

To avoid uncertainties caused by the initial solution, a fully-developed turbulent velocity field is generated by means of a separate large-eddy simulation (LES), which is interpolated on the numerical grid of the DNS. Afterwards, two flows through the domain (after the turbulence generating grid) are solved before sampling is started. In the case of LES, the same computational domain and boundary conditions are used as in the DNS study, while a numerical grid of approximately six million cells is utilized. Closure is obtained by means of the (wall-adapting local eddy-viscosity) subgrid scale model (WALE) by [39]. Furthermore, an universal equilibrium stress model based on the wall function of [40] is applied at the walls for the turbulent viscosity terms in order to bridge with a single cell the thin viscous sublayer, which is not fully resolved in the LES.

All the essential features of the investigated cases are summarized in the next subsection.

3.3. Summary of the Case Studies

Important features of the experimental and numerical investigations of the inclined impinging jets are listed in Table 1. It should be noted here that in the case of PIV measurements, all three inclination angles are investigated, while in the DNS, only the 45°-configuration is examined.

Table 1. Summary of the experimental and numerical studies with respect to the inclination angle.

Configuration	Experiment	DNS
0°-inclination	• fluid: dry air at $T = 298.15$ K and $p = 1$ atm. • flow: $Re = 5000$ based on nozzle outlet diameter. • results: mean and rms velocities, production of turbulent kinetic energy.	
45°-inclination	• fluid: dry air at $T = 298.15$ K and $p = 1$ atm. • flow: $Re = 5000$ based on nozzle outlet diameter. • geometry: 40 mm jet-to-plate spacing. • results: mean and rms velocities, production of turbulent kinetic energy.	• fluid: dry air at $T = 298.15$ K and $p = 1$ atm. • flow: $Re = 5000$ based on nozzle outlet diameter. • geometry: 40 mm jet-to-plate spacing. • results: mean and rms velocities, budget terms of turbulent kinetic energy, turbulence structures, wall shear stress.
90°-inclination	• fluid: dry air at $T = 298.15$ K and $p = 1$ atm. • flow: $Re = 5000$ based on nozzle outlet diameter. • geometry: 40 mm jet-to-plate spacing. • results: mean and rms velocities, production of turbulent kinetic energy.	

4. Experimental Results

In this section, general flow features of the inclined impinging jets are analyzed by means of two-component PIV measurements. Thereby, mean flow properties and features of the turbulence dynamics as apparent in the impinging flows are examined and compared for different impingement angles. Mean and rms velocity profiles close to the nozzle exit, which may be used as realistic turbulent inflow conditions for numerical simulations, are provided in Appendix B.

4.1. Mean Flow Properties

Figure 4 presents contour plots of the time-averaged magnitude velocity at the mid-plane section of the jet. Results of the low-speed PIV measurements are shown for the 0°-, 45°- and

90°-inclination configurations, respectively. Notice that an additional coordinate system is introduced with η representing the wall-normal direction, and ζ is the direction along the wall.

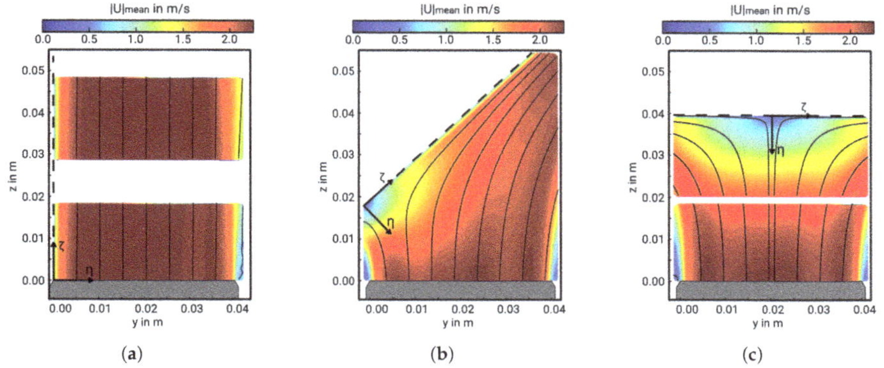

Figure 4. Contour plots of the time-averaged magnitude velocity at the mid-plane section of the jet obtained by low-speed PIV measurements. (—) streamlines of the flow field; (- -) position of the wall. (**a**) 0°-, (**b**) 45°- and (**c**) 90°-configuration.

Regarding the 0°-inclination angle as shown in Figure 4a, the stream of dry air leaves the nozzle and flows along the solid surface. Streamlines are parallel to the wall, and a characteristic turbulent boundary layer is formed in the vicinity of the solid surface, where the effects of viscosity become significant. In contrast, mean flow features and streamline patterns are inherently different in the 45°- and 90°-configurations (Figure 4b,c). Here, the jet can be seen to decay and spread in the y-direction, then impinges on the plate, next deflecting and splitting into two main streams along the plate. Thereby, streamlines are highly curved, especially around the stagnation region.

Next, profiles of the time-averaged wall-parallel and wall-normal velocity components are examined. Results are shown for wall-normal traverses at $\zeta/D = 0$, 0.25, 0.5 for the 45°- and 90°-configurations and at $\zeta/D = 0$, 0.25, 1 for the 0°-configuration. Profiles are normalized by the bulk velocity of the jet U_{bulk} and the hydraulic diameter of the nozzle D.

As can be clearly observed in Figure 5, general flow features of the 0°-, 45°- and 90°-configurations are fundamentally different. In the case of 0°-inclination, the flow field resembles a boundary layer flow where the mean flow is predominantly parallel to the wall, with the mean velocity $U_{mean,\zeta}$ varying mainly in the η-direction. Thereby, the velocity component in the wall-normal direction $U_{mean,\eta}$ is almost zero. In contrast, regarding the 90°-configuration, $U_{mean,\zeta}$ is initially zero at the stagnation point ($\eta/D = 0$). Further downstream, a wall-jet is formed as the flow deflects in ζ-direction, and the fluid is subject to a strong acceleration. Thereby, the velocity component in wall-normal direction $U_{mean,\eta}$ decreases with decreasing wall distance. Flow features of both the 0°- and 90°-configurations can be retrieved in the 45°-inclination case. Initially, as the flow deflects in ζ-direction, a wall-jet begins to form in the ζ-direction (Figure 5a) with a steep velocity gradient of $U_{mean,\eta}$ in the η-direction (Figure 5b), similar to that found in the 90°-configuration. Further downstream, the mean flow is predominantly in the ζ-direction, and the velocity component in wall-normal direction $U_{mean,\eta}$ vanishes. At this stage, the flow field resembles that of a 0°-configuration.

It turned out that the mean flow features and streamline patterns are inherently different in the 0°-, 45°- and 90°-configurations. The 0°-configuration resembles a boundary layer flow, while the stagnation region and wall-jets are predominant in the 90°-configuration. Flow features of both configurations (0° and 90°), particularly boundary layer flow properties, stagnation point and wall jet interactions, prevail in the 45°-inclination case.

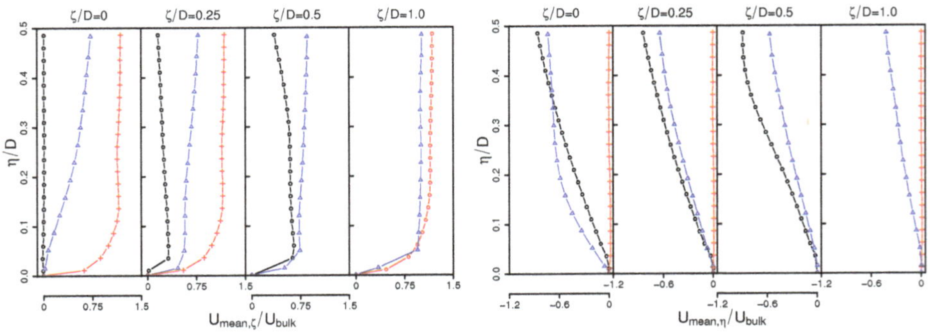

(**a**) (**b**)

Figure 5. Time-averaged velocity components in wall-parallel (**a**) and wall-normal (**b**) directions for different wall-normal traverses. (+) 0°-configuration; (△) 45°-configuration; (○) 90°-configuration.

4.2. Turbulence Dynamics

The effects of the inclination angle on turbulence dynamics are analyzed in Figure 6 in terms of magnitude root-mean-square (rms) velocity $|U|_{rms} = \sqrt{U^2_{x,rms} + U^2_{y,rms}}$ and production of turbulent kinetic energy $P_{tke} = \overline{u'_i u'_j} \frac{\partial \overline{U_i}}{\partial x_j}$ for several wall-normal traverses. Notice that in order to simplify the calculations of P_{tke}, the symmetric condition $\frac{\partial \overline{U_i}}{\partial x} \equiv 0$ in the span-wise direction x is used.

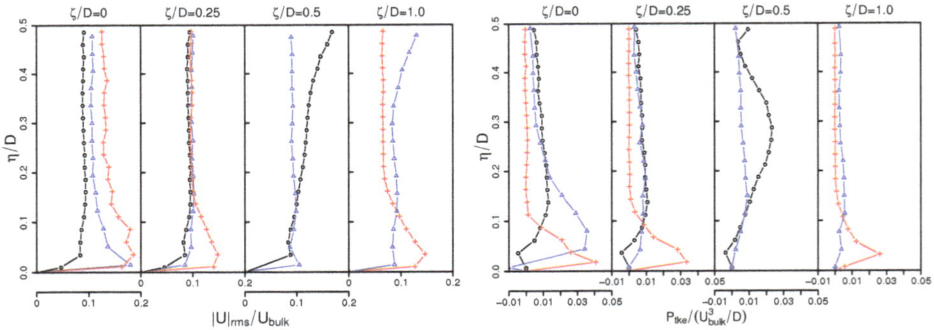

(**a**) (**b**)

Figure 6. Magnitude rms velocity (**a**) and production of turbulent kinetic energy (**b**) for different wall-normal traverses. (+) 0°-configuration; (△) 45°-configuration; (○) 90°-configuration.

As is visible in Figure 6a, there is approximately self-similarity far away from the wall. Thereby, normalized rms velocities are all in the same order of magnitude and essentially uniform. In contrast, close to the wall, the effects of the inclination angle are more dominant. It appears that the 0°-configuration contains the most vigorous turbulent activity, while rms velocities are small for the other inclination angles especially in the case of 90°. Regarding the production of turbulent kinetic energy in Figure 6b, it is interesting to observe that P_{tke} appears negative at the stagnation point in the 45°- and 90°-inclination configurations, while it is always positive in the case of 0°. Far away from the wall, values of P_{tke} are essentially smaller and more uniform. Therefore, it can be concluded that the flow dynamics are especially affected by the impinging region rather than by the far field. These findings are in good agreement with observations in fully-developed jets impinging normally on a solid surface (see, e.g., [8,21]) and hold obviously also for impinging flows that are not fully developed and impinge at a particular angle other than 0°.

To summarize: The turbulence flow dynamics differ considerably for the different inclination angles under consideration. Root-mean-square velocities decrease with increasing inclination angle in the vicinity of the wall and are fairly small at the stagnation region. Furthermore, the production of turbulent kinetic energy P_{tke} appears negative at the stagnation point in the 45°- and 90°-inclination configurations, while it is always positive in the case of 0°.

5. Numerical Results

After examining the general flow features of the inclined impinging jets with respect to the inclination angles using PIV measurement technique, DNS is utilized now to complement the experimental results of the 45°-configuration. Relying on the experimental findings discussed above, this inclination angle is selected since it includes all the flow features predominant in turbulent impinging flows, namely boundary layer flow properties, stagnation point and wall jet interaction. First, numerical results are compared with the experiment in order to validate the DNS and vice versa. Then, budget terms of the turbulent kinetic energy, turbulence structures and wall shear stresses are analyzed.

5.1. Comparison with Experimental Results

Figure 7 shows predicted time-averaged wall-parallel (a) and wall-normal (b) velocity components in comparison to the PIV measurements at several wall-normal traverses ($\zeta/D = 0$, 0.25, 0.5, 1.0). In order to supplement the measurements, profiles at $\zeta/D = -0.5$ are additionally provided.

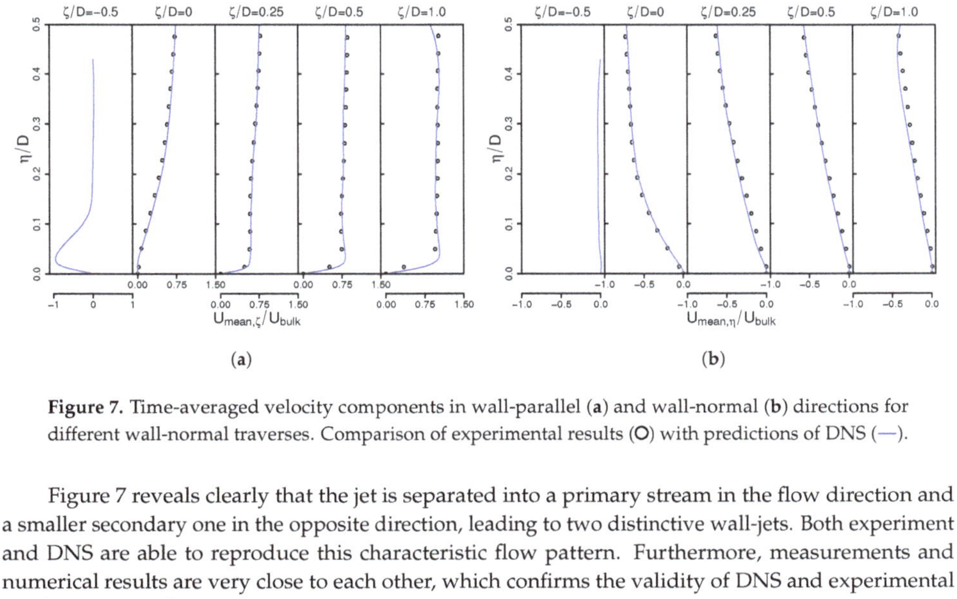

Figure 7. Time-averaged velocity components in wall-parallel (**a**) and wall-normal (**b**) directions for different wall-normal traverses. Comparison of experimental results (O) with predictions of DNS (—).

Figure 7 reveals clearly that the jet is separated into a primary stream in the flow direction and a smaller secondary one in the opposite direction, leading to two distinctive wall-jets. Both experiment and DNS are able to reproduce this characteristic flow pattern. Furthermore, measurements and numerical results are very close to each other, which confirms the validity of DNS and experimental results in terms of mean flow.

A comparison of rms velocity components is depicted in Figure 8. Again, DNS results at $\zeta/D = -0.5$, where experimental data are not available, are provided.

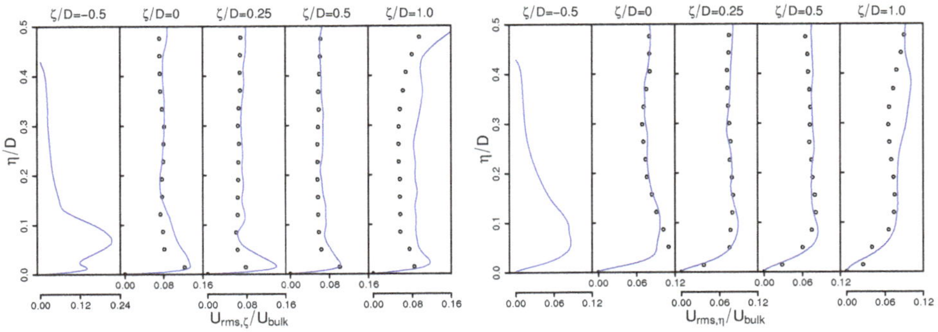

(a) (b)

Figure 8. Comparison of predicted wall-parallel (**a**) and wall-normal (**b**) rms velocity components with experimental results. PIV measurement (O); DNS results (—).

As it is apparent in Figure 8, the most vigorous turbulent activity appears in the vicinity of the wall, associated with high rms velocities that decline with increasing wall distance and remain approximately constant for $\eta/D > 0.2$. Thereby, it can be seen that rms velocities of the experiment and DNS differ slightly near the wall and close to the boundary of the PIV window ($\zeta/D = 1.0$). These discrepancies may be caused by reflections of the laser light from the solid wall and low particle seeding density at the boundary of the PIV window, respectively. Note that measurements of the fluid flow close to a solid wall are very challenging and associated with uncertainties. Nevertheless, deviations of numerical and experimental results are fairly small for such circumstances, which leads to the conclusion that the present DNS is appropriate to describe the turbulent flow field in the 45°-inclined impinging jet configuration.

5.2. Budget of Turbulent Kinetic Energy Transport

It was observed in the experiment (Section 4.2) that the production of turbulent kinetic energy P_{tke} appears negative at the stagnation point for inclination angles other than 0°. This observation warrants a closer examination of budget terms of the turbulent kinetic energy equation at the stagnation point. Regarding isothermal turbulent flows, the balance equation of turbulent kinetic energy reads [41]:

$$\frac{Dk}{Dt} = P_{tke} + \epsilon_{tke} + \Pi_{tke} + D_{tke} + T_{tke},$$
$$P_{tke} = -\overline{u_i' u_j'}\frac{\partial \overline{U_i}}{\partial x_j}, \quad \epsilon_{tke} = -2\nu\overline{S_{ij}' S_{ij}'}, \quad \Pi_{tke} = -\frac{\partial}{\partial x_j}\overline{u_j' p'},$$
$$D_{tke} = 2\nu\frac{\partial}{\partial x_j}\left(\overline{u_i' S_{ij}'}\right), \quad T_{tke} = -\frac{\partial}{\partial x_j}\left(\frac{1}{2}\overline{u_i' u_i' u_j'}\right), \tag{6}$$

where $k = 1/2\overline{u_i' u_i'}$ denotes the turbulent kinetic energy, P_{tke} is production, ϵ_{tke} viscous dissipation, Π_{tke} pressure-related diffusion, D_{tke} viscous diffusion and T_{tke} turbulent velocity-related diffusion due to third order moments. S_{ij} is the symmetric part of the velocity gradient tensor. Figure 9 shows a contour plot of the turbulent kinetic energy (a) and the corresponding normalized budget terms (b) along the wall-normal direction at the stagnation point ($\zeta/D = 0$). Budget terms are normalized by U_{bulk}^3/D.

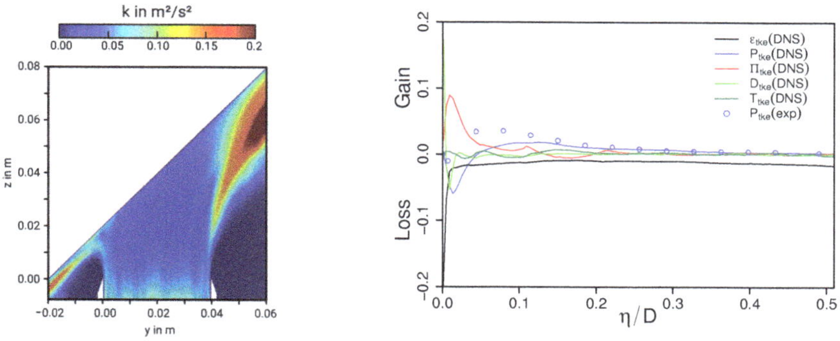

(a) (b)

Figure 9. Contour plot of the the turbulent kinetic energy k (**a**) and the normalized budget terms of k along the wall-normal direction at the stagnation point ($\zeta/D = 0$) (**b**). Budget terms are normalized using the bulk velocity U_{bulk} and the nozzle exit diameter D.

As might be expected, values of k are high at the shear layers and increase as the jet develops. Apart from the jet-edges, the turbulent kinetic energy remains relatively small, especially at the jet core. Surprisingly, values of k are also fairly small at the stagnation region, roughly in the range of the inflow turbulence. The reasons for such a behavior around the stagnation point becomes clearer by examining the budget terms of k in Figure 9b, that substantially differ from those of other turbulent wall-bounded flows. Far away from the wall, where k is relatively small, viscous dissipation is the dominant term, while other contribution terms in Equation (6) are small. This holds more or less up to $\eta/D \approx 0.2$. Then, in the vicinity of the wall, steep gradients prevail in the wall-parallel and wall-normal direction, leading to negative production of turbulent kinetic energy. Thereby, dissipation and viscous diffusion decrease, while pressure-related diffusion, which is usually negligibly small in wall-bounded flows, becomes notably large. Indeed, it is the pressure-related diffusion term Π_{tke} that balances the negative production of turbulent kinetic energy in the case of impinging flows. Finally, very close to the wall, dissipation is balanced by viscous transport, while the other terms in Equation (6) are zero, just as is the case for turbulent kinetic energy. A similar behavior of turbulent kinetic energy budget terms was reported in experimental studies of fully-developed jets impinging normally on a solid surface (see, e.g., [8]), which holds also in impinging flows that are not fully developed and impinge at a particular angle of 45°. Note that PIV measurements of P_{tke} from the present experimental study are also displayed in Figure 9b. Thereby, DNS and experimental results agree very well far away from the wall, while they differ in the near-wall region. As mentioned before, discrepancies may be caused by reflections of the laser light from the solid wall or limited resolution in the PIV measurements. Nevertheless, the principle physical behavior with negative production close to the wall and small values elsewhere are clearly retrieved by both experiment and DNS.

To sum up, the blockage of the impermeable wall has a considerable effect on the generation and destruction mechanisms of turbulent kinetic energy k. In particular, it turns out that values of k are relatively small at the stagnation region. Thereby, production of turbulent kinetic energy appears negative close to the wall, which is balanced predominantly by the pressure-related diffusion term.

5.3. Turbulence Structures

As figured out in the previous sections, turbulent impinging jets feature very complex flow properties including the presence of a stagnation point, shear layers and strong streamline curvature. Their impact on vortical and turbulence structures is addressed in this subsection, with the main focus on the near-wall region. First, in order to obtain a global perception of the flow structures, coherent vortices are identified by means of the Q-criterion [42]. It is defined as $Q = -1/2\left(g_{ii}g_{jj} - g_{ij}g_{ij}\right)$, where $g_{ij} = \partial U_i/\partial x_j$ is the velocity gradient tensor and Q its second invariant. $Q > 0$ represents the spatial region of a vortex and implies that irrotational straining is small compared with the vorticity [42]. Instantaneous isocontours of positive Q around the mid-plane section of the jet and in the vicinity of the impinging wall colored by the magnitude velocity are shown in Figure 10a,b, respectively. In order to illustrate the strength of flow circulation associated with the coherent structures, Figure 10c depicts the corresponding magnitude vorticity $|\omega|$ close to the wall.

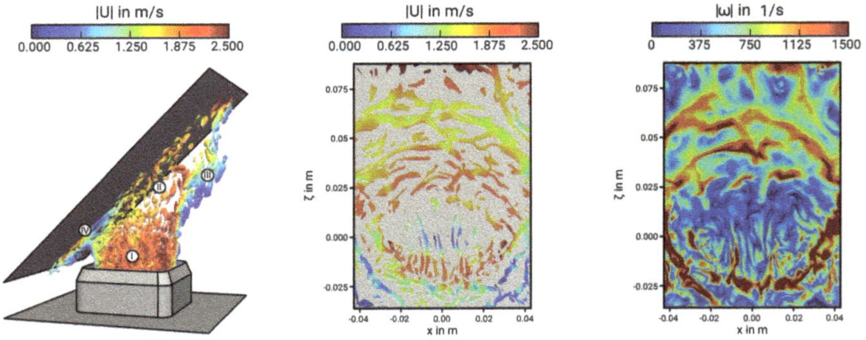

(a) (b) (c)

Figure 10. Instantaneous isocontour of positive Q around the mid-plane section (**a**) and in the vicinity of the wall (**b**) and snapshot of the vorticity magnitude close to the wall (**c**).

Considering Figure 10a, four main regions associated with the general flow pattern of the inclined impinging jet and the resulting nature of turbulent structures can be distinguished. First, the jet core region close to the nozzle exit (I) is dominated by relatively small turbulent scales of uniform shape that are generated by the turbulence grid inside the nozzle. These flow structures are carried along with the main flow and seem to dissolve further downstream. Secondly, the wall jet directed outward along the wall (II), where the flow interacts with the solid surface and outer shear layer. Here, flow structures are considerably larger and toroidally organized, especially in the vicinity of the wall. Thirdly, the shear layer on the side away from the impinging wall (III), which is triggered by the interaction of the jet with the ambient fluid. Thereby, large coherent roll-up vortices are created by the induced shearing that increase downstream. Finally the stagnation region (IV), where the fluid is subject to a strong acceleration and stretching, leading to thin streaks orientated in the flow direction. The resulting instances of the large-scale structures on the impinging wall are shown Figure 10b. Here, it can be seen as well that coherent fluid flow structures are extremely elongated and axisymmetric at the stagnation region. Thereby, the magnitude velocity is small, which suggests that these flow structures persist for a significant period of time until they slowly migrate away. Further downstream, turbulent structures are predominantly toroidally organized among many other random structures within the background. These structures have a considerably higher magnitude velocity and convect outward along the wall. A similar conclusion of the organization of vortical structures close to the wall can be drawn by examining the vorticity magnitude displayed in Figure 10c. Vorticity appears small at the stagnation region with isolated stretched nests of concentrated vorticity that are orientated in the flow direction. Further downstream, vorticity becomes circumferentially interconnected resulting in ring structures of strong flow circulation that serve as precursors of large coherent turbulent eddies in the near-wall region.

Next, anisotropy invariant maps are utilized in order to analyze the underlying physics of the turbulence associated with the fluid flow structures in the inclined impinging jet configuration. For this purpose, Figure 11 shows plots of the second and third invariants, II$= b_{ij}b_{ij}/2$ and III$= b_{ij}b_{jn}b_{ni}/3$, of the Reynolds stress anisotropy tensor defined as $b_{ij} = \overline{u_i'u_j'}/\overline{u_l'u_l'} - 1/3\delta_{ij}$ (see, e.g., [41]). Results are shown for wall-normal traverses at $\zeta/D = 0$ (a) and $\zeta/D = 1$ (b).

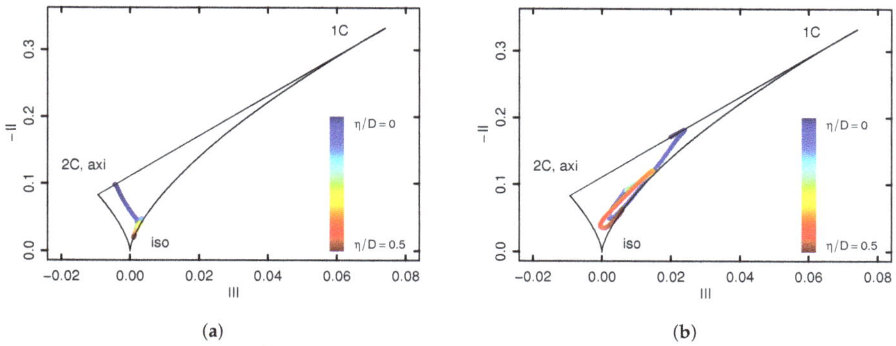

Figure 11. Non-linear anisotropy invariant maps at $\zeta/D = 0$ (**a**) and $\zeta/D = 1$ (**b**).

In line with the visual appearance of the turbulent structures in Figure 10a, it can be clearly observed in Figure 11a that the turbulence is in fact axisymmetric at the stagnation point with a negative third invariant and one small eigenvalue of the Reynolds stress anisotropy tensor. Away from the wall, anisotropic turbulence decays on a more or less straight trajectory directed towards the origin. After reaching the plane-strain limit (III= 0) at $\eta/D \sim 0.03$, which is located close to the jet shear layer facing the impinging wall, the turbulence tends towards the one-component direction. Subsequently, far away from the wall and close to the nozzle exit, the turbulence returns to the isotropic state. Here, turbulent structures have to a large extent a uniform shape. A rather different behavior of the turbulence appears farther downstream at the wall-jet region as depicted in Figure 11b. Very close to the wall, the turbulence is essentially two-component. Then, anisotropic turbulence reaches a peak at $\eta/D \sim 0.02$ and is close to being axisymmetric far away from the wall with III> 0. This behavior of the Reynolds stresses is very similar to that found in turbulent boundary layer flows, which confirms again the visual perception in Figure 10a.

Finally, characteristic turbulence length and time scales along the wall-normal traverse at the stagnation point ($\zeta/D = 0$) are provided in Figure 12. These data may be useful to evaluate turbulence models, especially in the context of RANS. Thereby, $L_t = k^{3/2}/\epsilon_{tke}$ is the turbulence length scale, $\eta_K = \left(\nu^3/\epsilon_{tke}\right)^{1/4}$ the Kolmogorov length scale and $\tau_{int,i}$ the Eulerian integral time scale of the velocity component i calculated using the temporal autocorrelation function of the recorded time series (see, e.g., [41]).

As it is expected, L_t is small very close to the stagnation point, increases immediately and then declines gradually with increasing wall distance. Similar, η_K is small close to the stagnation point, increases away from the wall and remains approximately constant for $\eta/D > 0.02$. Thereby, the energy containing turbulent length scales L_t are approximately ten to twenty times larger than smallest scales η_K. Regarding the integral time scales as shown in Figure 12b, they are indeed large at the stagnation point and decrease with increasing wall distance. This confirms that characteristic time scales of turbulent processes are relatively large around the stagnation point and that turbulent flow structures persist for a significant period of time at this region while they are convected slowly away.

By examining turbulent flow structures, it turns out that quasi-coherent thin streaks appear around the stagnation point. Thereby, the turbulence is essentially axisymmetric with large characteristic turbulence time scales. Further downstream, at the wall-jet region, the organization of the flow is predominantly toroidal, and the turbulence behavior is similar to that found in turbulent boundary layer flows.

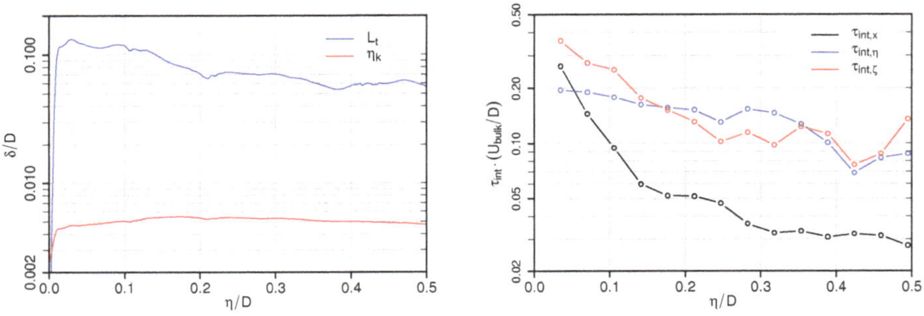

(a) (b)

Figure 12. Turbulent length (**a**) and time (**b**) scales as a function of wall distance. Turbulent length scale: $L_t = k^{3/2}/\epsilon_{tke}$; Kolmogorov length scale: $\eta_K = \left(\nu^3/\epsilon_{tke}\right)^{1/4}$; integral time scales: $\tau_{int,i}$.

5.4. Near-Wall Shear Stress

From the observations above, it is evident that the inclined impinging jet flow is characterized by a strong wall/flow interaction process. Close to the wall, Reynolds stresses vanish, and viscous shearing, exerted by the fluid on the impermeable wall, along with steep pressure gradients around the stagnation region dominate the fluid flow. Thereby, a thin but very important sub-layer is formed immediately adjacent to the solid surface. Usually, due to limited computing power, classical wall-function approaches are used in the context of LES and RANS to bridge with a single cell this very thin sub-layer where viscosity modifies the turbulence structure. However, it is well known that wall-functions based on the semi-logarithmic variation of the near-wall velocity do not apply under non-equilibrium flow conditions including impingement and steep pressure gradients, as is apparent in the inclined impinging jet (see, e.g., [43,44]). In such cases, advanced wall treatments are typically used, e.g., generalized wall functions [45], analytical wall-functions [46,47] or numerical integration of boundary layer equations [48], that produce the required value of the wall shear stress over the near-wall cell. Such values are difficult to obtain experimentally. It is therefore of particular interest to provide reliable reference data of wall shear stresses for validation purpose in order to appraise the assumptions made in the near-wall modeling approaches. For this reason and in order to explain the strong wall/flow interaction process in the inclined impinging jet flow, Figure 13 shows instances of the instantaneous (a) and time-averaged (b) absolute wall shear stress $|\tau_w|$ induced by the inclined jet on the wall.

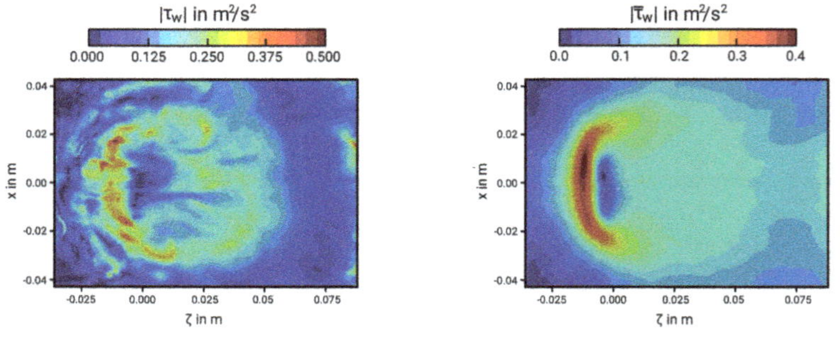

(a) (b)

Figure 13. Instantaneous (**a**) and time-averaged (**b**) instances of the absolute wall shear stress on the impinging wall.

Similar to fully-developed turbulent jets impinging perpendicularly on a solid surface, wall shear stresses are very low at the stagnation region and peak in its immediate vicinity (for a comparison, see, e.g., [5,49,50]). It is interesting to observe that, in the case of a 45° inclination angle, the wall shear stress is predominantly concentrated at the secondary opposed wall jet region ($\zeta < 0$), where the direction of the flow changes suddenly and the fluid is subject to a high acceleration in the wall-parallel direction. At the wall jet region in the main flow direction ($\zeta > 0$), values of $|\tau_w|$ are considerably lower (approximately half the peak value of the opposed wall jet region), and the peak is significantly smoother and tends to smear out in the main flow direction. Furthermore, it can be clearly observed that the fluctuations in the near-wall shear stress are more pronounced at the opposed wall jet region, indicating strong transient fluid flow processes at this region. These observations suggest that the wall/flow interaction process depends inherently on the impingement angle and is obviously more intensive in regions where the direction of the flow changes suddenly and the fluid is subject to high acceleration in the wall-parallel direction.

Finally, profiles of the wall shear stress in span-wise direction x and in wall-parallel direction ζ are provided in Figure 14a,b, respectively, which may be used for validation purposes in the context of LES and RANS.

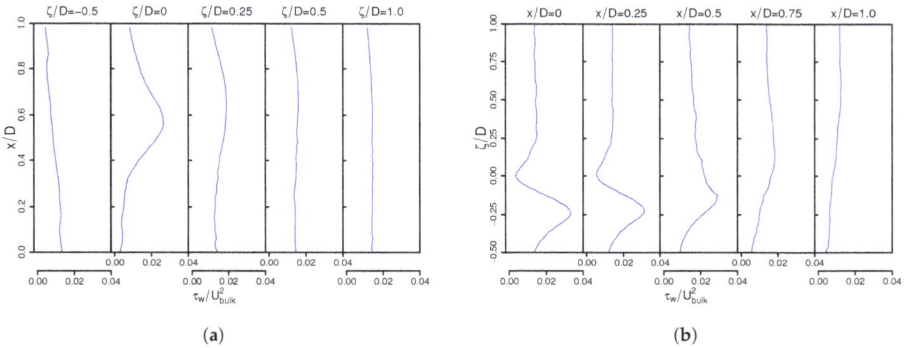

Figure 14. Profiles of the wall shear stress in span-wise direction x (**a**) and in wall-parallel direction ζ (**b**).

6. Concluding Remarks

Direct numerical simulation and particle image velocimetry have been applied complementarily in order to generate a database that allows characterizing the general flow features and turbulent flow properties of a highly turbulent, non-fully-developed jet impinging on a solid surface under different inclination angles (0°, 45°, 90°), with the main focus on the 45°-inclination. This comprehensive dataset includes near-wall turbulence statistics, budget terms in turbulent kinetic energy equation, anisotropy invariant maps, turbulent length/time scales and near-wall shear stresses that may be useful for the validation of near-wall modeling approaches of LES and RANS.

Some important observations found in the present study can be outlined as follows:

I Mean flow patterns are inherently different in the 0°, 45° and 90° jet configurations. In particular, the 0°-configuration resembles a boundary layer flow, while the stagnation region and wall-jets are predominant in the 90°-configuration. Flow features of both configurations, namely boundary layer flow properties, stagnation point and strong wall/jet interactions, prevail in the 45°-inclination case, which represents therefore a good generic benchmark test case for a wide range of technical applications.

II It turns out that the production of turbulent kinetic energy appears negative at the stagnation region in the 45°- and 90°-configurations, which is balanced predominantly by pressure-related diffusion. In the case of the 0°-inclination angle, the production term is always positive.

III By examining turbulent flow structures in the 45°-configuration, it turns out that quasi-coherent thin streaks appear around the stagnation point. Thereby, the turbulence is essentially axisymmetric with large characteristic turbulence time scales. Further downstream, at the wall-jet region, the organization of the flow is predominantly toroidal, and the turbulence behaves similarly to that found in turbulent boundary layer flows.

IV In the case of the 45°-inclination angle, near-wall shear stresses are very low at the stagnation point and primarily concentrated at the secondary opposed wall jet region. This suggests that the wall/flow interaction process depends inherently on the impinging angle and is obviously more intense in regions where the direction of the flow changes suddenly and the fluid is subject to high acceleration in the wall-parallel direction.

Acknowledgments: The authors gratefully acknowledge the financial support by the DFG (German Research Council) Sonderforschungsbereich/Transregio SFB/TRR 150 and the support of the numerical simulations on the Lichtenberg High Performance Computer at the University of Darmstadt. Andreas Dreizler was financially supported by Gottfried Wilhelm Leibniz-Preis (DFG).

Author Contributions: Florian Ries and Martin Rißmann performed the experiments, which were conceived of and designed by Andreas Dreizler, Benjamin Böhm and Martin Rißmann. Martin Rißmann, Yongxiang Li and Florian Ries evaluated and analyzed the experimental data. Florian Ries implemented the numerical approach and performed the DNS, while Yongxiang Li and Dario Klingenberg verified and validated the source code. Dario Klingenberg generated the numerical grid and provided analysis tools. Florian Ries and Yongxiang Li treated and exploited the numerical data and analyzed together with Amsini Sadiki and Kaushal Nishad the numerical results. Florian Ries wrote the paper, while Amsini Sadiki and Andreas Dreizler further supported improving the manuscript. Amsini Sadiki, Andreas Dreizler and Johannes Janicka contributed by providing materials and computing resources.

Conflicts of Interest: The authors declare no conflict of interest.

Appendix A. Code Verification

The method of manufactured solution (MMS) is applied to verify that the discretized governing equations, as implemented in the source code, are solved consistently. A manufactured solution is an exact solution to a set of PDE's that has been constructed by solving the problem backwards [51]. In this context, the analyst first selects a sufficiently differentiable function $U_i(x,t)$ to describe the desired evolution of the variables in space and time. This solution $U_i(x,t)$ does not necessarily satisfy the original set of PDE's, therefore a corresponding set of source terms are manufactured by applying the set of PDE's to $U_i(x,t)$ and added to the source code in order to balance the system [52]. The resulting set of PDE's including the source terms is then solved for different spatial and/or temporal resolutions. Finally, the order-of-accuracy in space and time is quantified and verified for the numerical approach, leading to a rigorous code verification in full generality. Further information about the concept and procedure of MMS can be found in [51–54].

Appendix A.1. Spatial Accuracy

A three-dimensional manufactured solution is chosen in the present study to quantify the spatial accuracy of the applied numerical approach. Thereby, the velocity field resembles a Taylor-Green vortex [55], which leads to the following smoothly varying solution for the velocity and pressure fields

$$\begin{aligned} U_x &= -cos\,(2\pi x) \cdot sin\,(2\pi y) \cdot sin\,(2\pi z)\,, \\ U_y &= 0.5 \cdot sin\,(2\pi x) \cdot cos\,(2\pi y) \cdot sin\,(2\pi z)\,, \\ U_z &= 0.5 \cdot sin\,(2\pi x) \cdot sin\,(2\pi y) \cdot cos\,(2\pi z)\,, \\ p &= \left(\frac{1}{3}x^3 - \frac{1}{2}x^2\right) \cdot \left(\frac{1}{3}y^3 - \frac{1}{2}y^2\right) \cdot \left(\frac{1}{3}z^3 - \frac{1}{2}z^2\right) \cdot 10^4, \end{aligned} \tag{A1}$$

where x, y, z are the coordinates in space. The solution is applied to the governing equations as introduced in Section 2.1 leading to non-zero source terms in the momentum equation (not shown here, due to its length), which are implemented in the source code. No source term appears in the continuity equation since the manufactured solution satisfies the divergence free condition. The kinematic viscosity is set to one to balance the order of magnitude of convection and diffusion terms equally in the momentum equation. A computational domain with $[0,1] \times [0,1] \times [0,1]$ for x, y and z is selected. Four equidistant computational grids consisting of 512, 4096, 32,768 and 262,144 control volumes are used for the verification study. Figure A1 shows a graph of the mean absolute errors (MAE) as a function of the mesh size for the magnitude velocity and kinematic pressure.

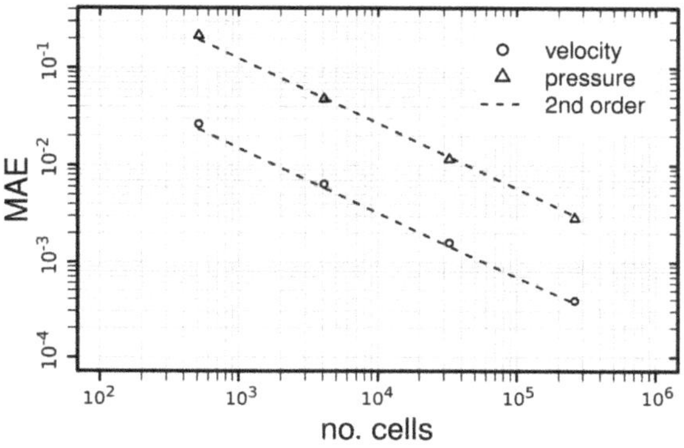

Figure A1. Mean absolute error (MAE) as a function of mesh size for the magnitude velocity and kinematic pressure.

As it is apparent in Figure A1, the mean absolute error (MAE) of the magnitude velocity and kinematic pressure both drop by a factor of four with each mesh refinement, thus matching the second-order slope and verifying that the numerical approach is of second-order accuracy in space.

Appendix A.2. Temporal Accuracy

Next, the temporal accuracy of the three-stages explicit Runge-Kutta scheme for time integration is addressed. For this purpose, the following manufactured solution is selected

$$U_x = sin\,(20\pi t)\,,\ \ U_y = -cos\,(20\pi t)\,,\ \ U_z = sin\,(20\pi t)\,,\ \ p = const., \tag{A2}$$

where t is the time. Note that a temporal analysis of the pressure is omitted, since no time derivative occurs in the Navier-Stokes equations. In accordance with the verification of the spatial accuracy, the solution is applied to the governing equations leading to non-zero source terms in the momentum equation and no source term in the continuity equation. For the analysis, the time step is gradually increased from $\Delta t = 2.5 \times 10^{-4}$ up to $\Delta t = 2 \times 10^{-3}$ by a factor of two and a time interval is chosen as $[0,1]$. The resulting graph of the mean absolute error norm with respect to the time step size is depicted in Figure A2.

As it can be clearly observed in Figure A2, the global errors decrease by a factor of four with each time step size refinement, verifying that the numerical approach is of second-order accuracy in time.

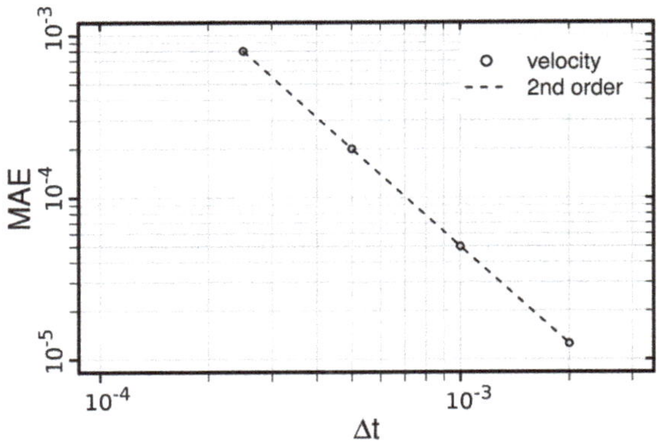

Figure A2. Mean absolute error (MAE) as a function of time step size for the magnitude velocity and kinematic pressure.

Appendix B. Inflow Conditions for Numerical Simulations

Profiles of mean and rms velocities measured at $z \approx 0.6$ mm downstream the nozzle exit are provided in Figure A3, which may be used as realistic turbulent inflow conditions for numerical simulations.

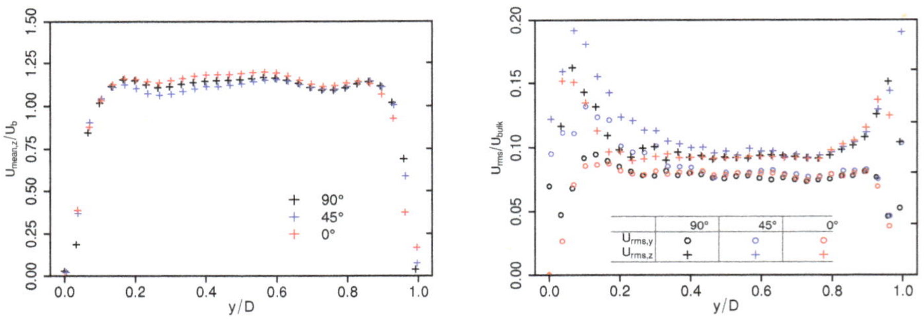

(**a**) (**b**)

Figure A3. Profiles of axial mean (**a**) and rms (**b**) velocities measured at 0.6 mm downstream the nozzle exit for the 0°-, 45°- and 90°-inclination configurations, respectively.

Thereby, the velocity profiles are not fully developed at the nozzle exit and the turbulent intensities are of ∼10–20%, which is approximately two times the intensity found in fully developed turbulent flows in a square duct at a similar Reynolds number (see e.g., [56,57]). Furthermore, it is apparent, that measured inflow mean and rms velocity profiles differ slightly for different inclination angles, especially at the side facing the solid wall.

References

1. Martin, H. Heat and mass transfer between impinging gas jets and solid surfaces. *Adv. Heat Transf.* **1997**, *13*, 1–60, doi:10.1016/S0065-2717(08)70221-1.
2. Jambunathan, K.; Lai, E.; Moss, M.A.; Button, B.L. A review of heat transfer data for single circular jet impingement. *Int. J. Heat Fluid Flow* **1992**, *13*, 106–115, doi:10.1016/0142-727X(92)90017-4.
3. Viskanta, R. Heat transfer to impinging isothermal gas and flame jets. *Exp. Therm. Fluid Sci.* **1993**, *6*, 106–115, doi:10.1016/0894-1777(93)90022-B.

4. Zuckerman, N.; Lior, N. Jet impingement heat transfer: Physics, correlations, and numerical modeling. *Adv. Heat Transf.* **2006**, *39*, 565–631, doi:10.1016/S0065-2717(06)39006-5.
5. Tummers, M.J.; Jacobse, J.; Voorbrood, S.G.J. Turbulent flow in the near field of a round impinging jet. *Int. J. Heat Mass Transf.* **2011**, *54*, 4939–4948, doi:10.1016/j.ijheatmasstransfer.2011.07.007.
6. Boughn, J.W.; Shimizu, S. Heat transfer measurements from a surface with uniform heat flux and an impinging jet. *J. Heat Transf.* **1989**, *111*, 1096–1098, doi:10.1115/1.3250776.
7. Cooper, D.; Jackson, D.C.; Launder, B.E.; Liao, G.X. Impinging jet studies for turbulence model assessment-I. Flow-field experiments. *Int. J. Heat Mass Transf.* **1993**, *36*, 2675–2684, doi:10.1016/S0017-9310(05)80204-2.
8. Nishino, K.; Samada, M.; Kasuya, K.; Torii, K. Turbulence statistics in the stagnation region of an axisymmetric impinging jet flow. *Int. J. Heat Fluid Flow* **1996**, *17*, 193–201, doi:10.1016/0142-727X(96)00040-9.
9. Behrouzi, P.; McGuirk, J.J. Laser Doppler velocimetry measurements of twin-jet impingement flow for validation of computational models. *Opt. Laser Eng.* **1998**, *30*, 265–277, doi:10.1016/S0143-8166(98)00030-X.
10. Fairweather, M.; Hargrave, G. Experimental investigation of an axisymmetric, impinging turbulent jet. 1. Velocity field. *Exp. Fluids* **2002**, *33*, 464–471, doi:10.1007/s00348-002-0479-7.
11. Geers, L.F.G.; Tummers, M.J.; Hanjalić, K. Experimental investigation of impinging jet arrays. *Exp. Fluids* **2004**, *36*, 946–958, doi:0.1007/s00348-004-0778-2.
12. Katti, V.; Prabhu, S.V. Experimental study and theoretical analysis of local heat transfer distribution between smooth flat surface and impinging air jet from a circular straight pipe nozzle. *Int. J. Heat Mass Transf.* **2008**, *51*, 4480–4495, doi:10.1016/j.ijheatmasstransfer.2007.12.024.
13. Satake, S.; Kunugi, T. Direct numerical simulation of an impinging jet into parallel disks. *Int. J. Numer. Methods Heat Fluid Flow* **1998**, *8*, 768–780, doi:10.1108/09615539810232871.
14. Chung, Y.M.; Luo, K.H. Unsteady Heat Transfer Analysis of an Impinging Jet. *J. Heat Transf.* **2002**, *124*, 1039–1048, doi:10.1115/1.1469522.
15. Hattori, H.; Nagano, Y. Direct numerical simulation of turbulent heat transfer in plane impinging jet. *Int. J. Heat Fluid Flow* **2004**, *25*, 749–758, doi:10.1016/j.ijheatfluidflow.2004.05.004.
16. Wilke, R.; Sesterhenn, J. Numerical Simulation of Impinging Jets. In *High Performance Computing in Science and Engineering '14*; Springer: Cham, Switzerland, 2015; pp. 275–287, doi:10.1007/978-3-319-10810-0_19.
17. Jainski, C.; Lu, L.; Sick, V.; Dreizler, A. Laser imaging investigation of transient heat transfer processes in turbulent nitrogen jets impinging on a heated wall. *Int. J. Heat Mass Transf.* **2014**, *74*, 101–112, doi:10.1016/j.ijheatmasstransfer.2014.02.072.
18. Behnia, M.; Parneix, S.; Durbin, P.A. Prediction of heat transfer in an axisymmetric turbulent jet impinging on a flat plate. *Int. J. Heat Mass Transf.* **1998**, *41*, 1845–1855, doi:10.1016/S0017-9310(97)00254-8.
19. Hanjalić, K.; Popovac, M.; Hadžiabdić, M. A robust near-wall elliptic-relaxation eddy-viscosity turbulence model for CFD. *Int. J. Heat Fluid Flow* **2004**, *25*, 1047–1051, doi:10.1016/j.ijheatfluidflow.2004.07.005.
20. Jaramillo, J.E.; Pérez-Segarra, C.D.; Rodriguez, I.; Oliva, A. Numerical study of plane and round impinging jets using RANS models. *Numer. Heat Transf. B Fundam.* **2008**, *54*, 213–237, doi:10.1080/10407790802289938.
21. Hadžiabdić, M.; Hanjalić, K. Vortical structures and heat transfer in a round impinging jet. *J. Fluid Mech.* **2008**, *596*, 221–260, doi:10.1017/S002211200700955X.
22. Uddin, N.; Neumann, A.O.; Weigand, B. LES simulations of an impinging jet: On the origin of the second peak in the Nusselt number distribution. *Int. J. Heat Mass Transf.* **2013**, *57*, 356–368, doi:10.1016/j.ijheatmasstransfer.2012.10.052.
23. Krumbein, B.; Jakirlić, S.; Tropea, C. VLES study of a jet impinging onto a heated wall. *Int. J. Heat Fluid Flow* **2017**, in press, doi:10.1016/j.ijheatfluidflow.2017.09.020.
24. Hall, J.W.; Ewing, D. On the dynamics of the large-scale structures in round impinging jets. *J. Fluid Mech.* **2005**, *555*, 439–458, doi:10.1017/S0022112006009323.
25. Hall, J.W.; Ewing, D. The development of the large-scale structures in round impinging jets exiting long pipes at two Reynolds numbers. *Exp. Fluids* **2005**, *38*, 50–58, doi:10.1007/s00348-004-0883-2.
26. Lee, J.; Lee, S.-J. Stagnation region heat transfer of a turbulent axisymmetric jet impingement. *Exp. Heat Transf.* **1999**, *12*, 137–156, doi:10.1080/089161599269753.
27. Dairay, T.; Fortuné, V.; Lamballais, E.; Brizzi, L.-E. Direct numerical simulation of a turbulent jet impinging on a heated wall. *J. Fluid Mech.* **2015**, *764*, 362–394, doi:10.1017/jfm.2014.715.
28. Wilke, R.; Sesterhenn, J. Statistics of fully turbulent impinging jets. *J. Fluid Mech.* **2017**, *825*, 795–824, doi:10.1017/jfm.2017.414.

29. Schmitt, M.; Frouzakis, C.E.; Tomboulides, A. Direct numerical simulation of multiple cycles in a valve/piston assembly. *Phys. Fluids* **2014**, *26*, 035105, doi:10.1063/1.4868279.
30. Yang, Q.; Zhang, Z.; Liu, M.; Hu, J.A. Numerical Simulation of Fluid Flow inside the Valve. *Procedia Eng.* **2011**, *23*, 543–550, doi:10.1016/j.proeng.2011.11.2545.
31. Chorin, A.J. Numerical Solution of the Navier-Stokes Equations. *Math. Comput.* **1968**, *22*, 745–762, doi:10.1090/S0025-5718-1968-0242392-2.
32. Issa, R.I. Solution of the implicitly discretised fluid flow equations by operator-splitting. *J. Comput. Phys.* **1986**, *62*, 40–65, doi:10.1016/0021-9991(86)90099-9.
33. Patankar, S.V.; Spalding, D.B. A calculation procedure for heat, mass and momentum transfer in three-dimensional parabolic flows. *Int. J. Heat Mass Transf.* **1972**, *15*, 1787–1806, doi:10.1016/0017-9310(72)90054-3.
34. Vuorinen, V.; Keskinen, J.-P.; Duwig, C.; Boersma, B.J. On the implementation of low-dissipative Runge–Kutta projection methods for time dependent flows using OpenFOAM. *Comput. Fluids* **2014**, *93*, 153–163, doi:10.1016/j.compfluid.2014.01.026.
35. Van der Houwen, P.J. Explicit Runge-Kutta formulas with increased stability boundaries. *Numer. Math.* **1972**, *20*, 149–164, doi:10.1007/BF01404404.
36. Hirsch, C. *Numerical Computation of Internal and External Flows: The Fundamentals of Computational Fluid Dynamics*, 2nd ed.; Butterworth-Heinemann: Oxford, UK, 2007; ISBN 9780080550022.
37. González Hernandéz, M.A.; Moreno López, A.I.; Jarzabek, A.A.; Perales Perales, J.M.; Wu, Y.; Xiaoxiao, S. *Design Methodology for a Quick and Low-Cost Wind Tunnel, Wind Tunnel Designs and Their Diverse Engineering Applications*; InTechOpen: London, UK, 2013; doi:10.5772/54169.
38. Grötzbach, G. Spatial resolution requirements for direct numerical simulation of the Rayleigh-Bérnard convection. *J. Comput. Phys.* **1983**, *49*, 241–264, doi:10.1016/0021-9991(83)90125-0.
39. Nicoud, F.; Ducros, F. Subgrid-scale stress modelling based on the square of the velocity gradient tensor. *Flow Turbul. Combust.* **1999**, *62*, 183–200, doi:10.1023/A:1009995426001.
40. Spalding, D.B. A Single formula for the "Law of the Wall". *J. Appl. Mech.* **1961**, *28*, 455–458, doi:10.1115/1.3641728.
41. Pope, S.B. *Turbulent Flows*; 11th Printing; Cambridge University Press: Cambridge, UK, 2011; ISBN 9780521598866
42. Hunt, J.C.R.; Wray, A.A.; Moin, P. Eddies, stream, and convergence zones in turbulent flows. In Proceedings of the 1988 Summer Program, Stanford University, CA, USA, June 1988; pp. 193–208.
43. Launder, B.E. Numerical computation of convective heat transfer in complex turbulent flows: Time to abandon wall functions? *Int. J. Heat Mass Transf.* **1984**, *27*, 1485–1491, doi:10.1016/0017-9310(84)90261-8.
44. Launder, B.E. On the computation of convective heat transfer in complex turbulent flows. *J. Heat Transf.* **1988**, *110*, 1112–1128, doi:10.1115/1.3250614.
45. Shih, T.-H.; Povinelli, L.A.; Liu, N.-S. Application of generalized wall function for complex turbulent flows. *J. Turbul.* **2003**, *4*, N15, doi:10.1088/1468-5248/4/1/015.
46. Craft, T.J.; Gerasimov, A.V.; Iacovides, H.; Launder, B.E. Progress in the generalization of wall-function treatments. *Int. J. Heat Fluid Flow* **2002**, *23*, 148–160, doi:10.1016/S0142-727X(01)00143-6.
47. Suga, K.; Craft, T.J.; Iacovides, H. An analytical wall-function for turbulent flows and heat transfer over rough walls. *Int. J. Heat Fluid Flow* **2006**, *27*, 852–866, doi:10.1016/j.ijheatfluidflow.2006.03.011.
48. Craft, T.J.; Gant, S.E.; Iacovides, H.; Launder, B.E. A new wall function strategy for complex turbulent flows. *Numer. Heat Transf. B Fundam.* **2004**, *45*, 301–318, doi:10.1080/10407790490277931.
49. El Hassan, M.; Assoum, H.H.; Sobolik, V.; Vétel, J.; Abed-Meraim, K.; Garon, A.; Sakout, A. Experimental investigation of the wall shear stress and the vortex dynamics in a circular impinging jet. *Exp. Fluids* **2012**, *52*, 1475–1489, doi:10.1007/s00348-012-1269-5.
50. Tu, C.V.; Wood, D.H. Wall pressure and shear stress measurements beneath an impinging jet. *Exp. Therm. Fluid Sci.* **1996**, *13*, 364–373, doi:10.1016/S0894-1777(96)00093-3.
51. Salari, K.; Knupp, P. *Code Verification by the Method of Manufactured Solutions*; SAND2000-1444; SciTech Connect: Washington, DC, USA, 2000.
52. Shunn, L.; Ham, F. Method of Manufactured Solutions Applied to Variable-Density Flow Solvers. In *Annual Research Briefs*; Center for Turbulence Research: Stanford, CA; USA, 2007; pp. 155–168.
53. Roache, P.J. *Verification and Validation in Computational Science and Engineering*, 1nd ed.; Hermosa Publishers: Socorro, NM, USA, 1998; ISBN 9780913478080.

54. Oberkampf, W.L.; Roy, C.J. *Verification and Validation in Scientific Computing*; Cambridge University Press: Cambridge, UK, 2010; ISBN 978-0-521-11360-1.
55. Taylor, G.I.; Green, A.E. Mechanism of the production of small eddies from large ones. *Proc. R. Soc. A Math. Phys.* **1937**, *151*, 499–521, doi:10.1098/rspa.1937.0036.
56. Joung, Y.; Choi, S.-U.; Choi, J.I. Direct numerical simulation of turbulent flow in a square duct: Analysis of secondary flows. *J. Eng. Mech. ASCE* **2007**, *133*, 213–221, doi:10.1061/(ASCE)0733-9399(2007)133:2(213).
57. Gavrilakis, S. Numerical simulation of low-Reynolds-number turbulent flow through a straight square duct. *J. Fluid Mech.* **1992**, *244*, 101–129, doi:10.1017/S0022112092002982.